Refrigeration, Air Conditioning, Range and Oven Servicing

Refrigeration, Air Conditioning, Range and Oven Servicing

Edited by
ROBERT SCHARFF

McGRAW-HILL BOOK COMPANY
NEW YORK
ST. LOUIS
SAN FRANCISCO
AUCKLAND
DÜSSELDORF
JOHANNESBURG
KUALA LUMPUR
LONDON
MEXICO
MONTREAL
NEW DELHI
PANAMA
PARIS
SÃO PAULO
SINGAPORE
SYDNEY
TOKYO
TORONTO

Library of Congress Cataloging in Publication Data
Main entry under title:

Refrigeration, air conditioning, range and oven servicing.

Includes index.
1. Refrigeration and refrigerating machinery —Maintenance and repair. 2. Household appliances, Electric—Maintenance and repair.
I. Scharff, Robert. II. Title.
TP492.7.R44 643'.6 75-29239
ISBN 0-07-055144-8

234567890 HDHD 78543210987

The editors for this book were Tyler G. Hicks and Lester Strong, the designer was Edward J. Fox, and the production supervisor was Teresa F. Leaden.

It was printed and bound by Halliday Lithograph Corporation.

Contents

Preface

The purpose of this book is to give the service technician the necessary information needed to maintain and repair refrigeration appliances—refrigerators, ice makers, freezers and water coolers—and air conditioning appliances—room air conditioners, dehumidifiers, humidifiers, electronic air cleaners, and central vacuum cleaning systems. Chapter 9 is devoted completely to electric ranges and ovens.

It is not necessary to break down the subject matter any further in the Preface; the extent of the completeness of information can be found in the Contents, Index, and of course, the text itself. The book is not a treatise on *appliance* engineering; rather it is intended to tell the service technician how to maintain and service refrigeration and air conditioning appliances, as well as electric ranges and ovens.

This book—as the three others in the series (*Basics of Electric Appliance Servicing, Small Appliance Servicing Guide,* and *Major Appliance Servicing*)—was prepared and written with the full cooperation of the Association of Home Appliance Manufacturers (AHAM) and its Service Committee under the chairmanship of S. E. Upton of the Whirlpool Corporation. While the AHAM and all its members were most cooperative in the preparation of this book, we would especially like to thank the following for their outstanding help: C. H. Ramlow of Kelvinator; David Moore of Amana; D. C. Shaffer and S. L. Gessaman of Frigidaire; Guy Turner, V. A. Miller and A. R. Sabin of Whirlpool; R. W. Rivett of Gibson; R. J. Kalember and A. F. Horn of General Electric; D. L. LeBlac of Corning; S. N. Ludwig of Fedders; W. H. Lewis and Gary Brown of McGraw-Edison Company; H. E. Dillmore and A. E. Last of Westinghouse; and Robert Holding of AHAM. A portion of the text, as well as many of the illustrations, was taken from the following home study courses: *Refrigeration Service Part I* and *Refrigeration Part II* (Whirlpool); *Refrigeration Theory* (Gibson); and *Silver Brazing* (Kelvinator). Other illustrations were taken from the following technical slide films: *Refrigeration Servicing* (Gibson); and *Basics of Room Air Conditioners, Repair Procedures for Refrigeration Sealed Systems, Electric Range Service, The Whirlpool Central Vacuum Cleaner* and *Understanding Electric Range Components* (Whirlpool).

In addition, we want to thank Janet Just, who did the typing of the manuscript; Ronald L. Graffius of G. J. Bear Company, who reproduced a portion of the illustrations; and Mary Puschak, who coordinated all the art for this book.

Refrigeration, Air Conditioning, Range and Oven Servicing

Refrigeration servicing

CHAPTER

1

As with all servicing endeavors, by familiarizing ourselves with the pertinent fundamental laws, we are able to better understand the physical phenomena taking place inside the refrigeration system, do a better job of service analysis, and arrive at the corrective answers more quickly. But if we were to start at the beginning of refrigeration, we would have to go back to a time when the icebox was in every home and the iceman was a very familiar figure. In those days it was necessary to have the iceman regularly deliver a big chunk of ice, and just as regularly the users had to empty the drain pan underneath. There was nothing mysterious about the cooling process. The ice was cold and eventually everything in the icebox became cold, too. Everyone knew that ice melted, so no one ever questioned the necessity of emptying the drain pan.

Modern mechanical refrigeration has changed this concept tremendously. Today we have forgotten the chore of emptying the drain pan. We no longer need someone to put ice into the icebox; instead, we can take ice cubes out almost any time we want—the refrigerator still stays cold.

HEAT AND THE REFRIGERATION PROCESS

Refrigeration may be defined as the *transfer or removal of heat.* In domestic refrigeration, heat is removed from the interior of the cabinet and given off to the air in the room. A study of refrigeration is a study of heat and heat transfer.

"Hot" and "cold" are relative terms used to indicate whether a particular item is above or below body temperature. The natural condition of a substance is extremely cold if heat is removed from it. Technically, refrigeration does not produce cold; instead, refrigeration removes heat which allows the item to become cold. By this established phenomenon of nature, we can say that cold is the absence of heat.

Heat

Heat is a form of energy known by its characteristics and effects; it is molecular motion. All substances are made up of tiny molecules which are in a state of rapid motion. As the temperature of a substance is increased, the molecular motion increases, and as the temperature decreases, the molecular motion decreases.

Heat always flows from a warmer substance to a cooler substance, that is, from higher temperatures to lower temperatures. Faster-moving molecules impart some of their energy to slower-moving molecules; therefore, the faster molecules slow a little and the slower ones move a little faster. Sometimes, however, when heat is added or removed, the molecules, instead of moving slower or faster, change their shape. The change in shape is caused by one or more of the atoms in the molecules shifting to a different position, and this shift in the atomic structure of the molecule may cause it to change from a gas to a liquid or vice versa. For example, if you remove enough heat, water becomes ice, or if you add enough heat, ice becomes water; and if you add still more heat, water becomes steam.

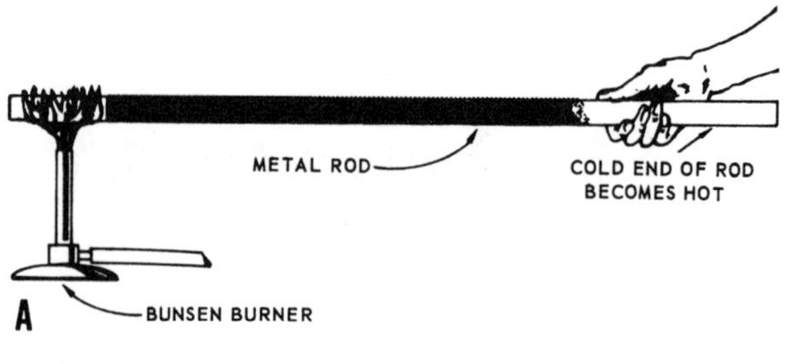

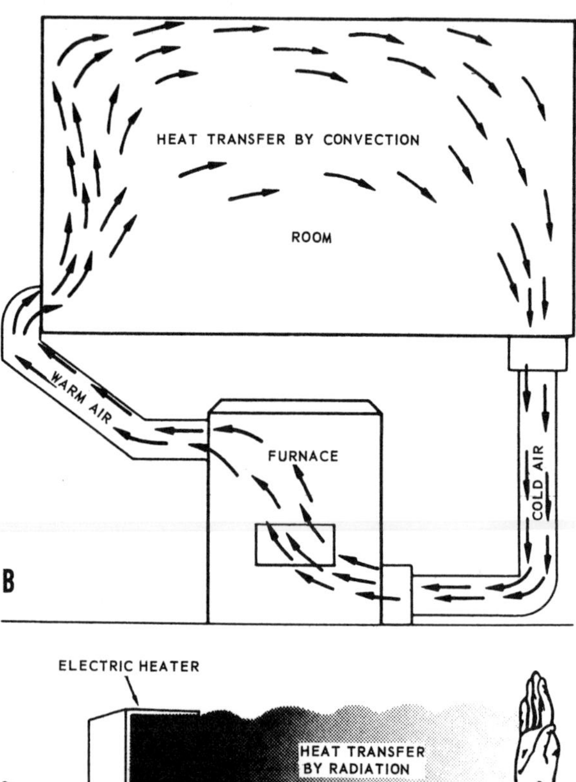

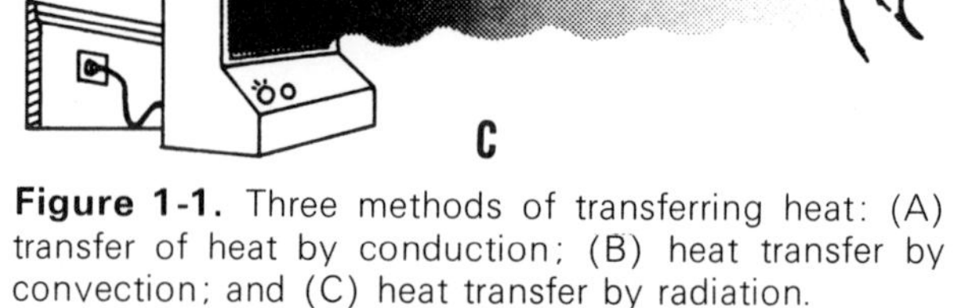

Figure 1-1. Three methods of transferring heat: (A) transfer of heat by conduction; (B) heat transfer by convection; and (C) heat transfer by radiation.

Heat transfer. Keep in mind that heat always travels from a warmer object to a cooler object. Heat is transferred in three ways (Fig. 1-1).

1. *Conduction.* The transfer of heat by contact is called conduction. If one end of a metal rod is heated in a fire, the heat travels along the rod, increasing the temperature of the other end by conduction. Every material has a conduction value, some high and others low. Materials which have a low heat conductivity are called *insulators.* However, no material completely stops or offers a perfect barrier to the passage of heat.
2. *Convection.* Heat transferred by moving a heated substance, such as air or water, is known as convection. Heat is transferred from a furnace throughout a house in a warm-air system by convection. Heat travels from the warmer items in a refrigerator to the cooler evaporator by convection, as a result of the circulation of air inside the refrigerator.
3. *Radiation.* The transfer of heat without heating the medium or substance through which the heat passes is known as radiation. The earth receives heat from the sun by radiation, but the space between the earth and the sun is not heated directly by the sun's rays.

Heat measurement. Two methods of heat measurement are commonly used, one to indicate the intensity, the other to indicate the quantity or amount of heat present. *Intensity* of heat, commonly known as *temperature,* is usually measured by a Fahrenheit thermometer. *Quantity,* or the amount, of heat is measured by the British thermal unit, commonly abbreviated Btu.

Temperature may be defined as the heat intensity or heat level of a substance. Measuring temperature does not tell the amount of heat in a substance, but it does give an indication of the degree of warmth or how hot a body is.

The molecular theory of heat states that temperature is an indication of the speed of motion of the molecule. It is important not to use the words "heat" and "temperature" carelessly. Temperature measures the speed of motion of one molecule, while heat is the speed of motion of the molecule plus the number of molecules (weight so affected). As an example of temperature and the quantity of heat, consider a small copper dish heated to 1,340°F; it does not contain as much heat as 5 lb of copper heated to 300°F, even though the small dish is much warmer. That is, its heat level is higher or its intensity of heat is greater but does not contain as much heat as the larger mass.

The British thermal unit is a measure of heat just as a mile is a measure of distance, a gallon is a measure of liquids, and a pound is a measure of weight. One Btu is the amount of heat required to change the temperature of 1 lb of water 1°F under all average conditions. If 1 Btu of heat is added to 1 lb of water, the temperature will be increased 1°F. If 1 Btu of heat is taken away from 1 lb of water, the temperature will decrease 1°F.

Heat effects. We know that heat changes the temperature of objects—that is, causes expansion and contraction—and can also change the form of substances, as when ice is changed to a liquid (water) by addition of heat.

Years ago, scientists discovered that ice can take on large amounts of heat while remaining at a constant temperature. As long as the ice remained, the temperature stayed at 32°F. They also found that it was possible for water to be in liquid form at 32°F. However, the 32°F water rapidly assumed the temperature of the surrounding air. They were able to measure the temperature rise of water with the use of a thermometer; this they called *sensible* heat. The heat required to change ice to water could not be measured with a thermometer; thus the name *latent* heat, meaning hidden heat, was used. Of course, like everyone else, the scientists were curious about what happened to the latent heat. Where did it go? At first they thought it was in the water that melted from the ice. But that was not exactly the right answer because, upon checking the water temperature as it melted from the ice, they found it was only a shade warmer than the ice itself. It was not nearly warm enough to account for all the heat the ice had absorbed. The only possible answer was that the latent heat had been used up in changing the ice from a solid to a liquid. Using this knowledge, we can define sensible and latent heat as follows.

1. *Sensible heat.* Heat which produces a change in temperature of a substance without changing the state of the substance is known as sensible heat. An example is water at 100°F being heated to 150°F. The 50°F change in temperature is easily read on a thermometer; therefore, it is known as sensible heat.
2. *Latent heat.* The amount of heat required to change the state of a substance without changing the temperature of the substance is known as latent heat. Latent heat, also called *hidden* heat, is the more important in the study of refrigeration.

To further explain the importance of heat in refrigeration, let us take a look at three states of matter: solids, liquids, and gases. All matter is in one of these three states, solids such as ice, liquids such as water, or gas such as steam or air. Matter can be changed from one state to another by adding or removing heat.

The amount of heat which must be removed from an object to change it from a liquid to a solid, such as water to ice, is known as the *latent heat of fusion.* The amount of heat which must be added to an object to change it from a liquid to a gas, such as water to steam, is known as the *latent heat of vaporization.* It is important to remember that latent heat changes the state of a substance without changing its temperature. Using the diagram in Fig. 1-2, we can easily follow the effects of heat on water.

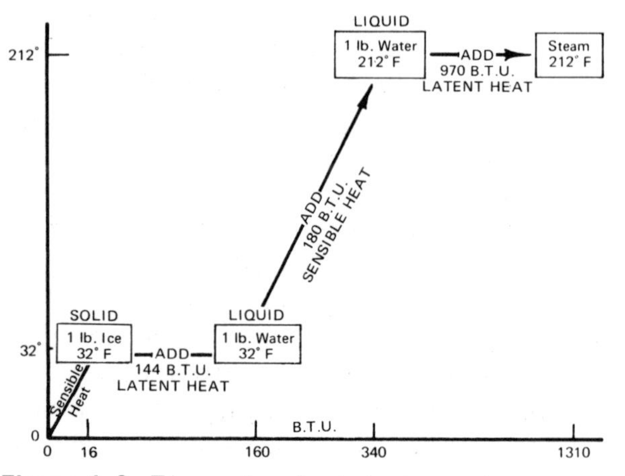

Figure 1-2. Temperature-heat chart.

If we have 1 lb of ice at 32°F and add 144 Btu of latent heat to it, we can change it to 1 lb of water, still at 32°F. We have learned that latent heat cannot be measured with a thermometer, and know that it changes the state of a substance, in this case, ice to water. Now we can see that if we add 180 Btu of sensible heat (that is, heat that can be measured with a thermometer) to this 1 lb of water at 32°F, we shall have 1 lb of water at 212°F. By adding an additional 970 Btu of latent heat to the 1 lb of water at 212°F, we will have changed the water from liquid at 212°F to steam at 212°F.

It is important to note the comparatively large amount of heat required to change the state of a substance, for example, water to steam or ice to water. The latent heat of vaporization of refrigerants used in domestic refrigeration will vary under certain conditions and is different from the latent heat of water, but as in the case of water, its latent heat is relatively large as compared with the amount of heat required to change its temperature.

In a modern refrigerator, freezer, air conditioner, or dehumidifier, liquid refrigerant is piped under pressure to the evaporator coil. In the evaporating coil the pressure is greatly reduced, and the refrigerant boils (changes to a gas) and absorbs considerable heat from the coil. This produces a low temperature and cools the evaporating coil. This refrigerant gas is pumped out of the coil and compressed by the compressor. It then flows into the condenser which is located outside the refrigerator where the heat that was absorbed in the evaporator coil is released ("squeezed out") to the surrounding atmosphere, the refrigerant returns to liquid form again, and the cycle is repeated.

Pressure-Temperature Relation

The temperature at which a gas will condense (change from a gas to a liquid) is the same as the boiling or vaporization temperature of the liquid when the pressure is the same. For example, on the temperature-heat chart it can be noticed that it takes 970 Btu of latent heat to change 1 lb of water at 212°F to 1 lb of steam at

212°F. If we remove 970 Btu from the 212°F steam, on the other hand, we will then have 212°F water.

As a gas is compressed, its temperature is raised. A gaseous refrigerant returning to the compressor from the cooling coil will probably be at or slightly below room temperature. This same gas, as it leaves the compressor and enters the condenser, will be at a much higher temperature. Little or no heat has been added. However, the total heat of the gas is now squeezed into a much smaller space, and a much higher temperature results. This is called *heat of compression*, although technically no heat has been added. Knowing this, we can say that when the pressure is increased, the temperature is also increased, or if the pressure is decreased, the temperature is decreased. An example of this is water which boils at 212°F. If we took the same quantity of water to the top of a mountain where the atmospheric pressure is less, the water would boil at a temperature less than 212°F. An example of increased pressure is a pressure cooker. By increasing the pressure on the water by 15 lb, it is feasible to raise the temperature of the water to 250°F before its boiling point is reached. Knowing this, we can understand how a pressure cooker saves so much time in preparing food. (Much higher temperatures can be reached.)

Now that you have a better understanding of the temperature and pressure relationship, you can see that by controlling the pressure in the evaporator, we can control the temperature at which the refrigerant boils. You can also see that the pressure in the condenser determines the temperature at which the gas will condense (return to a liquid).

In one part of the refrigerating system (evaporator), liquid is changed to a gas, absorbing heat; in the other part of the refrigerating system, gas is changed back to a liquid by pressure and removal of heat. The pressures are naturally different in these two parts of the system. The low pressure is in that part where the liquid changes to a gas, the high pressure is in that part where the gas changes to a liquid. These two parts are called the *low* and the *high* side of the refrigerating system.

A good question here would be, On a modern refrigerator, why is a condenser so warm to the touch? Remember that it was stated earlier that heat travels only from a warmer to a cooler object. We know that the refrigerant boils in the evaporator because of the lower pressure created there by the compressor. The compressor then draws this gas back, compresses it, and pumps it into the condenser. The compressor must be capable of applying enough pressure to the gas to raise its temperature above that of the room, thus allowing the heat in the condenser to flow to the cooler surrounding air.

REFRIGERANTS

Any fluid, when used as a cooling medium, may be termed a *refrigerant*; however, for our purpose a refrigerant is defined as a chemical fluid which can be changed easily from a liquid to a gas and from a gas to a liquid, in compression systems, to utilize the latent heat of vaporization in obtaining the desired cooling effect. Such a fluid is a *primary refrigerant* and should not be confused with nonvolatile fluids, such as brine or water, which are not true refrigerants but are merely carriers of sensible heat. In fact, the principles involved in the physical changes necessary to produce refrigeration are the same for all chemical refrigerants used in compression systems.

In order to intelligently service refrigerating systems, it is essential that the service technician understand the physical changes of the refrigerant in the system and be familiar with the pressure-temperature relation of the refrigerant, as well as its chemical characteristics.

The refrigerant commonly used in today's household system is dichlorodifluoromethane, commonly known as Freon-12, which has the chemical formula CCl_2F_2. Today, the Freon family of refrigerants originally designated as Freon-12, Freon-22, etc., are known under various trade names simply as Refrigerant-12, Refrigerant-22, etc., or merely as R-12, R-22, etc., or F-12, F-22, etc.

The Freon refrigerants are colorless and almost odorless, and their boiling points vary over a wide range of temperature. Those in general use are nontoxic, noncorrosive, nonirritating, and nonflammable under all conditions of usage. (*Caution*: Even though nonflammable, Freon, when exposed to open flame, will produce phosgene gas which is highly toxic.) They are generally prepared by replacing the chlorine or hydrogen molecules with fluorine. Chemically they are inert, and are thermally stable up to temperatures far beyond conditions found in actual operation.

The two most used refrigerants today are Freon-12 and Freon-22.

1. *Freon-12* (*R-12*). R-12 has a boiling point of −21.6°F and is extensively used as a refrigerant in air-conditioning systems as well as in refrigerators and freezers. The health hazards resulting from exposure to R-12 when used as a refrigerant are remote. Freon is in a class of specially nontoxic gases. Vapor in any proportion will not irritate the skin, eyes, nose, or throat, and being odorless and nonirritating, it offers no possibility of causing panic should it escape from a refrigeration system. A pound of R-12 liquid expands to 2.8 ft^3 of vapor at 68°F. R-12 is a stable compound capable of undergoing, without decomposition, the physical change to which it is commonly subjected in service, such as vaporization, compression, and high temperatures.
2. *Freon-22* (*R-22*). R-22 has a boiling point of −41.4°F and is used as a refrigerant in industrial and commercial low-temperature refrigerating systems to −150°F and also in window-type and free-standing room coolers and central air-conditioning units. R-22 is also used in locker and processing plants where lower temperatures result in quicker freezing of foods, and in countless other low-temperature industrial applications.

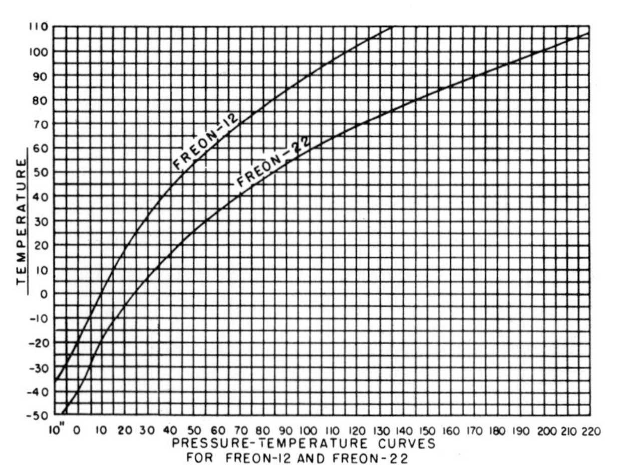

Figure 1-3. Pressure-temperature curves for Freon-12 (R-12) and Freon-22 (R-22).

The relations between pressure and corresponding temperatures in Fahrenheit degrees of R-12 and R-22 are given in the chart (Fig. 1-3). For example, at a pressure of 87 lb/in^2, the temperature of R-12 will be 82°F. If the pressure is increased to 130 lb/in^2, the temperature will increase to 107°F, etc.

Operating pressures will vary with the temperature of the condensing unit, amount of condenser surface, operating back pressure, condition of the condenser surface, extent of superheating of the refrigerant gas, and other factors. *Note*: Superheat is that condition in which a gas is heated to a temperature higher than it would normally be at the pressure under which it is existing. That is, the superheat in a refrigerant vapor is sensible heat, which raises the temperature of the vapor above its saturation temperature. Superheat usually is expressed in degrees rather than in Btu. The amount of heat, in Btu, required to raise the temperature of 1 lb of vapor 1°F will depend upon the specific heat of the refrigerant vapor.

In refrigeration work, the degree of superheat in a refrigerant vapor is of utmost importance, since it affects the capacity of the evaporator, controls the liquid flow through the capillary tube, is used for subcooling liquid by use of a heat exchanger, affects the capacity of the compressor, and affects the quantity of cooling medium required over the condenser.

It is important to remember that all refrigerants contain a certain amount of moisture. The manufacturers of refrigerants have set definite specifications limiting the moisture content of the refrigerants. The moisture content in all factory-charged systems is well below the specified limits.

If excess moisture is present, Freon refrigerants will hydrolyze, that is, react with the water to form hydrochloric acid or hydrofluoric acid. The conditions encountered in refrigerating compressors, over long periods of operation and at elevated temperatures of the discharge vapor, cause an acid reaction when the moisture content is above the maximum allowable quantity.

Freeze-up is the most troublesome effect of excessive moisture in Freon systems. The solubility of water in R-12 is relatively low, and the solubility, in percent of water by weight, is rapidly reduced as the temperature of the refrigerant is lowered. The small quantity of water, in excess of that which will dissolve in the refrigerant, at a given temperature, will freeze out as ice. The ice usually forms at the metering device, that is, at the outlet of the capillary tube where the liquid suddenly expands into the larger-size tubing of the evaporator.

At 30°F, R-22 will hold approximately 25 times more water in solution than will R-12. At −20°F, R-22 will hold approximately 51 times more water in solution than will R-12. This indicates fewer freeze-ups with R-22 than with R-12. However, when the moisture content of R-12 is below the freeze-out point, there is not sufficient moisture to cause corrosion, whereas R-22 with moisture below the freeze-out point may still contain enough moisture to cause damaging acid reactions.

Oil sludge may result from the acids formed by moisture. Moisture or acid, together with heat, dissolves metal particles from brass or copper parts. These copper particles are carried in solution with the oil and are deposited in the compressor. This copper plating will first be noticeable at the discharge valve where the highest temperature exists.

Excessive moisture in a refrigeration system may occur from one or more of the following:

1. Poor evacuation of the system
2. Faulty method of charging systems
3. Defective workmanship on refrigerant piping
4. Mechanical failure, causing leaks
5. Moisture on replacement parts
6. Wet oil or refrigerant
7. Faulty service

REFRIGERATING SYSTEM

The essential elements required to accomplish the refrigerating process and make up the refrigerating system (Fig. 1-4) are the following:

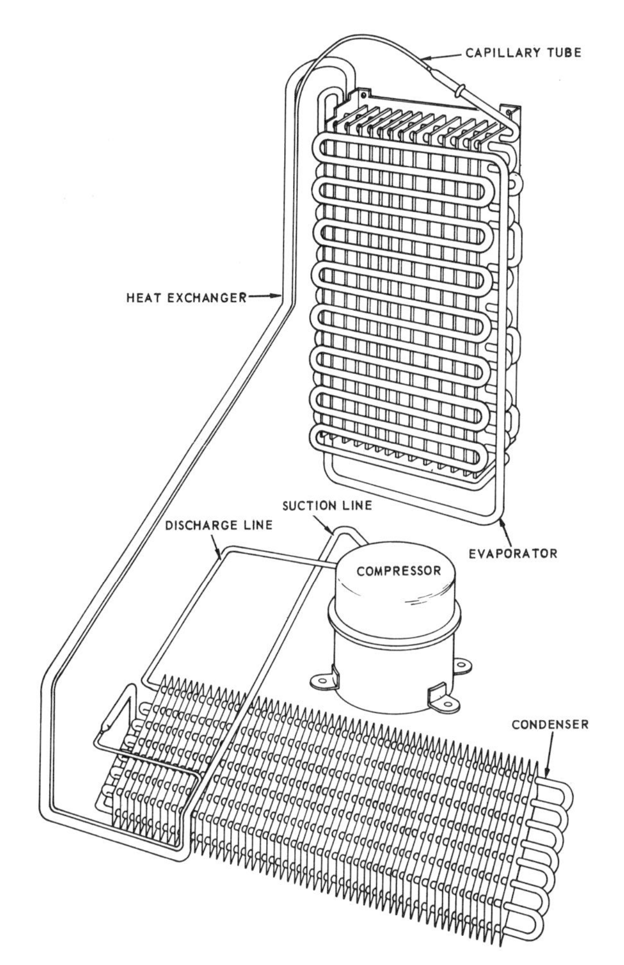

Figure 1-4. Basic refrigeration system.

1. Evaporator or cooling unit
2. Compressor or pump
3. Condenser
4. Liquid-metering device (i.e., capillary tubes)

The evaporator, sometimes referred to as the *cooling unit, chilling unit, freezing unit,* or *frozen-food chest,* normally is located in the top portion of the food compartment. It may extend across the entire top of the cabinet and frequently has a separate door. The evaporator absorbs heat from the compartment faster than heat can enter through the walls, from food, and from other sources. Thus, it lowers the temperature of everything inside the compartment.

As heat always flows from a warmer to a colder substance, heat from foods and liquids in the cabinet will flow to the evaporator, thereby reducing the temperature of the stored foods and liquids. The evaporator is maintained at a lower temperature than the space refrigerated and so absorbs heat from it. This causes the air to fall, as cold air is heavier than warm air, and forces the warm air to rise. This action sets all the air in motion; therefore, heat is carried to the evaporator by natural air circulation, that is, by convection. This circulation may be aided by a fan.

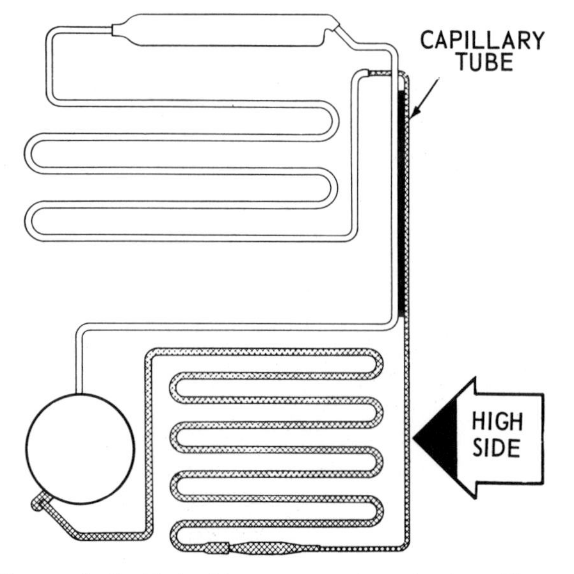

Figure 1-5. High-side pressure in a basic refrigeration system.

The condensing unit assembly comprises a compressor, an electric motor for driving the compressor, the condenser, the mounting base for these components, and, in forced-convection units, a condenser fan driven by a separate fan motor. In a hermetically sealed system, the motor is direct-connected to the compressor and both are sealed inside a steel dome, with a permanent supply of oil.

Two tubes are attached to the evaporator and normally enter the cabinet through a hole in the back wall. One tube is the suction line and conveys vaporous refrigerant from the evaporator to the compressor. The other tube is the liquid line and conveys liquid refrigerant from the condenser to the evaporator. These two tubes usually are soldered together to form a heat exchanger. The cold suction gas cools the hot liquid refrigerant before it enters the evaporator, thus increasing the efficiency of the system.

The pressure at the inlet end of the small capillary tube connected to the condenser outlet line is at high pressure. If you were to look back along the line toward the compressor, you would note that this high pressure exists in the condenser, the compressor discharge line, and all the way back to the discharge valve of the compressor pump. This much of the system, from the discharge valve on the compressor pump to the capillary tube, is called the *high side* because it is under high pressure (Fig. 1-5).

The inside of the evaporator is at low pressure because vapor is being sucked out of it by the pump. This low pressure exists also in the suction line, in the condenser dome, and right up to the intake valve on the compressor pump. This part of the system between the capillary tube outlet and the intake valve on the pump, in a path through the evaporator, is called the *low side* because it is under low pressure (Fig. 1-6).

The pressure on the high side of the system is called the *head pressure* and that on the low side is called the *suction pressure,* or *back pressure.* Remember these names and what they mean because no words are used more often when talking about mechanical refrigeration.

We have both liquid and gas in the low side,

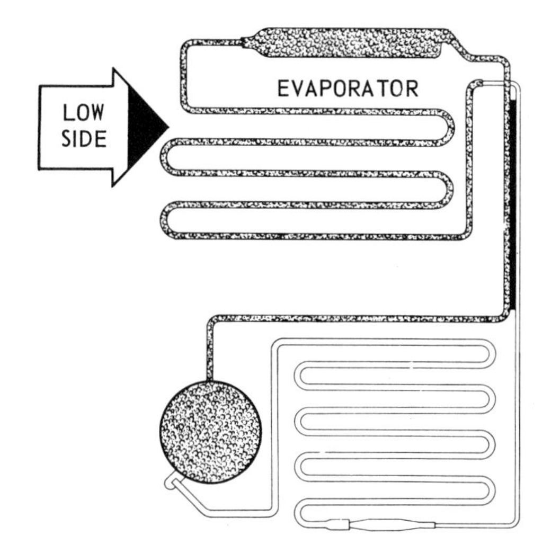

Figure 1-6. Low-side pressure in a basic refrigeration system.

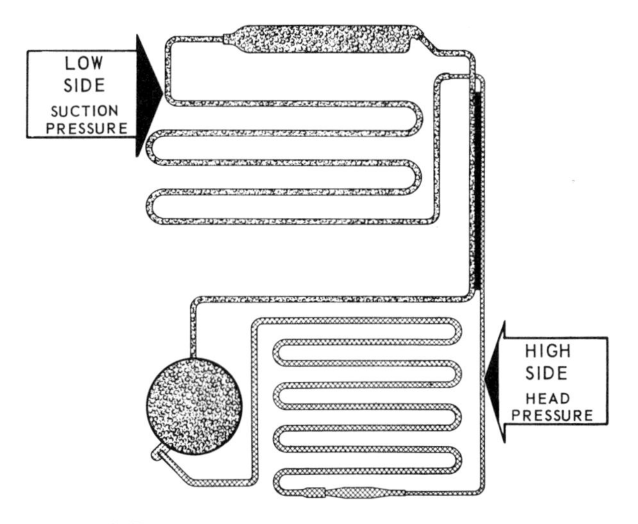

Figure 1-7. High-side and low-side pressure working together in a basic refrigeration system.

and we have both liquid and gas in the high side. Therefore, you cannot correctly say that either side contains all gas or all liquid. It is not a matter of gas or liquid that distinguishes between the low side and the high side—it is the *pressure.*

The capillary tube has head pressure at one end and suction pressure at the other end, when the compressor is in operation. The pressure difference and the inside diameter of the tube determine the rate of flow of liquid refrigerant from the high side to the low side.

The capillary tube also serves to "unload" the high side when the compressor is not running; that is, it permits the pressure to equalize between the high side and the low side. This makes for easier starting of the compressor motor as both the suction and discharge sides of the pump are at the same pressure. Since less starting torque is required, a split-phase type of compressor motor can be used instead of a more complex capacitor type, which has high starting torque characteristics.

Since the service technician must have a detailed knowledge of the components of the refrigeration system, let us take a closer look at them.

Compressors

The function of the compressor is to establish a pressure difference and thus cause the refrigerant to flow from one part of the system to the other. This is the only responsibility of a compressor in a refrigerating system. It is not intended to be a pump just for circulating the refrigerant; rather, its job is to exert pressure for two reasons. Pressure makes the vapor hot enough to cool off in warm room air. At the same time, the compressor raises the refrigerant's pressure above the condensing point at the temperature of the room air so it will condense. It is this difference in pressure between the high and low sides that forces liquid refrigerant through the capillary tube and into the evaporator.

Modern refrigeration compressors are usually of the hermetically sealed type, either rotary or reciprocating, although numerous early refrigeration systems used open-type compressors. The open type is one in which the motor is connected to the compressor by means of a belt or coupling. In the sealed type, the motor and compressor are directly connected to the same shaft and sealed in the same compartment.

Reciprocating compressors. The reciprocating-type compressor is simply a cylinder with a piston moving back and forth in it. As in an

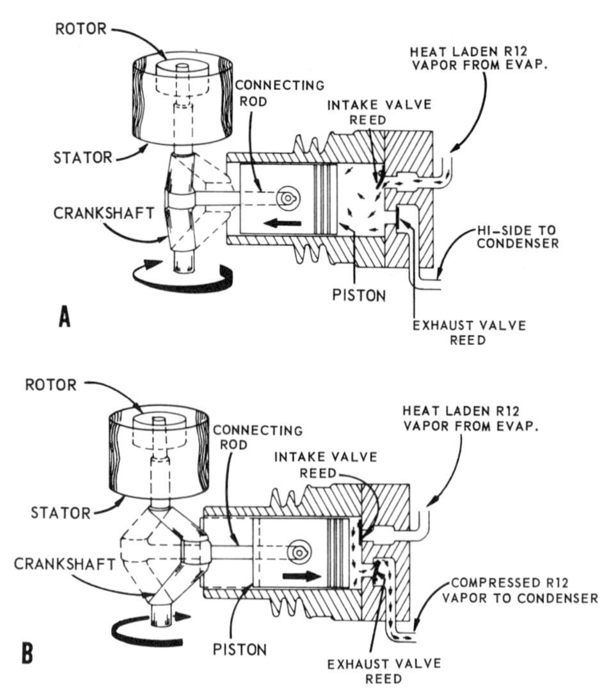

Figure 1-8. The operation of a typical reciprocating compressor.

automobile engine, a connecting rod and crankshaft are part of the mechanism.

In practice, an electric motor drives a "rotor" on the end of the crankshaft. As the crankshaft revolves with the rotor, it works through the connecting rod to move the piston back and forth in the cylinder (Fig. 1-8). As the piston is drawn back, it draws refrigerant vapor from the evaporator into the cylinder of the compressor. A reed-like valve in the cylinder head acts as a check valve. It permits refrigerant vapor to flow into the cylinder, but it closes and blocks the path as soon as any vapor turns to flow out of the cylinder.

By the time the piston has reached the back of its stroke, it has drawn a full load of vapor from the evaporator into the cylinder. Then the piston starts on its forward stroke, pushing the entrapped vapor ahead of it. The vapor cannot return to the evaporator because the intake valve is closed, blocking its path. However, there is another reed-like valve in the cylinder head. This one is arranged so that vapor can get out of the compressor cylinder. All vapor that is pushed through this valve is directed through tubing to the condenser. Like the intake valve, this exhaust valve is a one-way door. It lets vapor out of the compressor cylinder, but blocks the way to any of that same vapor that wants to return. Having reached the end of its forward stroke, the piston is ready to start back again to repeat the cycle by drawing more vapor from the evaporator.

The servicing of refrigerating systems using hermetically sealed compressors does not differ in any important respect from servicing those systems in which the compressor and motor are mounted separately. Because the unit is completely sealed and tested at the factory, trouble is seldom found with the compressor-motor assembly but in other parts of the system. The motor, located above the compressor, operates in a vertical position, whereas the compressor operates horizontally. The construction of the motor-compressor permits operation of the compressor in oil, simplifying lubrication problems. The dome is under low pressure in a reciprocating compressor. Just as there are many makes and models of automobiles, all the same in basic principles of operation but differing considerably in detail, so it is with reciprocating compressor units. Not only do they range in size and number of cylinders, but as a rule the compressors used in domestic refrigerators are either one- or two-cycle affairs.

Rotary compressors. The rotary type of compressor is very simple in its operation. Yet, in some respects, its simplicity makes its operation slightly more difficult to understand. Actually, the only moving parts in a rotary compressor consist of a steel ring, an eccentric or cam, and a sliding barrier.

Both the ring and the cam are housed in a steel cylinder. The steel ring is somewhat smaller in diameter than the cylinder, and it is so situated off-center that one point on the outer circumference of the ring is always in contact with the wall of the cylinder. This, of course, leaves an open, crescent-shaped space on the opposite side between the ring and the cylinder wall. An electric motor rotates the cam. As it (the cam) spins, it carries the ring around with it, imparting a peculiar rolling motion to the ring. The ring literally rolls on its outer rim around the wall of the cylinder.

Bearing in mind the way the ring rolls around in the cylinder, let us now see how it can compress a vapor. First, we shall drill a hole in the wall of the cylinder and use it as an entry to permit refrigerant vapor from the freezer to flow into the crescent-shaped space. By rotating the cam just a fraction of a complete turn, the ring will almost immediately cover up the drilled hole. As soon as that happens, the refrigerant vapor is trapped in the crescent-shaped space. There is no place it can go. However, if we drill another hole near the other end of the crescent-shaped space, the entrapped vapor will have a way of getting out. We can connect a pipe to this exit port and lead it to the condenser. Now, as the cam continues to rotate and roll the ring around in the cylinder, it will push the crescent-shaped wedge of refrigerant ahead of it, compressing the vapor as it pushes it out the exit port to the condenser. But, as you probably have already noticed, the ring cannot roll very far before it uncovers both the inlet and the exit ports. Obviously, we need some means of directing the refrigerant vapor out through the exit port and, at the same time, blocking its passage to the inlet port.

The most simple and the most effective way of doing that is to put up a barrier between the two ports. Of course, this barrier must be sort of flexible or sliding because one end of it is constantly moving back and forth as the ring rolls around in the cylinder. So we shall cut a slot in the wall of the cylinder and make it deep enough that the barrier can slide all the way in. Then we shall add some springs that will tend to push the sliding barrier out and hold it snugly against the ring no matter what position the ring is in. Now, as the ring rolls around in the cylinder, our movable barrier will follow every movement of the ring. With our movable barrier in place, the refrigerant vapor entrapped in the crescent-shaped space can go only one way as the ring rolls around and pushes the vapor ahead of it. The only way it can go is out through the exit port—the movable barrier blocks its passage to the inlet port.

That then, is the simple cycle of operations in a rotary type compressor. As the ring rolls

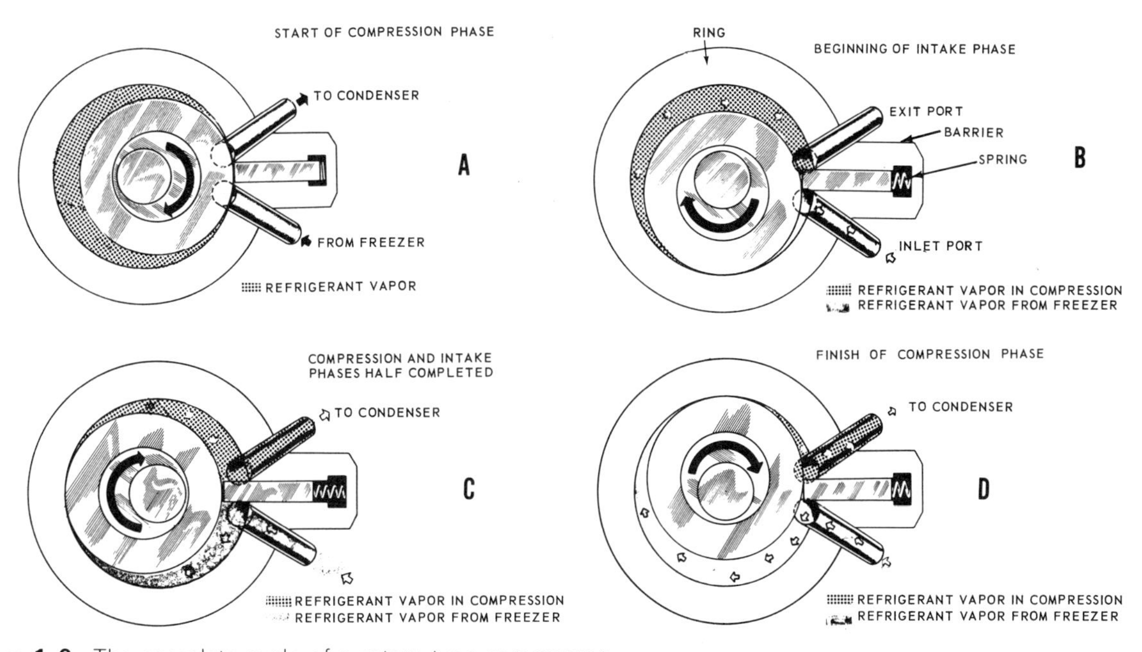

Figure 1-9. The complete cycle of a rotary-type compressor.

around in the cylinder, its point of contact literally "runs in a circle" around the cylinder wall. All refrigerant vapor ahead of that point of contact is pushed toward the movable barrier which deflects it out the exit port. Meanwhile, once the point of contact has passed the inlet port, a fresh charge of refrigerant vapor is drawn from the evaporator into the compressor.

As with the reciprocating-type compressor, numerous variations of design are to be found in different makes of rotary compressors. None of them, though, is simpler or has fewer moving parts than the rotary compressor just described. Also, the overall efficiency of a rotary compressor is higher than that of the reciprocating type. Rotary compressors are also more compact and can produce a deeper vacuum than a reciprocating compressor. It is for this reason that a rotary compressor is used as a vacuum pump for evacuating a refrigeration system for servicing. However, a reciprocating compressor can handle larger volumes of refrigerant.

Oil within the rotary compressor not only lubricates the moving parts, but it also serves to cool the compressor. The dome of a rotary compressor is under high pressure.

Condensers

Since the refrigerant leaves the compressor in the form of high-pressure vapor, some method must be found to change the vapor back into a liquid. It is the function of the condensing unit to condense the vapor to a liquid so that it can be reused in the refrigeration cycle.

As the refrigerant vapor is pumped into the condenser by the compressor, its temperature and pressure increase. This high temperature allows an effective transfer of heat to the surrounding room air from the surface of the condenser. A large portion of the heat transferred to the room air is the latent heat which was absorbed by the refrigerant in the evaporator. The loss of heat to the room air is sufficient to condense the refrigerant vapor to a liquid.

Condensers are of three main types: air-cooled by natural air circulation, air-cooled by means of a forced fan draft, and indirectly air-cooled through being in thermal contact with (and inside the exterior wall of) the cabinet.

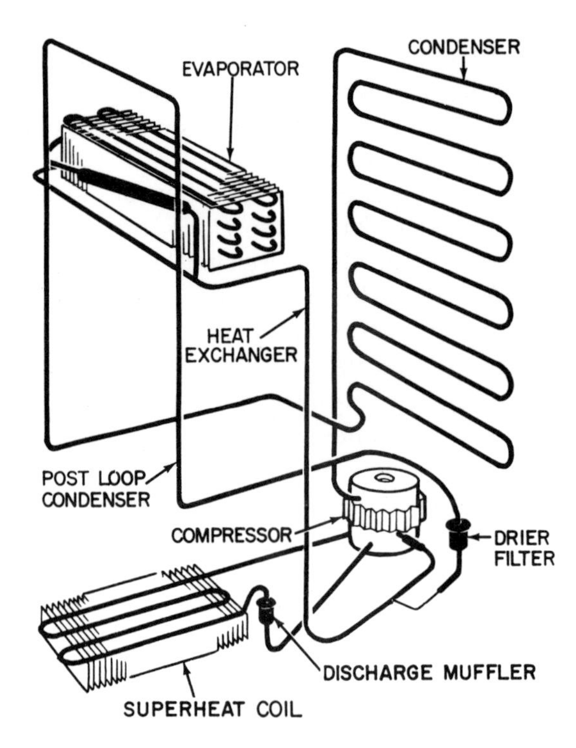

Figure 1-10. Typical static condenser system.

Static condensers. The most common condenser system is the static type shown in Fig. 1-10. This type is usually physically larger than a forced-draft condenser, and is constructed of either finned tubing or tubing interlaced with wire. It is located on the back of the refrigerator cabinet and relies on the thermal expansion of the surrounding room air to create a draft (convection currents) over the tubing.

Forced-draft condensers. Forced-draft condensers are usually constructed with finned tubing. As shown in Fig. 1-11, a fan is used to move air over the condenser. The fan is wired into the circuit in such a way that it runs only during the time the compressor is running. This type of condenser is located in the machine (compressor) compartment and will have the necessary air baffles to direct the air over the finned tubing.

Warm-wall condensers. A warm-wall condenser is actually another type of static con-

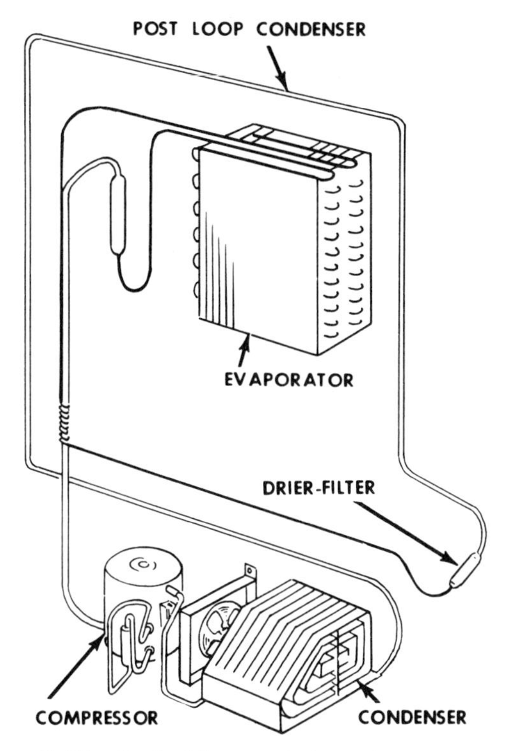

Figure 1-11. Typical fan-cooled condenser system.

denser. This type has the condenser tubing fastened directly to, and in thermal contact with, the inside of the outer wall of the cabinet. Heat is transferred to the cabinet wall from the tubing and from the wall to the surrounding room air. The outside of the cabinet will feel warm to the touch when this type of condenser is used and is in operation.

Capillary Tubes

The capillary tube is essentially a metering device used as a part of the refrigerant circuit. It consists normally of a miniature tube, the length and bore of which depend on the size of the condensing unit and the kind of refrigerant used. The refrigerating system must be carefully engineered, for not only must the capillary be matched to the characteristics of the compressor, for the operating conditions which the system is most likely to encounter, but also both the condenser and evaporator must be specifically designed for use with a capillary tube. The bore or inner diameter is very small, and the length varies greatly from a few inches up to several feet. Since the capillary tube offers a restricted passage, the resistance to the flow of refrigerant is sufficient to build up a high enough head pressure in the condenser to change the vapor to a liquid. The operating balance is obtained by properly proportioning the size and length of the tube to the particular system on which it is to be used. In other words, the capillary tube allows the high-side pressure to unload, or equalize with the low-side pressure, during the OFF cycle, and thus permits the compressor to start under a NO LOAD condition.

The inside diameter of the tube must, in any event, be such as to keep the tube full of liquid under normal operating conditions. Because of the tiny size of the tube bore, it is important to keep the refrigerant circuit free from dirt, grease, and any kind of foreign matter, since these obstructions may close up the tube and thus make the system inoperative. If the capillary tube becomes plugged, the evaporator will defrost, and the unit may run continuously.

Evaporators

The function of the evaporator in a refrigerator is to absorb heat from the air surrounding it, the heat being introduced by food placed in the refrigerator, penetrating through the insulation, and entering the refrigerator when the door is opened. Evaporators used in present-day designs are of the direct-expansion type. They are simple to construct, low in cost, and compact, and they also provide a more uniform temperature and rapid cooling. The evaporator consists simply of metal tubing or passages through which the refrigerant flows. Figure 1-12 illustrates various types of evaporators.

In operation, when the liquid refrigerant leaves the capillary tube and enters the larger tubing of the evaporator, the sudden increase in tubing diameter creates a low-pressure area which causes the boiling point of the refrigerant to drop, allowing a more rapid absorption of heat units. In the process of passing through the evaporator, the refrigerant will absorb heat from the surrounding cabinet air and will gradually

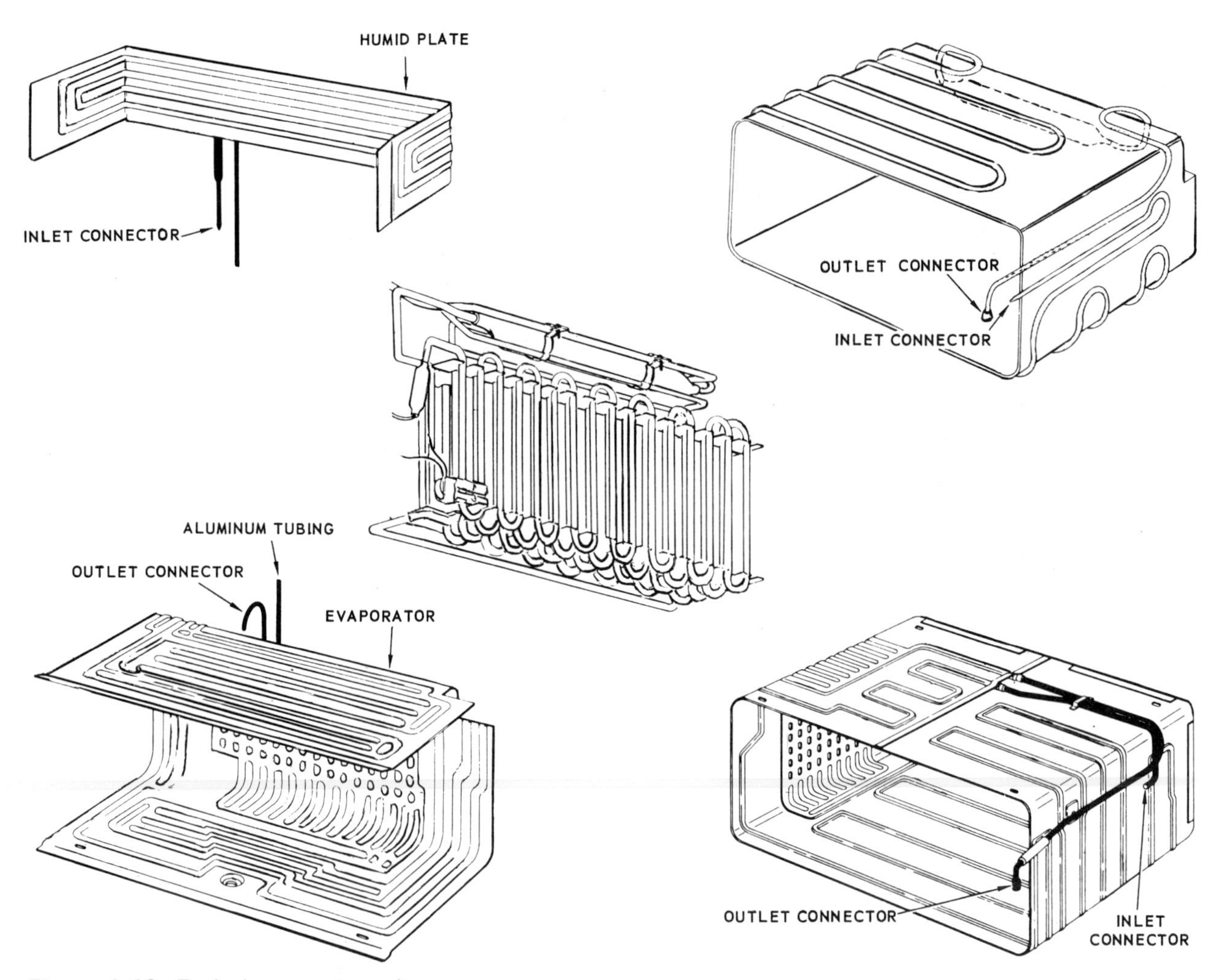

Figure 1-12. Typical evaporator units.

change from a liquid to a mixture of liquid and vapor, and finally to a vapor.

Additional Components

The previous discussion described the basic or fundamental components necessary for any refrigeration system. Other components are sometimes added for various reasons. They include:

1. Filter-driers
2. Heat exchangers
3. Accumulators
4. Precoolers

Filter-driers. A filter-drier, sometimes a drier-strainer, is located in the liquid line either at the inlet or outlet end of the condenser. Its purpose is to filter, trap minute particles of foreign materials, and absorb any moisture which may be in the system. Fine mesh screens filter out foreign particles, and the desiccant absorbs the moisture.

Each time the refrigerant system is entered, the filter-drier *must* be replaced. If the system did not contain a filter-drier originally, one *must* be installed. The filter-drier, as already mentioned, may be found (or may be installed) in either the high or low side of the system. However, the drying agent used will be different, depending on which side the filter-drier is installed.

Heat exchangers. The heat exchanger is formed by soldering a portion of the capillary tube to the suction line. The purpose of the heat

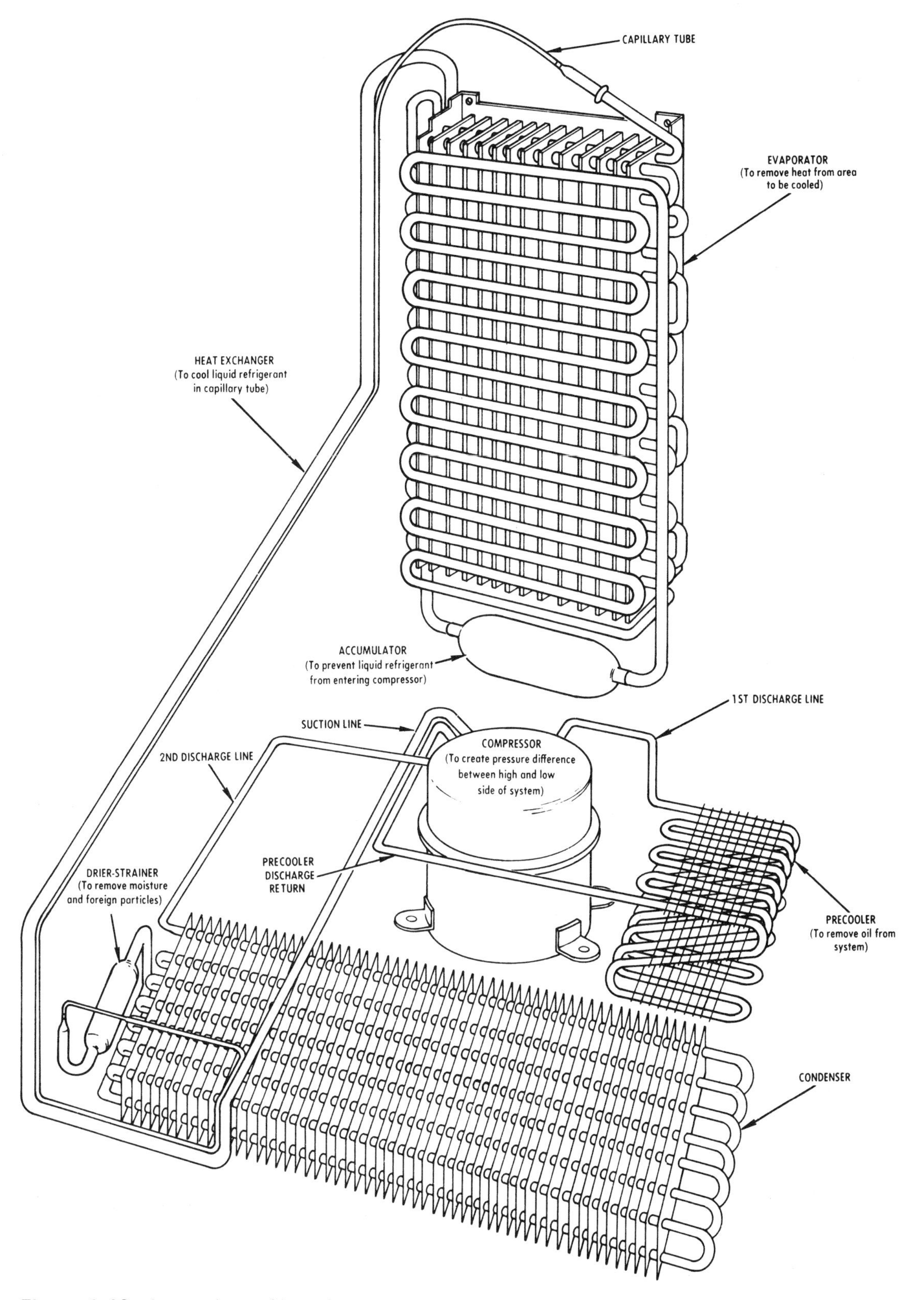

Figure 1-13. A complete refrigeration system.

exchanger is to increase the overall efficiency of the system by using the cold suction gas leaving the evaporator to cool the warm liquid refrigerant passing through the capillary tube to the evaporator. If the hot liquid refrigerant from the condenser were permitted to flow uncooled into the evaporator, part of the refrigerating effect of the refrigerant in the evaporator would have to be utilized to cool the incoming hot liquid down to evaporator temperature.

Accumulators. The accumulator is added as part of the evaporator in some systems, and is a large cylindrical vessel designed to hold any liquid refrigerant which may not have changed to a vapor in the evaporator. It is in this manner that any liquid refrigerant remaining in the low side of the system is prevented from entering the suction line and causing damage to the compressor. The accumulator must therefore be positioned so that the outlet is always higher than the inlet.

Precoolers. With the use of rotary high-side dome compressors, the lubricating oil may become vaporized and leave the compressor as a high-temperature vapor in quantities large enough to either damage the compressor or cause oil restrictions further along in other parts of the system. To minimize the amount of oil circulated through the system, a precooler, or oil-cooler coil, is installed. As a rule, the precooler coil is connected between the first discharge port and the discharge return port on the compressor. Thus, the oil vapor condenses and is returned to the compressor.

Refrigeration Control Devices

In order to be of service, the refrigeration unit must function without attention; that is, it must be made fully automatic in its operation. This is accomplished by the use of certain control devices.

There are two basic categories of control devices—refrigerant and electrical controls. Refrigerant controls regulate the flow of refrigerant within the system. The capillary tube is the most commonly used refrigerant control. Other refrigerant controls that are sometimes used are

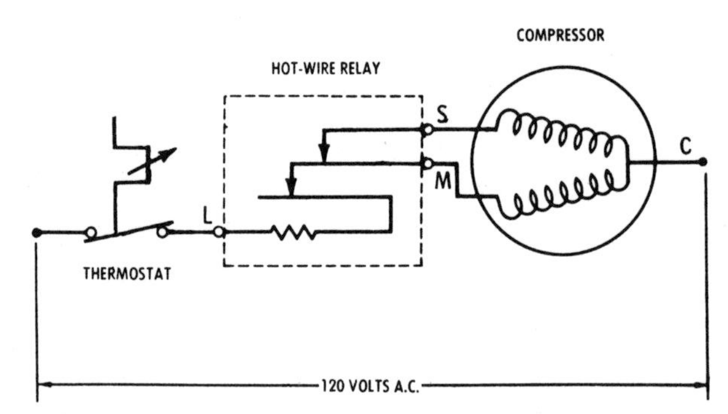

Figure 1-14. A simplified wiring diagram of a refrigerator.

automatic expansion valves and thermostatic expansion valves. These valves, like the capillary tube, are merely refrigerant-metering devices. Their use in recent times has decreased because they rely on mechanical action which can be a source of trouble. A discussion of these valves will not be made because of their limited use. Instead, we shall concern ourselves here with only the various electrical controls.

Electrical control devices can all be considered switches, and consist of two major categories:

1. Operating controls
2. Protective controls

Operating controls consist of:

1. Thermostats
2. Relays
3. Defrost controls
4. Humidistats

Protective controls consist primarily of thermal cut-outs.

The electrical wiring may differ considerably from unit to unit, depending on the protective feature of the motor-compressor as well as the defrosting circuit or circuits. All refrigerator systems are made up of a wiring harness, which includes the line cord, compressor leads, light switch, control leads, and heater-wire leads (where used). In their simplest form, the electric components consist of the following:

1. Temperature control (thermostat)
2. Motor-compressor
3. Relay

Thermostats. The most common method of temperature control presently employed in household refrigeration units is thermostatic control. There are two basic types of thermostats: gas-bulb and bimetal thermostats. With the gas-bulb type, a highly expansive gas is sealed into a metallic bulb which is fastened to the evaporator and connected to a bellows by means of a small-diameter tube (Fig. 1-15). The bulb and tube are charged with a highly volatile (readily vaporizable) fluid. As the temperature of the bulb increases, gas pressure in the bulb-bellows assembly also increases, and the bellows pushes the operating shaft against the spring pressure. The shaft operates the toggle or snap mechanism. Consequently, the travel of the shaft finally pushes the toggle mechanism off-center, and the switch snaps closed, starting the motor-compressor. As the motor-compressor runs, the control bulb is cooled (owing to the lowering temperature of the evaporator), gradually reducing the pressure in the bulb-bellows assembly. This reduction of pressure in the bellows allows the spring to push the shaft until it has finally traveled far enough to push the toggle mechanism off-center in the opposite direction, snapping the switch open and stopping the motor-compressor. The control bulb then slowly warms up until the motor-compressor again starts, and the cycle repeats itself. The temperature at which the switch is turned on and off is adjustable by means of a selector knob which applies a greater or lesser force against the bellows, changing the effect of the internal pressure existing in the bellows. This selector knob controls the cabinet temperature and is adjusted by the user.

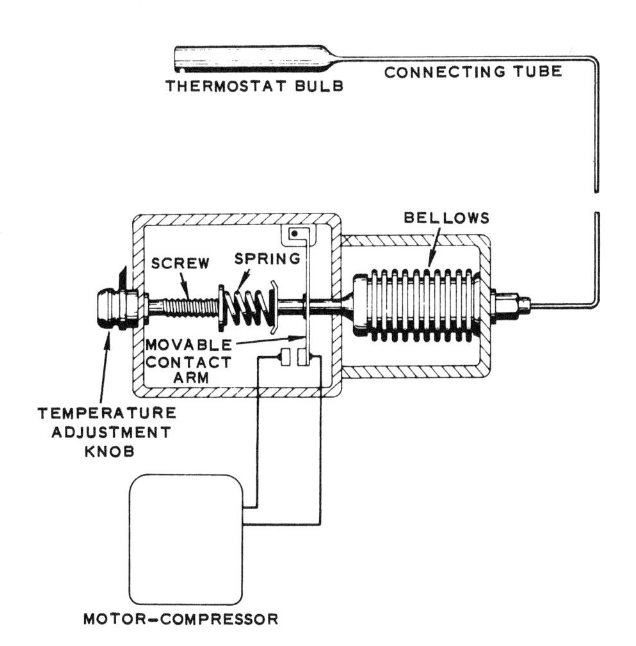

Figure 1-15. A typical refrigerator thermostat assembly.

The bimetal type of thermostat, as described in Chap. 6 of *Basics of Electric Appliance Servicing*, operates by the expansion and contraction of metals due to temperature changes.

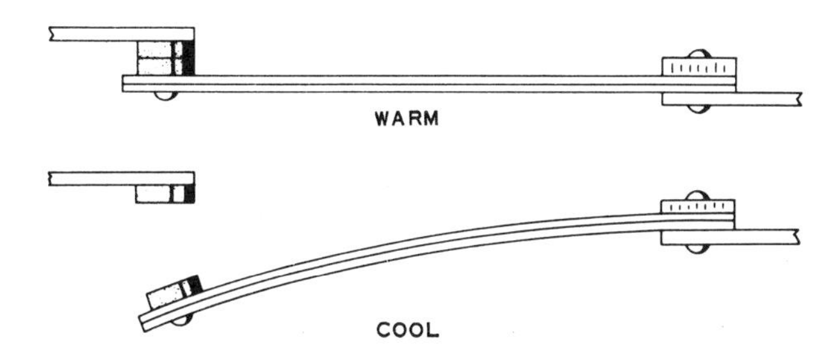

Figure 1-16. A simple bimetal thermostatic control.

Two metals, having different coefficients of expansion, are welded together to form a bimetallic blade or unit, and this blade is securely anchored at one end. An increase in the temperature of the surrounding air will cause the metals in the blade to expand, but at different rates (Fig. 1-16). This difference in rates causes the blade to bend or warp, closing the contact points and turning on the motor-compressor of the refrigerating unit. As the surrounding air cools, because of the refrigerating action of the system, the blade warps and opens the contacts, thus stopping the motor compressor. This cycle repeats itself over and over, and thus prevents the temperature from rising or dropping too far from the desired setting. Bimetal thermostats are usually designed for control of heating and cooling in air-conditioning units, refrigeration storage rooms, etc.

Dehumidifiers use a control device, called a

humidistat, to start and stop the compressor, just as a thermostat starts and stops the compressor in a refrigerator. The difference, however, is that the humidistat senses moisture content in the air rather than temperature (see Humidistat in Chap. 5).

Motor-compressor. As stated earlier in the chapter, the motor-compressor is usually a split-phase induction-type motor, the basic type used for most fractional horsepower appliances. It has the advantage of relatively low cost, and its principle of operation is simple. It has two stator windings, a running winding and a starting winding.

The start winding is made into two or four coils in series, depending on the motor speed. These windings are physically and electrically positioned several degrees from the running winding. A smaller wire is used in the start winding; however, it has more turns than the running winding. Because of the counter-emf in the greater number of coils in the start winding, the current will build up slower; consequently, the magnetic effect of these windings will be retarded in relation to the running windings. This will cause a torque on the rotor which causes it to start turning. The start winding is disconnected as soon as the motor reaches approximately 75 percent of its running speed. Remember that split-phase motors have relatively low starting torque; therefore, the high- and low-side system pressures must be equalized before the motor-compressor will start.

Starting relays. As we know, the motor-compressor employs both a starting and a running winding. The running winding is energized during the complete cycle of operation, whereas the starting winding is energized only during the starting period.

The starting relay is designed to remove the starting (phase) winding and/or the start capacitor from the circuit after the motor has reached a predetermined speed. If these circuits are not removed when the motor reaches approximately three-fourths running speed, the motor will draw excessive current and could be damaged. The three basic types of relays in use today are

1. Hot wire
2. Current-operated
3. Potential (voltage-operated)

The *hot-wire relay* is generally used in low-horsepower applications. The current drawn by both the run and start windings passes through a resistance-type wire (see Fig. 1-14). As the current draw increases, the wire expands, allowing one set of contacts to open. If, for any reason, the current draw is excessive, the wire continues to expand, causing both sets of contacts to open. Of the three types of relays in use today, the hot-wire type is the only one that furnishes built-in overload protection.

The *current-operated type of relay* has a coil connected in series with the run winding of the compressor, as can be seen in Fig. 1-17. When the compressor attempts to start, it draws heavy current through both the run winding and the relay coil. This strong current flow through the relay coil creates a magnetic field strong enough to cause the start contacts to lift and close, energizing the start winding in the compressor. When the motor reaches approximately three-fourths running speed, the current flow through

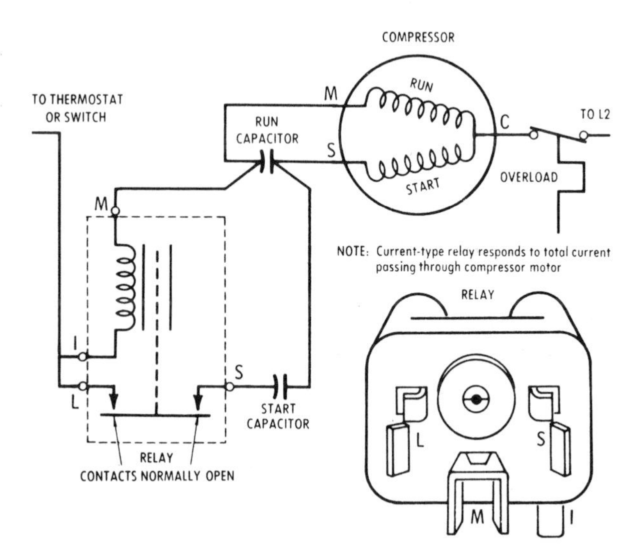

Figure 1-17. Current-operated relay connections.

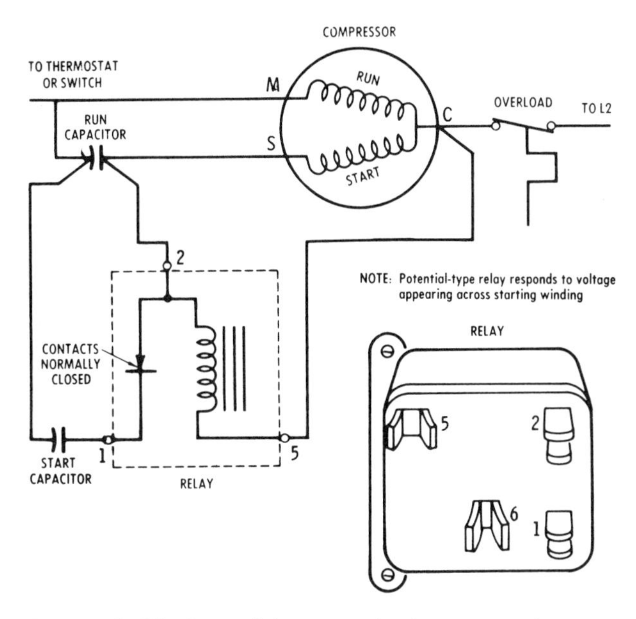

Figure 1-18. Potential-operated relay connections.

the relay coil decreases (because of back-emf in the motor) and the start contacts fall open. *Note:* The current-operated relay must be used with an overload protector and must be mounted in an upright position so that the contacts can fall freely to the open position.

The *potential-operated relay* is generally used in high-horsepower applications where the need for greater power and starting torque exists. Note the two capacitors in Fig. 1-18. The start capacitor is of a large value while the run capacitor is smaller in value. The relay coil is wired in parallel with the start winding of the compressor, and its contacts are normally closed. When voltage is applied to the circuit, the voltage across the relay coil is low, and the relay contacts stay closed. Current flows through both the start and run capacitors, and a high starting torque is developed in the motor. As the motor picks up speed, the voltage seen by the relay coil increases, causing a larger current to flow through it. When the motor reaches approximately three-fourths running speed, sufficient current will flow through the relay coil causing the magnetic field developed to open the start contacts. This removes the start capacitor from the circuit. The relay contacts remain open throughout the rest of the running cycle. When the motor is started, the relay contacts are closed, and open only after the high current of starting has passed. This action extends the life of the starting relay contacts. This type of relay must also be mounted in an upright position or it will not function.

The noticeable difference between the current-operated and voltage-operated relays is the wire size used for the coils. Heavy wire is used for the current-operated relay because all the current flowing through the run winding must flow through it. Fine wire is used for the voltage-operated type because the current that flows through it is small by comparison.

Automatic defrost controls. During normal operation, moisture will collect on the surface of the evaporator and freeze (from frost), impairing the efficiency of the system. Various automatic defrost methods have been developed to remove this frost, most of which use electric heaters to melt the frost accumulation. Two of the more commonly used methods are the cycle defrost and the timed defrost.

The cycle-defrost system (Fig. 1-19) makes use of a wide-differential thermostat which has a cut-in point (contacts closed, compressor running) above freezing. The above-freezing cut-in point allows the evaporator to completely de-

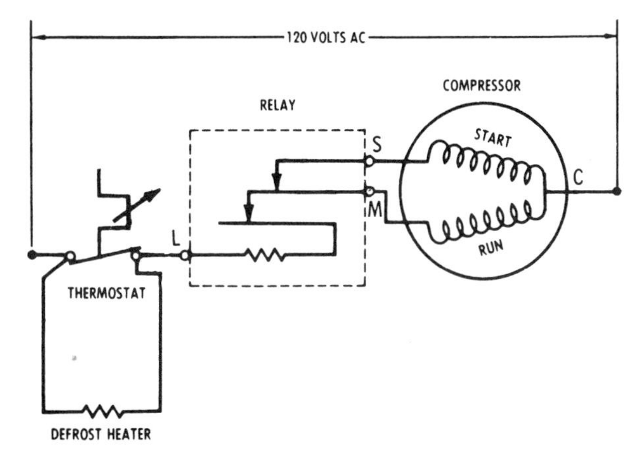

Figure 1-19. A wiring diagram of a cycle-defrost system.

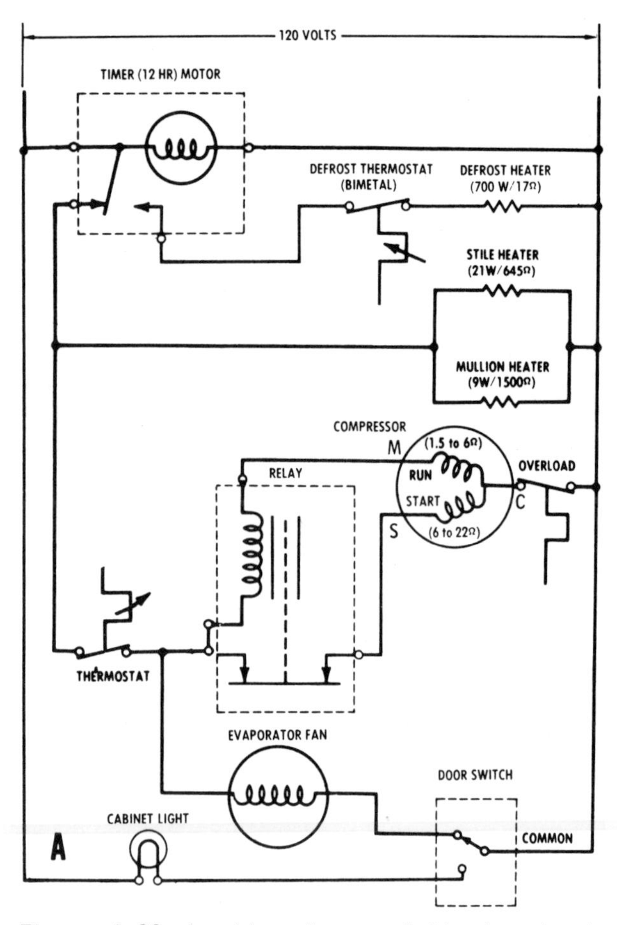

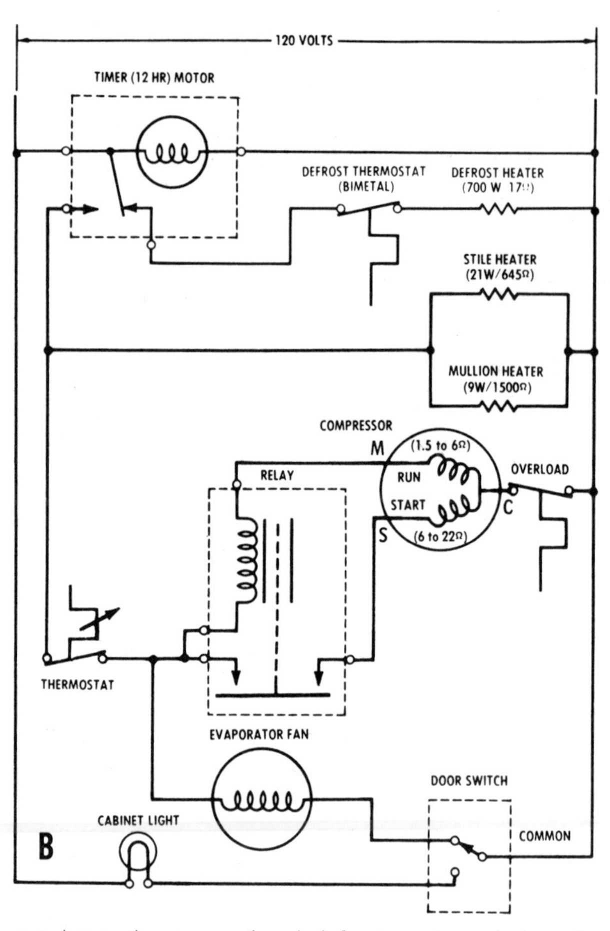

Figure 1-20. A wiring diagram (left) of a timed-defrost system during the cooling cycle, while at the right we have the same timed-defrost system during the defrost cycle.

frost before the next running cycle. The defrost heater (which is in direct contact with the evaporator) is wired across (in parallel with) the thermostat contacts so that when the contacts are open, the defrost heater operates. The heater operates on every off cycle of the compressor.

The most popular defrost system uses a clock timer to initiate the defrost cycle. The number of defrost periods varies from one to four in 24 h, depending on the particular timer used. The timer contacts initiate either the defrost cycle or the cooling cycle. When the timer is in the cooling cycle, the thermostat controls the OFF-ON periods of the compressor. Note in Fig. 1-20 that the stile and mullion heaters are energized during this time, preventing frost buildup in those areas. Note that the fan inside the cabinet is also running to circulate the cabinet air. If the cabinet door is opened during this period, the fan will shut off and the interior light will turn on.

When the timer is in the defrost cycle, the thermostat cannot turn the compressor on. In other words, the thermostat has no control of the compressor when the defrost timer is in the defrost position.

The defrost cycle terminates approximately 20 min after being turned on. The defrost heater is wired in series with a bimetal thermostat whose contacts will open at some predetermined temperature, thereby disconnecting the heater. The length of time it takes for the contacts of the bimetal thermostat to open is determined by the amount of frost on the evaporator. The time remaining on the defrost timer (before the compressor can be turned on) ensures complete removal of moisture from the evaporator.

Start and run capacitors. Start capacitors are generally in the high-capacity (high microfarad) range and are used as a form of booster to start the motor. They are wired in series with the start windings. Since these are high-capacity dry-type electrolytics, they must not be left in the circuit more than 5 s. If left connected for longer periods, they will rapidly become overheated, with damage or failure the possible result.

Run capacitors are generally of the the low-capacity type and are oil-filled. They also serve as an aid in starting the motor, but are left in the circuit permanently to improve the efficiency of operation of the motor. Although called run capacitors, they are connected in series with the start windings.

Some start and run capacitors have a high-value resistor connected across their terminals. This is a bleed resistor, and its purpose is to remove any residual charge that may be left in the capacitor after the unit shuts off. Many run capacitors have a built-in fuse and must be discharged through a resistor to prevent damage to the capacitor. *Note*: The term *condenser* is often used in place of *capacitor*; both terms mean the same thing. When used in this context, *condenser* does not mean the condenser through which refrigerant flows.

Overload protectors. In series with the motor windings is a separate bimetallic overload protector, held in place on the compressor by a spring clip. The short wire lead on the overload protector connects to the common terminal on the compressor. Should the current in the motor windings increase to a dangerous point, the heat developed by a passage of current through the bimetallic disk will cause it to deflect and open the contacts. This breaks the circuit to the motor windings and stops the motor before the damage can occur. The dome-mounted overload protector provides added protection for the compressor motor because, in addition to protecting against excessive current, it also protects against excessive temperature rise.

After an overload or a temperature rise has caused the overload protector to break the circuit, the bimetallic disk cools and returns the contacts to the closed position. The time required for the overload switch to reset varies with room temperature and compressor dome temperature. The overload protector is specifically designed with the proper electrical characteristics for the compressor motor and its application. Any replacement must be exact.

REFRIGERANT CYCLE

A thorough understanding of the cycle of operation, that is, what takes place inside a refrigerator, is necessary before a correct diagnosis of any service problems can be made. Only by a thorough study of the fundamentals will one be able to master the field of refrigeration.

A cycle, by definition, is an interval or period of time occupied by one round or course of events in the same order or series. The domestic refrigerating system embodies a perfect cycle of operation. The refrigerant in the system is continually changing its physical form from liquid to gas in the evaporator, and from gas to liquid in the condenser. The total heat added to the refrigerant in the evaporator, plus the heat added during compression, which is mechanical energy converted into heat energy, is exactly equivalent to the amount of heat removed from the refrigerant in the condenser by the room air. Let us review the various components of a refrigerator system and see how they perform in the total refrigerant cycle.

Since the evaporator is the place where the desired refrigeration effect takes place, it is logical that we start at this point to describe the path of refrigerant flow through the system. You should keep in mind that heat transfer is always from the higher temperature level to the lower temperature level, and that heat travels from the food compartment to the evaporator surface by natural-convection air currents set up in the food compartment, or by forced convection by a fan. It then travels by conduction through the evaporator shell or tubes to the refrigerant.

If a transfer of heat is to take place between the air passing over the evaporator surface and the refrigerant in the evaporator shell or tubes, a temperature difference between the air and the refrigerant must be created. When the system is not in operation, the temperature and pressure of the liquid refrigerant in the evaporator are determined by the temperature of the air surrounding the evaporator. For example, if the room temperature is 70°F, the food compartment air will be at 70°F and the refrigerant in the evaporator will be at 70°F.

The outlet of the evaporator tubing usually terminates in an "accumulator" or chamber, in which the vaporized liquid refrigerant collects, and from the top of which the gas can be drawn off by the compressor pump, through the suction line joining the accumulator with the suction side of the pump.

When the compressor pump starts, some R-12 gas is removed from the evaporator shell or tubes, causing a lower pressure in the evaporator. As the pressure is lowered, the temperature and boiling point also are lowered, creating a temperature difference between the refrigerant and the cabinet air. This temperature difference starts a flow of heat from the cabinet air, surrounding the evaporator, to the liquid refrigerant. Continuous action by the compressor, in drawing off gas, maintains the low pressure and low boiling temperature in the evaporator and continuous boiling of the liquid refrigerant. Heat continues to flow into the refrigerant from the cabinet air. The latent heat of vaporization of the refrigerant is utilized. For every ounce of refrigerant changed to a gas, a definite number of heat units, or Btu, are taken out of the food compartment.

To evaporate or boil the greatest quantity of liquid refrigerant in the evaporator, it is necessary to keep all the interior surfaces of the evaporator shell or tubes wet with liquid refrigerant. Doing so provides maximum evaporator surface efficiency and classifies the evaporator as a "flooded" type.

Vaporized refrigerant leaves the evaporator at saturation temperature, which is the same temperature as the liquid in the evaporator. Some drops of liquid may be entrained in the gas leaving the evaporator. These drops of liquid vaporize in the suction line, making the suction line cold. This serves no useful purpose except to help the service technician determine the proper refrigerant charge in the system. The technician can feel the suction line and tell whether liquid is flooding back. An excessive quantity of liquid refrigerant entrained in the gas leaving the evaporator will lower the capacity of the compressor pump, in pounds of gas handled. Certainly, any quantity of entrained liquid should be sufficiently small to evaporate before the suction line leaves the insulated cabinet.

The suction gas, after leaving the evaporator, enters the heat exchanger where the cold suction gas comes in contact with the warm liquid refrigerant from the condenser. Since the liquid is much higher in temperature than the gas, there is a transfer of heat from the liquid to the gas. The transfer of heat from the liquid refrigerant to the suction gas gives two results: (1) It cools the warm liquid before it enters the evaporator, and (2) it further superheats the suction gas before it enters the compressor.

Cooling of the liquid is very advantageous because the cooler it is upon entering the evaporator, the smaller the loss of refrigeration required to cool the remainder of the liquid down to boiling temperature in the evaporator. The lost refrigeration consists of the "flash gas," or that portion of the liquid entering the evaporator which flashes immediately into gas, pulling heat out of the remainder of the liquid, to cool the remainder of the liquid down to boiling temperature. The net result is increased efficiency, or refrigerating capacity, of the system.

The superheated suction gas leaves the heat exchanger and flows through the suction line into the compressor dome. The compressor pump (a reciprocating type is used in this example) draws the gas off the upper portion of the dome where it will be most free of oil and liquid-refrigerant particles.

The gas enters the compressor cylinder on the suction stroke, through the compressor suction intake valve. When the piston reaches the end of the suction stroke and starts on the compression stroke, the suction intake valve closes, and the piston starts compressing the gas. As the piston continues its compression stroke, the pressure in the cylinder increases until it is greater than the pressure in the discharge line. The compressor exhaust valve then opens and the compressed gas flows into the discharge line. When the piston has reached the end of its compression stroke, the exhaust valve closes to seal off the discharge line from the cylinder.

The discharge gas passes through the discharge line to the condenser. This gas is highly superheated, and its density has been raised by compression. In other words, it is now a high-pressure, high-temperature gas.

The purpose of the condenser, as previously detailed, is to remove sufficient heat to condense the high-pressure, high-temperature gas into a liquid. To remove heat from the gas, its temperature must be raised higher than the temperature of the cooling medium, room air. This is accomplished by the compressor, which continues to pump gas into the condenser under high pressure and density. Each ounce of gas contains a definite number of heat units. As more ounces, and, therefore, more heat units are added to the fixed volume of gas in the condenser, the temperature of the gas increases until there is sufficient temperature difference between it and the cooling medium, room air, to establish a flow of heat from the gas to the room air. The farther the superheated gas travels through the condenser tubes, the more it is cooled by the room air, until finally its saturation temperature is reached. Further heat removal then results in the gas changing into a liquid. The liquid, still under high pressure, passes through the condenser outlet line to the small-diameter liquid line, or capillary tube. Just prior to entering this tube, the liquid passes through a refrigerant strainer to remove any free particles of scale, dirt, or other foreign matter that might plug the small opening in the capillary tube.

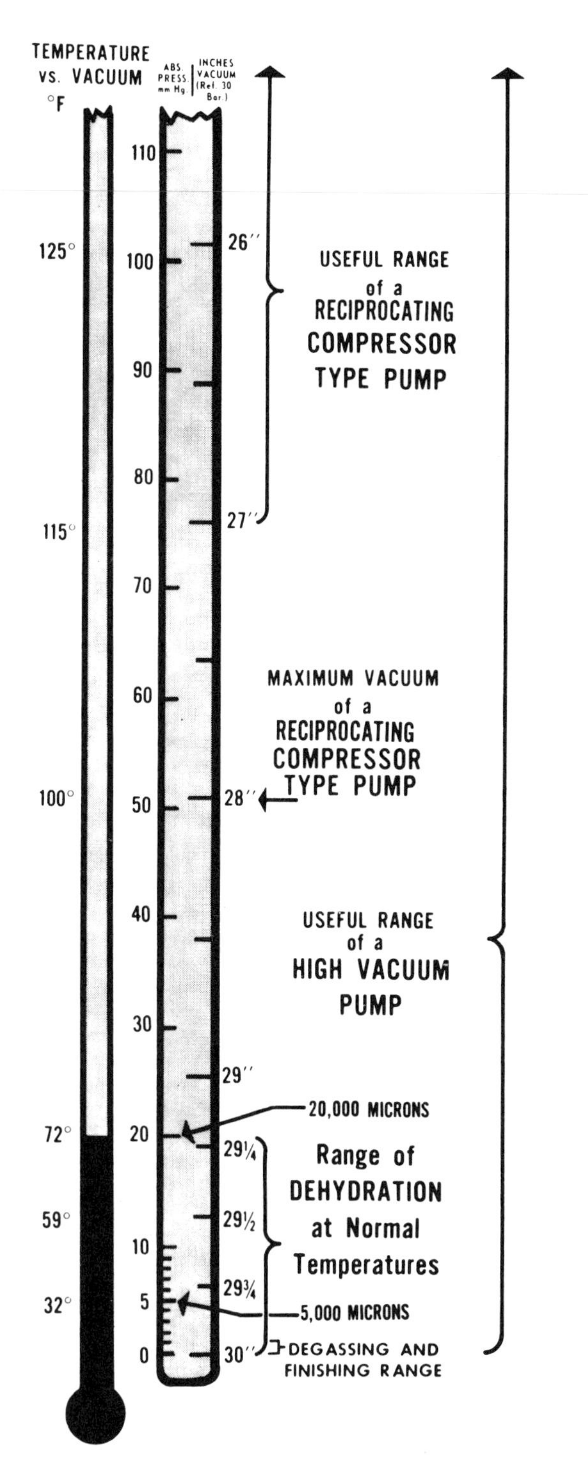

Figure 1-21. Emphasis in refrigerator servicing must be placed on the boiling of water in a vacuum, a process called *dehydration*. Any moisture that is found in a refrigeration system will cause troubles, either as corrosion, freeze-up, or both. For this reason a system must be completely dehydrated before putting in fresh refrigerants.

The high-pressure liquid enters the capillary tube and flows through it into the evaporator. In flowing through the capillary tube, its pressure is gradually reduced until it emerges from the capillary tube at evaporator suction pressure. The drop in pressure occurring in the capillary tube is due to several factors. The small inside diameter of the capillary tube causes frictional resistance to the flow of liquid. The length of the tube, the viscosity of the liquid, and the pressure difference between the ends of the tube are other factors. These factors are used in determining the size and length of the tube to be used.

Because the capillary tube is soldered to the suction line to form a heat exchanger, the liquid refrigerant in the capillary tube is subcooled by the cool suction gas. The subcooling of the liquid reduces its heat content, and therefore, its temperature—before it enters the evaporator. Thus, the cycle is completed, with the refrigerant liquid back in the evaporator, at low boiling temperature, ready to take on a new quantity of heat and start repetition of the complete cycle. This cycle is continuous as long as the compressor is in operation.

SERVICE TOOLS

Refrigeration servicing requires some special tools. Here are the more important ones needed:

High-vacuum pump. A good high-vacuum pump is essential in the repair of a refrigeration system. There are three functions in which it is needed.

1. To remove old refrigerant from the refrigeration system
2. To remove air and other noncondensible gases from the system
3. To remove moisture after the repair work has been completed

If air and moisture are left in the sealed refrigeration system during repairs, they will increase the condensing temperatures. When this happens, a chemical reaction of the moisture, oil, and refrigerant becomes accelerated, and finally oxidation or breakdown of the oil occurs, resulting in carbon deposits throughout the sealed system.

The presence of moisture, even without air, causes harmful acids to form from reaction of the moisture with the refrigerant. Most refrigeration repair shops have proper equipment for evacuating refrigeration systems. They have learned from costly experience that a contaminated system not only results in unpaid (repeat) service calls but also creates unsatisfied customers, who in turn condemn the shop's workmanship as well as the product itself.

The minimum requirement of any vacuum pump used in refrigeration is that it must be capable of obtaining 500 microns (μm) (29.98 in of vacuum). Because of this, we do not recommend using a rebuilt rotary-type refrigerator as a vacuum pump, which in all too many cases would do good to pull 28 in of vacuum. Remember that water boils at 212°F under atmospheric pressure at sea level; if we reduce the pressure on this water, it will boil at a lower temperature. For example, at 28 in vacuum water boils at 101°F; bring the vacuum to 29.98 in, and water will then boil at 80°F. By the way, a reciprocating-type compressor is less capable than the rotary-type of pulling at least 29.8 in of vacuum and should never be considered a vacuum pump.

Because the vacuum pump is used to remove dirt, water, and acids from the refrigeration system, these foreign materials could contaminate the pump, and in time may impair its efficiency. To make certain that this does not occur, the oil in the vacuum pump should be changed at frequent intervals. Always change the oil after evacuating a burned-out or wet system. This oil change should be made according to manufacturer's directions. Incidentally, a burned-out system is one in which the insulation on the compressor motor has charred because of a short-circuit or an overload. Since the motor windings are in the same housing as the refrigerant, the burned material will contaminate the refrigerant and oil in the system. On the other hand, a wet system is one in which moisture-laden air has entered, usually through the low side because it is under vacuum. The moisture contaminates the refrigerant and oil in the system.

Vacuum gauge. A good high-vacuum pump is of little use if you do not have and use a good vacuum gauge, capable of reading a high vacuum. The average compound (bordon type) service gauge is not accurate, and the scale is too small to read.

A gauge rugged enough for field use and yet very accurate is a thermocouple vacuum gauge, which reads in microns. To use it, follow the manufacturer's instructions most carefully.

Equally as accurate, but costing less, is the closed-end mercury manometer with scales reading in inches and tenths of an inch (Fig. 1-22). Properly used, the manometer will indicate whether a leak or moisture exists in the system. If there is a *leak* in the system, the manometer reading will continue to change until it reaches atmospheric pressure. If there is *moisture* in the system, the reading will change only to a point, and then it will stabilize.

Vacuum in inches of mercury is determined by noting the difference between the mercury level in the two tubes, then subtracting this difference from 30 (considered to be a perfect vacuum). Using the drawing here as an example, note that the level on the left side (the closed end) is higher than the level on the right. This should always be the case when a reading is taken. In our example, the left side reads 0.3 in above zero and the right side 0.3 in below zero. The difference, therefore, is 0.6 in. Subtracting this difference from 30 in, we have a vacuum reading of 29.4 in of mercury. As shown in Fig. 1-22, the top of each mercury column is actually convex; this can lead to an inaccurate reading. Read either the top or bottom of the hump, but be certain to read both columns in the same way.

Care of the manometer requires that the storage or shipping carton should be used to keep the mercury in the tube during shipment or when not in use. It is made up of a rod threaded on one end, a nut, a sleeve, and a rubber seal. The keeper is inserted into the open tube until the seal almost touches the mercury. When the nut is tightened, it forces the sleeve down against the rubber seal. The rubber will flatten out against the sides of the tube and seal the mercury in. This keeper must be removed before the manometer can be used. When in use, the manometer must be in an upright position to give accurate readings.

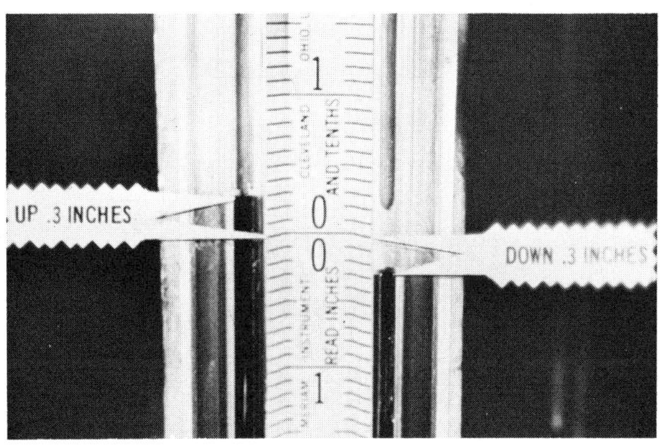

Figure 1-22. Typical mercury manometer (top) and detailed view of the manometer scale (bottom).

Gauge manifolds. The gauge manifold (Fig. 1-23) is the common connection point for the hoses from the sealed system under repair, and for the hose to the refrigerant supply and the vacuum pump. The hand valves open or close the valve ports leading to the hoses directly below them.

The unit on the left is a compound gauge;

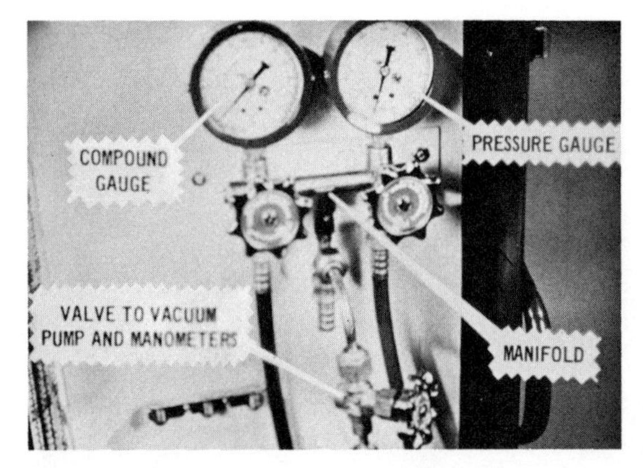

Figure 1-23. Typical gauge manifold.

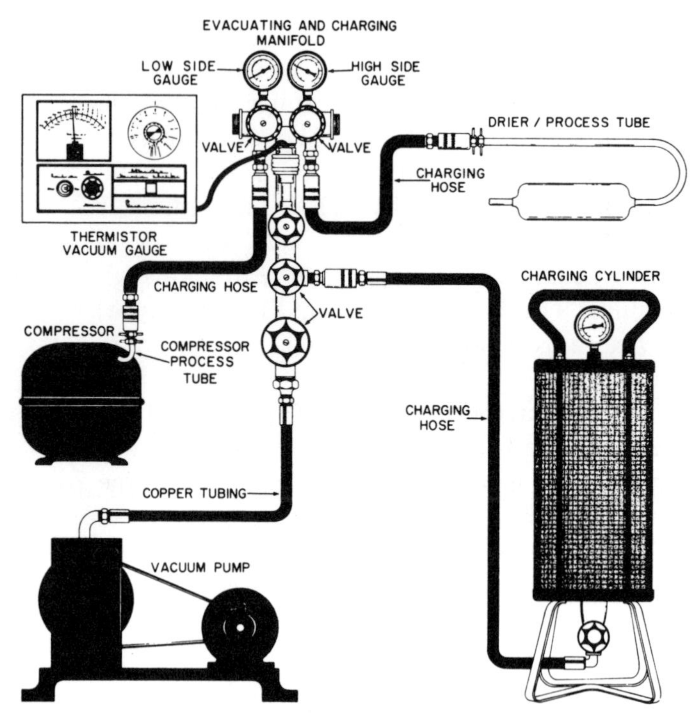

Figure 1-24. An illustration showing the evacuation and charging system connections.

that is, it will read either vacuum or pressure. The gauge on the right side will read pressure only. Each gauge will read only the pressure in the outlet hose directly below it. A bypass port around the hand valve is built into the manifold body. The valve merely opens (or closes) a passageway between the hose and the common manifold.

To properly connect the gauges to a system requiring repair, the hose leading from the compound gauge (left-hand side) should always be connected to the low side of the system at the suction line where either pressure or vacuum can be read. The hose from the pressure gauge on the right side of the manifold should always be connected to the high side of the system at the second discharge line from the compressor where only pressure will be read.

Calibrated charging cylinder. Most of our present-day hermetic systems use capillary tubes for control of the refrigerant. This is a good feature that prolongs the life of the complete system, but these systems require closer tolerance of the refrigerant charge. Therefore, the use of a calibrated-volume charging cylinder (Fig. 1-25) is the most popular method of charging refrigerant into hermetic systems. A must, however, is a cylinder with a feature to compensate for variation in volume because of a difference in pressure and temperature of refrigerant.

Most calibrated charging cylinders (often called a "dial-a-charge cylinder") comprise a

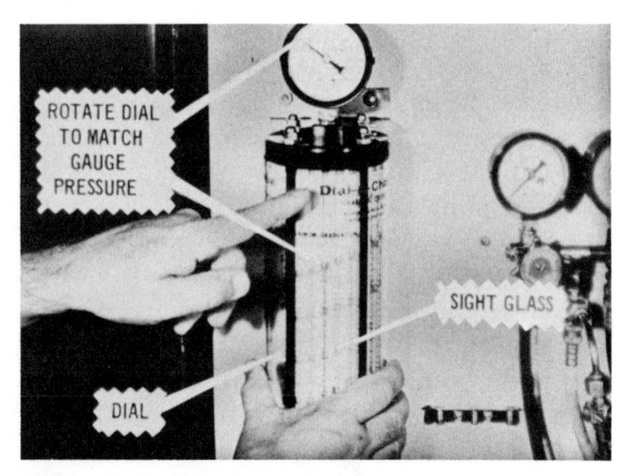

Figure 1-25. Typical calibrated charging cylinder.

tank equipped with a vertical sight glass for determining the liquid level in the cylinder. A plastic dial encircling the assembly is imprinted with graduation marks calibrated in ounces. The volume of refrigerant in the tank will change with a change in temperature. Temperature determines pressure, so note the pressure of the gauge on the tank. Then by rotating the calibrated dial until the proper refrigerant scale corresponding to the pressure is lined up with the sight glass, the liquid level can be converted into ounces of refrigerant in the tank.

To fill the charging cylinder, attach a hose from the refrigerant supply drum or cylinder to the valve. (Always purge air from the hose by passing some refrigerant vapors through it before making the last fitting tight.) Invert the supply drum and open both the supply valve and the cylinder valve. (The reason the supply drum or cylinder should be inverted is to be sure that liquid, not vapor, enters the cylinder.) Watch the liquid level as it fills. Always fill the cylinder with 8 to 10 oz more than you need for the repair, but not beyond the top calibration mark. If the refrigerant ceases to flow into the cylinder because of pressure inside, release the vapor pressure by operating the valve at the top of the charging cylinder. Replace the cap after bleeding.

After charging the cylinder, shut off the cylinder valve and place the supply drum upright. Grasp the charging hose near the cylinder valve and slide your hand toward the drum. The heat from your hand will help drive liquid in the hose back into the drum. Do this several times and then close the supply valve. Now, when you carefully disconnect the hose, there will be a minimum of lost liquid. When the job is completed, be sure to place the supply drum or cylinder upright.

Refrigerant hoses. Refrigerant hoses for evacuation and recharging of sealed systems must be in good condition and leak-tight. Some hoses come equipped with valve core depressors. Hose adaptors with core depressors are available for use with hoses not equipped with depressors. Rubber seal rings are inserted in the ends of hoses to provide a good seal. They sometimes become damaged and need replacement. You should always carry some spare seals in your toolbox. Connections should be made finger-tight. Do not tighten with pliers.

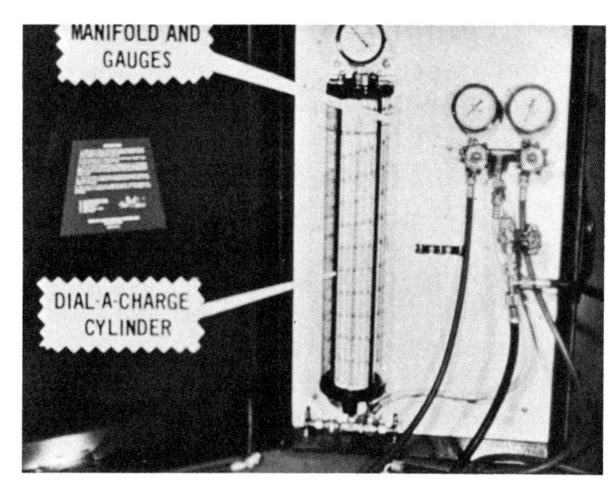

Figure 1-26. Typical charging board which contains the basic charging equipment used in a sealed system repair.

Brazing equipment. The term *hermetically sealed* describes a system that is welded together. Equipment must be available to open and reseal the refrigerant circuit. This requires the hot flame produced by oxygen and acetylene brazing and welding torch. Temperatures developed by Prest-O-Lite and propane-type torches are not adequate to provide the concentration of heat necessary to make a leakproof joint.

All solder joints, except for one possible exception, are silver-brazed. More on silver brazing is given later in this chapter.

Access valves. Access valves, often called *piercing valves*, have been developed which can easily be installed on either copper or steel tubing. They can be installed almost anywhere on the system and without loss of refrigerant even if the system is under pressure. After installation, they become a permanent part of the system. The valve consists of a saddle nut, line-piercing valve body, valve core, and sealing cap. For entering a system, the saddle nut is silver-brazed to the tube at the point of entry.

Saddle nuts come in several sizes to fit the tube to be tapped. Normally, one is placed on the high side of the system, usually on the second discharge line, and one on the suction line on the low side of the system. The line-piercing valve body with the core in it is then screwed into the saddle nut. As the body is screwed into the saddle nut, the hardened point cuts a hole in the tube while, at the same time, the tapered body presses against the sides of the hole to form a seal between the tube and the valve body. The valve is kept closed by the spring-loaded valve core. Saddle nuts are counterbored to resist solder flow into the center opening.

To open the valve, the core stem must be depressed with a depressor-type connector, or removed with a core-removal tool. The core serves only as a temporary stop for servicing purposes. For a permanent seal, the sealing cap must be screwed on finger-tight. These valves should be located so that the following requirements can be met.

1. Heat of brazing will not affect existing joints.
2. Core tools can be installed and process hoses connected.
3. Capped valve will not cause interference problems when the unit is reinstalled in the cabinet.

Caution: When installing access valves on a system charged with refrigerant, locate the saddle nuts a safe distance from existing brazed joints and apply only enough heat to flow silver solder when brazing the nuts to the tubing. Excessive heat may cause rupture of tubing or an existing joint, resulting in spewing of oil and refrigerant.

Valve kits (Fig. 1-27) are usually available for four sizes of tubing: $\frac{1}{4}$, $\frac{5}{16}$, $\frac{3}{8}$, and $\frac{1}{2}$ in. The technician should stock a supply of kits. They are not included in compressor mounting or repair kits and must be ordered separately. Two kits will be needed for all sealed system entries except for some room air conditioners.

Valve-core removal tool. The access-valve-core remover and replacer tool is recommended for opening and closing the access valve. Two are required, one for the low side and one for the high side of the system. This tool is equipped with a female fitting that screws onto the access-valve body. It opens the valve by unscrewing the core and withdrawing it from the valve body by means of a knob on a keyed shaft. That is, the long stem engages the valve core, unscrews it, and pulls the valve core back into the body of the tool. Then the full flow of vapors can pass unobstructed through the access valve for faster evacuation and charging.

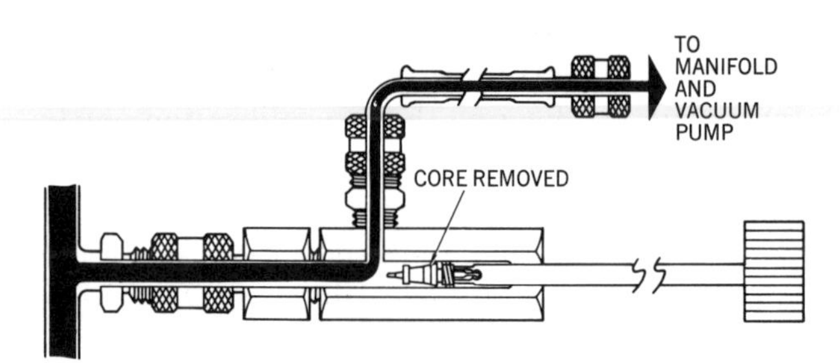

Figure 1-28. A cross-sectional view of a core-removal tool installed on an access valve.

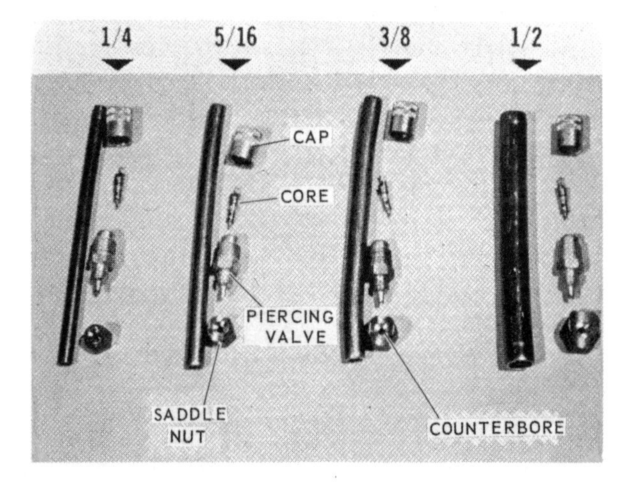

Figure 1-27. The contents of a piercing valve kit.

To close the valve, insert the core into the valve body and screw it into place. This tool must be kept clean and oiled. With most core tools, the stem seal is lubricated with a wick. A few drops of refrigeration oil in the wick periodically will keep the seal leakproof. Keep a supply of replacement O rings handy in order to replace the damaged rings when necessary.

If access-valve removal tools are not available, a hand valve on the high-pressure connection is suggested. It is equipped with a valve depressor. The hand valve may be closed at the end of the charging operation for clearing refrigerant liquid out of the charging hoses.

Leak detectors. A good leak detector is essential in order to ensure that a tight system will be held for many years after the repair has been made. Every system should be checked *before* and *after* final charging. There are three common methods of ensuring a leakproof system.

The most common leak detector in use is the halide torch. When properly used and adjusted, it will detect a refrigerant leak of 1 oz/year. This torch, using acetylene or propane fuel, burns with a clear blue flame until the sensing hose passes the area of a refrigerant leak. A small leak causes the blue flame to turn green; a large leak causes the flame to turn bright purple.

Electronic leak detectors are frequently used in leak testing. They are very sensitive and will detect a leak as small as $\frac{1}{2}$ oz/year. Always follow the instructions from the manufacturer of the tester being used. The tip of the probe should be placed slightly below, or as near as possible to, the suspected leak. Allow 3 to 5 s for the unit to react.

In recent years, lightweight, battery-operated, transistorized electronic leak detectors have entered the market. These gun-type detectors need less warming up and control adjustment, but they do not have quite the sensitivity of standard electronic detectors. When employing the battery-operated-gun detector, follow the manufacturer's directions for usage.

Neither the halide torch nor the electronic leak detector will operate in the presence of urethane foam insulation. The expander material in foam insulation contains some of the ingredients of refrigerants. The leak detector will sense it and react as it would with a leak.

Where urethane foam insulation is present in the refrigeration unit, leak testing is done with a soap bubble solution. This solution is sold commercially or can be made up with liquid dishwashing detergent. Simply apply the solution to the area being tested and watch for bubbles to form. When testing is completed, remove all bubble solution from the tubing. If this is not done, corrosion will occur. By the way, the bubble method of leak detection is not restricted to foamed-urethane areas.

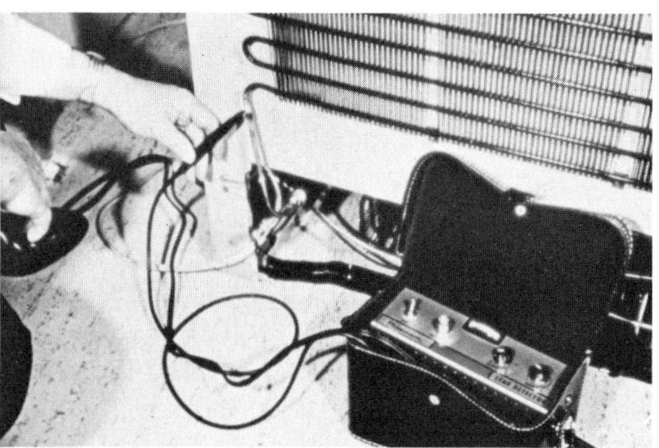

Figure 1-29. A propane or halide torch (top), an electronic leak detector (center), and transistorized leak detector gun (bottom).

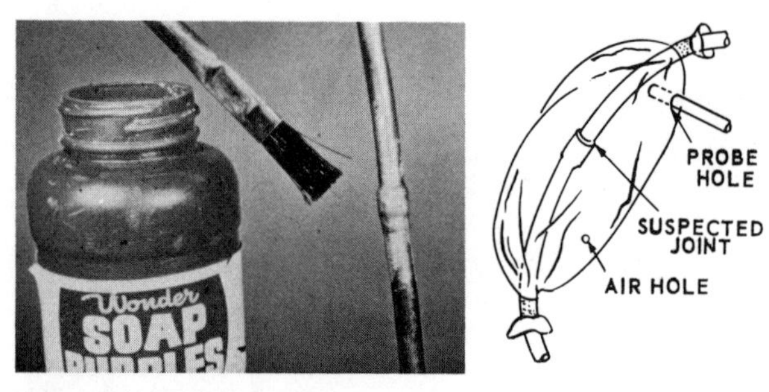

Figure 1-30. Detecting a leak by applying a soap solution (left), and by using a leak detection envelope (right).

Test-charging hose. A convenient way to leak-test a unit that has lost its charge is to introduce a test charge of refrigerant gas into the high and low sides of the system directly from the refrigerant supply cylinder or drum by means of a test-charging hose (Fig. 1-31). Such a hose can be made by assembling three hoses together by connecting one end of each to a common T fitting. To use the assembly, connect the other end of one hose to the outlet valve of a refrigerant supply cylinder. Connect one of the other hoses to the high-side and one to the low-side access valves on the unit. With the supply cylinder sitting with the outlet valve up, open the cylinder valve and both access valves. This will charge both the high and low sides of the unit with gas for leak testing. The valves should be left open while testing to maintain gas pressure in the system.

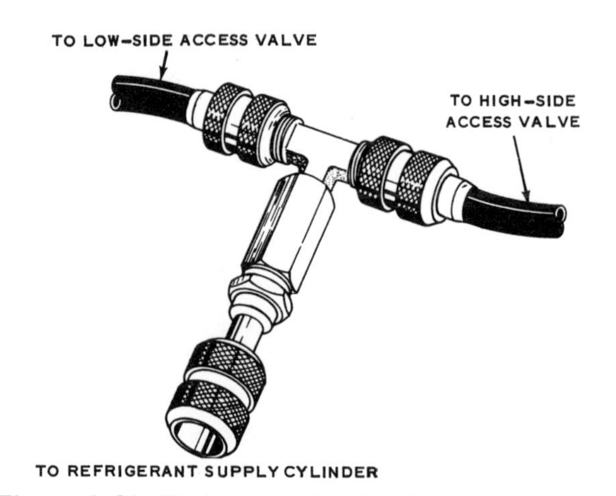

Figure 1-31. Typical test-charging hose setup.

Heat gun. The heat gun (Fig. 1-32) is an excellent source of heat for speeding up the evacuation process, particularly in the case of a system into which moisture has entered (wet system). This gun produces a large volume of air heated to temperatures over 500°F. A 250-W heat lamp equipped with spring clamp may be used instead, but more time is required. With either method, care must be exercised in applying heat in the presence of plastic to avoid damage to plastic parts.

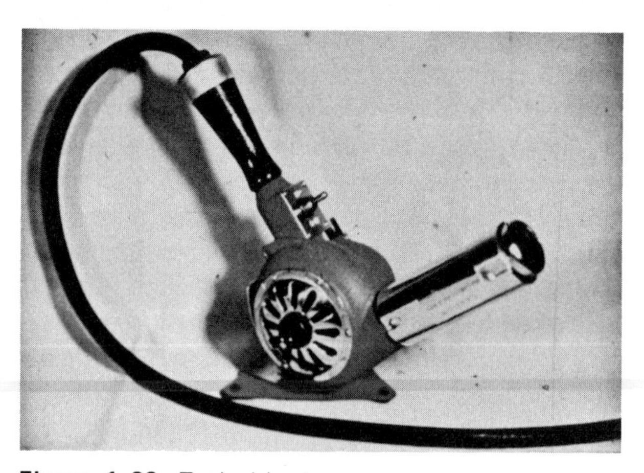

Figure 1-32. Typical heat gun.

Starting test set. When servicing a single-phase unit having a hermetic motor-compressor that will not start, you can save a lot of time and expense by "proving" the compressor. If the compressor motor starts through the test set, then the cause of failure lies in the external electric components. In other words, a starting test (Fig. 1-33) can be used to test a compressor without disconnecting the cabinet wiring.

The starting or wired direct test set can also be used to facilitate dehydration of the compressor by using it to wire a 250-W heat lamp in series with the compressor running winding to generate heat within the compressor (see Fig. 2-41). The heat lamp in series with the running winding reduces the voltage applied to the motor enough that the motor cannot run.

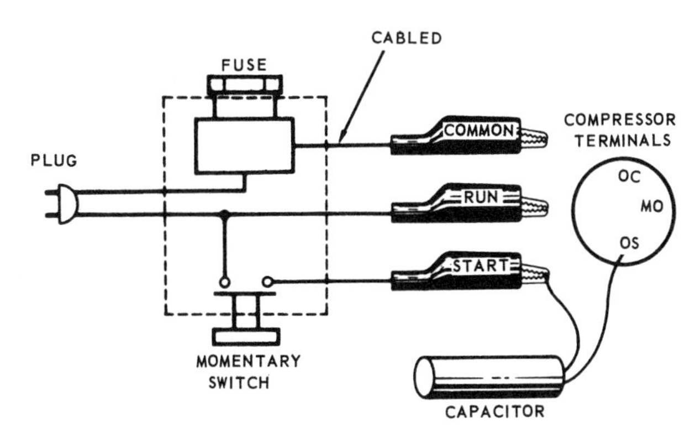

Figure 1-33. Typical starting test set.

However, sufficient current will flow through the motor winding to generate considerable heat, thus driving moisture from the compressor assembly.

Purging hose. A purging hose is used to control refrigerant being removed from a unit. The old refrigerant, if not directed outside the work area, will contaminate the air. This could make leak testing difficult. Also, the refrigerant, while nontoxic to human beings, can be dangerous to small animals, particularly birds. Therefore, 15 to 20 ft of purging hose is necessary so that the refrigerant can be discharged out of doors. When discharging a unit, it is advisable to purge both high and low sides of the system. A convenient way to do this using the test-charging hose is to connect one hose to the high-side and one to the low-side access valves. The third hose is connected to the purging hose leading out of doors. Opening both high- and low-side access valves will permit the unit to discharge down to atmospheric pressure. This will reduce the amount of contaminates in the system which would otherwise be drawn through the vacuum pump and contaminate it.

Tube-working tools. The following tools are needed for cutting, sizing, and forming tubing to do an efficient job of refrigeration system repair:

1. *Tube cutter* capable of cutting copper or steel tubing up to $\frac{1}{2}$ in outside diameter (OD).
2. *Tube bender* for $\frac{1}{4}$, $\frac{5}{16}$, $\frac{3}{8}$, and $\frac{1}{2}$ in OD tubing.
3. *Swaging tool set* consisting of holding block with $\frac{3}{16}$, $\frac{1}{4}$, $\frac{5}{16}$, $\frac{1}{2}$, and $\frac{5}{8}$ in swaging tools.
4. *Triangular file* for scoring capillary tubes.

Oil-changing equipment. Where there may be moisture in the system, changing oil in the unit compressor is recommended. To change compressor oil, a container for catching and measuring the oil taken out of the compressor is needed. A 1-pt plastic measuring cup graduated in ounces and fractions of an ounce will serve the purpose. For recharging the compressor with oil, a 1-pt plastic squeeze bottle with a nozzle that can be inserted into the second discharge return stub is ideal.

Refrigerant supply cylinder. A cylinder of refrigerant approved by the Interstate Commerce Commission is needed for filling the charging cylinder and for use in leak checking. This cylinder or drum should be equipped so that it can be set with the outlet valve down for dispensing liquid, as well as up for dispensing gas. Disposable cylinders or drums in 10-, 20-, 25-, and 30-lb sizes are available and are recommended for ease of handling and as a means of saving both refrigerant and time, as opposed to large cylinders requiring local refrigerant transfer to smaller cylinders for portable use. Incidentally, many technicians now purchase their refrigerant in disposable drums. The drums or cylinders are

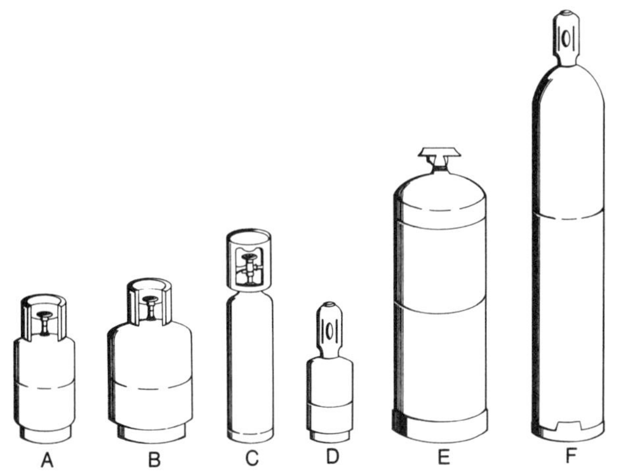

Figure 1-34. Typical refrigerant cylinders: A through D are service cylinders; E is a "throwaway" type; and F is a supply cylinder.

equipped with valve cores. Hand valves are needed for entering the drum or cylinder.

It is of the utmost importance to handle cylinders of refrigerant with care and to observe the following precautions.

1. Never drop cylinders or permit them to strike each other violently.
2. Never use a lifting magnet or a sling (rope or chain) when handling cylinders.
3. When caps are provided for valve protection, keep them on the cylinders except when the cylinders are in actual use.
4. Never overfill cylinders. *Caution*: Never recharge throwaway-type cylinders or drums. To do so may result in serious injury or death. Whenever refrigerant is discharged from or into a cylinder, immediately thereafter weigh the cylinder and record the weight of the refrigerant remaining in the cylinder.
5. Never mix refrigerants in a cylinder.
6. Never use cylinders or drums as rollers, for supports, or for any purpose other than to carry refrigerants.
7. Never tamper with the safety devices in valves or cylinders.
8. Open cylinder or drum valves slowly. Never use wrenches or tools except those provided or approved by the manufacturer.
9. Make sure that the threads on regulators or other unions are the same as those on cylinder valve outlets. Never force connections that do not fit.
10. Regulators and pressure gauges provided for use with a particular gas must not be used on cylinders containing other gases.
11. Never attempt to repair or alter cylinders or valves.
12. Never store cylinders or supply drums near highly flammable substances such as oil, gasoline, or waste.
13. Always wear safety goggles when working with refrigerants.
14. Never heat cylinders or drums with an open flame to remove refrigerant.

SILVER BRAZING AND REFRIGERATION SYSTEMS

Silver brazing is the joining of metals by means of heat, using a filler metal consisting of silver, copper, and a small percentage of other metals. The flow temperature of this filler metal alloy must be above 800°F to classify as silver brazing.

The domestic refrigeration industry in recent years has used silver-based alloys to bond components and tubing. Current regulations governing application and installation codes ban the use of lead- and tin-based "soft solders" in direct contact with a refrigerant. This ban is due to the corrosion, tensile strength, and vibration factors encountered with a refrigeration system. However, one area where soft solder is used currently in the household system is in the bonding of the suction line and the capillary tube to form a heat exchange. Quite frequently a corrosive atmosphere is encountered in the home that accelerates deterioration of the bond between these parts. The current recommendation is to use a soft solder with zero lead content when the heat exchange requires repair. This solder is commonly called 95-5; it has a tin content of 95 percent and antimony content of 5 percent.

When a repair has been made to the heat exchange on a single-coil system, the insulating sleeve must always be replaced, sealed along the length with duct tape or equivalent, and the ends tied to seal the cold surfaces from water vapor. When the soldered area is kept dry, electrolysis is prevented and the bond will not deteriorate.

To achieve the temperatures necessary to flow the silver alloys, acetylene has been employed over the years either in an air-mix torch assembly or, for higher temperatures, in the widely used oxygen-acetylene system. The refrigeration industry has used, as previously mentioned, air-acetylene flames. Generally speaking, this involves techniques developed by Prest-O-Lite, an automotive-based vendor that developed an

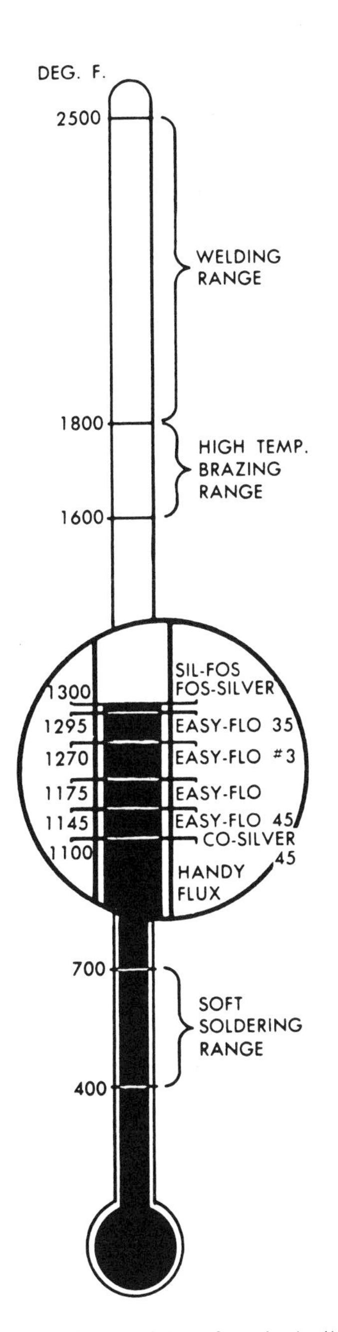

Figure 1-35. Melting points of typical silver solders.

acetylene-air headlight system for the motor car before Kettering developed the battery-powered automotive electric system.

Acetylene gas is very unstable and can explode spontaneously at pressures above 15 lb/in^2. This is the reason many low-pressure gauges on acetylene regulators are "red lined" above this pressure. The simpler single-stage regulators as used on Prest-O-Lite systems are designed so that peak pressures are below this level.

Anyone who has used this equipment has observed that "full" tank pressures are in the order of 250 lb/in^2. To achieve a volume of gas safely under high pressure, a procedure that involves the absorption of acetylene gas into liquid acetone has been used. The acetylene tank has a nonflammable porous mineral inside that acts as a "sponge" for the acetone. To prevent excessive loss of acetone and thus reach a critical point, the following safety procedures should be employed.

1. Never use an acetylene cylinder when it is lying down; if hauled lying down, always set it upright at least $1\frac{1}{2}$ h before using.
2. Never open a Prest-O-Lite needle valve more than $1\frac{1}{2}$ turns.
3. Do not use high-volume tips on small cylinders since they cannot deliver the volume of gas required safely.
4. Always shut off acetylene cylinders at the tank when operation is finished so that if the regulator leaks, dangerous pressures will not be developed in the hose. The more expensive dual-stage regulators minimize this hazard, but they also should always be shut off after use.
5. Any cylinder under pressure presents a potential rupture hazard, and acetylene cylinders that are subjected to severe shock can spontaneously explode. They should always be transported with safety caps in place (where provided) and secured in such a manner that they will not roll or turn over. Never subject them to excessive temperatures.

While discussing cylinders under pressure, a discussion of oxygen is also in order. Cylinders of oxygen, when "full," carry pressures in the 2,000-lb/in^2 range. Should the valve stem be broken, these cylinders become missiles capable of rupturing masonry walls. The lowering of ignition temperatures due to high oxygen content of the atmosphere upon tank rupture can also lead to severe fires. This factor has resulted in severe accidents. The sign "No Smoking" around an oxygen tent in a hospital is an indication of this danger.

Procedure for a good brazed joint. Since most service work involves the repair of an existing system, certain procedures should be followed to ensure good results and safe conditions.

1. Always be aware of surrounding activity. Do not open a refrigeration system in the presence of a gas flame. When Freon vapors pass through a flame, hydrochloric acid fumes are produced with high corrosive effect on any materials contacted. These fumes have been erroneously labeled *phosgene*. Phosgene gas, which has no odor, is produced under these conditions in rather low volume with the end result that the acid fumes act as a warning agent to the operator in a manner similar to the way acrilan acts in regard to natural gas.
2. Always provide for adequate ventilation. Oxygen from the air is consumed by the torch, and normal fumes of brazing are toxic to a degree.
3. Plan the procedure you will follow when installing a component or effecting a repair. Whenever possible do not remake an "old" joint. Have all tools and materials on hand before you start.
4. Clean the area around the joint to be made before opening a system. The metals to be joined should be thoroughly cleaned by using steel wool, wire brush, or a "hard" abrasive. Sandpaper should not be used because of the silicon particles which would fuse as glass when heat was applied. When a repair is involved to an existing system, it is good planning to clean the metal surfaces to be brazed before opening the system. This procedure minimizes the possibility of foreign material and air getting into the system. When you remove oil around the joint, remove it with a solvent such as trichloroethylene.
5. Use diagonals to cut the process tube or connecting lines as a normal procedure so that release of internal pressures can be controlled. This procedure will eliminate most of the oil foaming and oil discharge at the repair area when compared with opening the system with a tubing cutter or tubing saw. *Note*: Where steel process tubes are involved, a tubing saw will give one a little more length of stub to work with.
6. Tubing may be joined using commercial swedging tools, or the next larger size of tubing can be used as a coupling.

The service technician may find that tubing will vary to the extent that proper clearance will not be obtained on a sleeved joint. This is the result of using thinner-walled tubing and the difference in specification between an inside and an outside tolerance. It is good practice to score

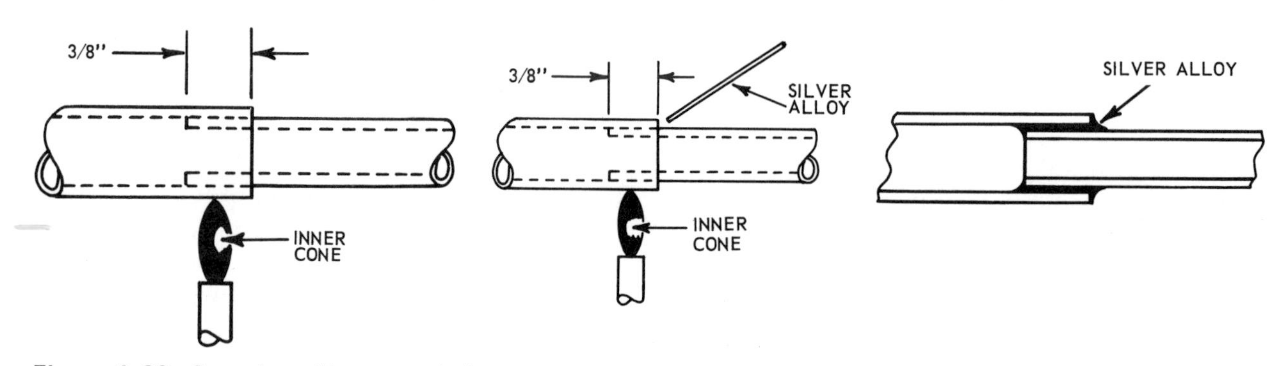

Figure 1-36. Steps in making a good silver solder joint.

tubing with a tubing cutter and then break it rather than cutting it off completely. This minimizes the burr one has to remove before making the joint. The clearance between the two parts to be joined should not be more than 0.005 in and should be uniform; either part being out of round would cause problems. Sizing may be accomplished by using the proper swedge or flare block opening. Punch-type swedging tools are used with a flare block to form swedged joints in the field.

In most cases it will be found that where $\frac{3}{16}$-in OD tubing, for example, is joined together, a short piece of $\frac{1}{4}$-in OD tubing used as a coupling will effect a more satisfactory repair. When a small tube is to be inserted into a larger one (capillary tube), the joint should be sized, as shown in Fig. 1-36, using diagonal pliers. Cleaning has already been mentioned if paint, oxides, etc., were removed previously. Any parts handled in sizing should always be recleaned.

The best known silver alloy solders are Easy-Flo and Sil-Fos. If the service technician makes silver-brazed joints infrequently, the use of Easy-Flo 45 could materially assist in making good joints. There are several characteristics of Easy-Flo 45 that make this so. The flux is a positive indicator as to when the filler rod should be applied to the joint with this material. When the flux makes the transition to a clear liquid, the metal has reached the flow point for Easy-Flo 45. Easy-Flo 45 does not have the characteristic of liquating (the separation of metals as to their different flow points) which is involved with Easy-Flo 35 and Sil-Fos. The size of the rod or wire alloy should be no larger than $\frac{1}{16}$ in in diameter; $\frac{3}{64}$ in or smaller would be preferable in refrigeration repair work. The smaller rod size does not "cool" the metal to be joined when it melts as is the case where the rod has a greater mass. This factor alone would have a tendency to eliminate overheating problems and reduce the amount of metal used per joint. The thin edge of the mechanical joint can pull the smaller rod into it by capillary action, eliminating to a great extent the flow of the filler metal outside the joint. Incidentally, never use phosphor silver alloys in the 5-percent silver range since this material produces essentially a "cast" type of joint that is subject to fracture by vibration.

Here are the steps necessary to obtain a good brazed joint.

1. Select the proper heat source and size of tip to deliver proper flow temperatures for the joint to be made. Heat requirements are such that the air-acetylene torches produce flame temperatures of 2800°F and can be used to make the joints required for domestic refrigeration repair. The Prest-O-Lite system using either the original tips or the more recently developed Harris automatic ignition torches produce a relatively large secondary flame area. When space is not a problem, the volume of flame acts as a shield from the atmosphere and assists one in producing a good joint. The high air volume designs, such as Uniweld's Twister, produce a hotter, more compact, flame which closely approaches that produced by an oxygen-acetylene torch. Where large tube sizes are involved, as would be encountered when changing air-conditioner compressors, the oxygen-acetylene torch is most often used. Propane household torches normally sold in hardware and discount stores are not recommended, as previously stated, for use as a flame source for silver soldering because the temperature reached is inadequate.
2. Arrange tubing so that one has proper clearance from surrounding objects, anticipate where the flame or heat can cause damage, and direct the flame or protect area with asbestos or "heat sink" as indicated. A strip of highly absorbent cloth soaked with water and applied as a "bandage" is very effective in stopping heat flow down a tube.
3. Wet a shop towel and have it lying close to the working area.
4. If you use Easy-Flo or other similar material, you should use a suitable flux. (*Caution*:

The proper use of any flux in refrigeration applications is very important because of the corrosive nature of the material.) The joint should be completely assembled and the flux applied to the outside of the tube. Should a tubing size of a diameter above $\frac{3}{8}$ in be involved (as in an air conditioner), partially assemble the joint about halfway, coat both parts with flux, and insert to the full depth of the joint. Where possible, rotation of the free end will ensure a more uniform coating of flux.

5. The assembly should always be properly supported and aligned so that stresses set up by brazing temperatures will not cause the joint to move or pull apart.
6. The proper heat source and size of the tip used is dictated by the type and location of the joint to be made. The flame should bring the joint to the flow temperature of the filler metal as quickly as possible while at the same time minimizing the likelihood of flux burnoff or overheating. A "soft" flame slightly reducing in nature is desirable when an oxygen-aceteylene torch is used. A low-velocity flame should be used when possible as it reduces the chances of flux burnoff and overheating. The torch orifice should be sufficiently large to envelop the tube circumference with flame regardless of the heating gas used. A flame of this size will shield the work and reduce the formation of oxides. Start heating the male member of the joint about $\frac{1}{2}$ to 1 in away from the fitting; this will expand the metal for a tighter fit and better filler metal flow later. Use the secondary flame to evaporate the water out of the flux if you use an Easy-Flo rod. In a short while a white powder coating will form. Examine this coating to see that complete coverage of the joint has been achieved. If not, add more flux and repeat the above procedure.
7. Once a coating has been established, increase the temperature of the joint by moving the tip of the primary flame closer to the work, concentrating most of the heat on the female member but sweeping the flame steadily back to the male member to achieve an even temperature in the joint. Avoid overheating the fitting face (female), as the thin edge can be oxidized very quickly. *Caution*: As a general rule, regardless of the alloy or heat source used, be very careful to keep the inner blue cone (center flame) from oxidizing the work or causing flux burnoff. When you are learning, keep the cone at least $\frac{1}{2}$ in away from the joint. After you develop skill, you can hold it so that the tip just touches. As the temperature rises because of this rapid heat flow, the flame should be pulled back so that the secondary flame is used to maintain flow temperature.

   The flame envelope should always surround the joint once it is heated and until such time as you are sure a good seal has been accomplished. The flux will go from the white powder to a milky appearance and then to a clear fluid as the temperature rises (approximately 1100°F). If Easy-Flo or Easy-Flo 45 is used, the "clear" liquid indicates that flow temperature has been reached. Where Easy-Flo 35 or Sil-Fos is the filler metal, an additional temperature rise of 200°F is required to achieve flow temperature. The joint will show red under normal room lighting when proper temperature has been reached. Apply the alloy to the fitting face with a brushing motion to allow you to "feel" for temperature and prevent the rod from sticking to the joint.

   We would recommend that the rod (as shown in Fig. 1-37) be applied to the "back" side of the joint as the molten metal will always flow toward the heat.
8. Once sufficient alloy has been melted into the joint, move the torch toward the base of the fitting and around it in an arc to ensure that good capillary action has been created and the alloy covers the joint area. One will quite frequently apply the alloy before the joint has reached a temperature where it will flow. This situation will

"weld" the rod to the joint; be patient and allow the flame to bring the temperature up to the flow point rather than pull or break the rod.

9. When you are satisfied with the joint, pull the flame away. Cool the tubing and joint down to room temperature with a wet rag, as soon as possible. The longer this metal is exposed to the atmosphere at temperatures in excess of 200°F, the more corrosion will result. In addition, one is likely to be burned by hot tubing as the repair work proceeds. Contrary to some opinions, no adverse effect will be produced with either steel or copper tube joints by rapid cooling.
10. These instructions have been based on the presumption that a system containing Freon is being repaired. Should this be "new" work or a system that has been evacuated, an internal shield should be supplied using a minute flow of Freon or inert gas such as nitrogen to minimize internal oxidation.
11. If a flux has been used, the wet cloth or towel used to cool the joint will remove most of the residue. One should continue mechanical cleaning until all traces of flux are removed to prevent corrosion from creating a leak around the joint area. There will be times when one must break and remake a joint; in this case always flux the area before disconnecting and flux again before remaking. This is true when copper-to-copper joints and Sil-Fos are used which normally would not require a flux.

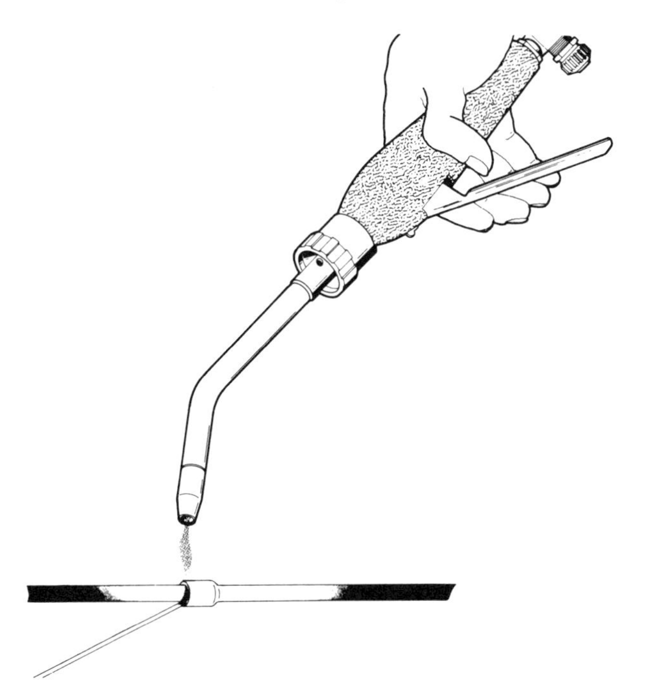

Figure 1-37. Applying a rod to the "back" side of the joint.

While the whys and wherefores of brazing techniques have been discussed in some detail, some additional mention should be made as to the hazards of overheating a silver-based alloy. When the alloy is overheated, the cadmium vaporizes or "boils off," first creating a "brown" smoke which is very poisonous. These fumes, when breathed, can result in internal hemorrhaging and cause a serious health problem. There is no occasion where temperatures in the vaporizing range should ever be reached, and if copper is ever melted at the joint it means you overheated at least 700°F. *Note*: When remaking old joints, always use the same metal rod.

Generally speaking, there are six essentials for making a good silver-brazed joint:

1. Good fit and proper clearance
2. Clean metal
3. Proper fluxing
4. Assembling and support
5. Heating and flowing the alloy
6. Final cleaning

In service work you should also take time to plan in advance the procedures you will use to change or install the individual part. Many times parts have been preassembled into a component and installed in this manner into the final assembly. Because of this factor many times a joint can not be made in the field as it was in production. A window air-conditioner compressor is an example of this. The compressor, condenser, evaporator, and connecting lines are assembled as a "unit"; this component in turn is mounted to the base, and the fans, fan motor, shrouds and controls, etc., are then installed.

This makes the suction and discharge lines on the compressor very difficult to remake with the compressor in place on the base mounting.

We would recommend that extension lines be formed and four joints made so better access can be achieved. Make the suction and discharge connections on the compressor before installation and swedge the discharge and suction lines at a suitable location to achieve easy brazing. Always make vertical joints with the female fitting at the bottom wherever possible. This procedure couples the capillary flow of the molten metal with the flow of gravity.

Second choice would be a horizontal joint; torch patterns previously described would apply and the alloy should be applied to the top face of the joint. Avoid wherever possible a vertical "overhead" joint; clearance of the fitting and the application of the heat are very critical as the flow pattern of the alloy depends entirely on capillary attraction. The torch pattern would be similar to other types of joints previously described. When flow temperature has been reached, all heat must be applied to the upper female fitting, and the alloy rod touched to the fitting face only; otherwise alloy will flow down the outside away from the joint. The heat should be concentrated at the base of the upper fitting and the flame rotated to create capillary flow. With an experienced operator, the probability of leakproof overhead joints will be about 70 percent on the first attempt.

Refrigerators

CHAPTER

2

The most common adaptation of mechanical refrigeration is the household refrigerator. In its usual form, it may be defined as an insulated cabinet, shaped and arranged for the storage of perishable foods, and from which heat is removed from the interior by mechanical means to produce a storage temperature most satisfactory for food preservation and to provide a supply of ice cubes—the entire operation being controlled automatically.

CUSTOMER EDUCATION

As was stated in the other books of this series, when entering the home of a customer on a call, the service technician is usually provided with an excellent chance to answer questions about the operation and use of the appliance. In addition to replying to direct questions, the service technician can offer information which will increase the value of the appliance to the customer, and perhaps avoid a future service call.

The following points of information will give you a good background for dealing with situations which require customer instruction. For instance, if thoughtful consideration is given to the food preservation function of a modern combination refrigerator-freezer, it is not surprising that units run as long as they do. A running period of 75 or 80 percent is not considered excessive. In extremely hot and humid weather, running time will, in some cases, approach 100 percent.

The customer should be diplomatically informed that the refrigerator unit will run more than it will be off, and if food preservation is to be safe and effective, this kind of operation must be considered normal. Also remember that each time a refrigerator door is opened, some of the heavy, cold air slides out of the cabinet onto the floor. This creates a low-pressure area within the cabinet which draws the warm air in the room into the cabinet. The rapid increase of the air temperature inside the cabinet causes the cold control to energize the refrigerator unit. The unit then runs until the cabinet air temperature has been lowered to a degree corresponding to the setting of the cold control. The greater the number of door openings, the longer the unit will operate, and the greater will be the power consumption.

The customer should therefore be advised to keep door openings at an absolute minimum. One way to do this is to remove all the items from the refrigerator at one time when a meal is being prepared, instead of opening the door each time something is needed. Hot, humid weather imposes a greater load upon the refrigerator unit. During humid weather, moisture will accumulate on the cold inner walls of the refrigerator cabinet. This is quite normal, and is similar to the formation of moisture on the outside of a glass which is filled with a cold liquid. Although this condition is normal during periods of high humidity, it is annoying, and can be minimized by opening the refrigerator door less frequently.

All home freezers and combination refrigerators will pull down to 0°F. This can be accomplished by setting the control colder if there are no poor door seals or other maladjustments to impair efficiency. However, it is not mandatory to have 0°F if food is kept for less than a year. Foods can be kept for 3 to 4 months at +10°F without harm and from 6 to 9 months at +3 to +8°F. Even at 0°F or less, food kept more than a year will deteriorate in appearance, taste, and nutritive value.

Good air circulation is also vital to the efficient refrigeration and preservation of food. If the air is restricted from circulating to all parts of the cabinet, the food in the lower area will not be refrigerated sufficiently. It is therefore important that packages and other items of food be placed in the refrigerator in such a way as to allow for sufficient air circulation. Large bulky items, especially square packages, that are pushed against the rear wall of the cabinet, prevent the cold air from circulating downward to the lower shelves and vegetable drawers. Space should be left between packages, and nothing should be placed too close to the cabinet walls. It is especially important that the refrigerator cabinet never be overloaded. In addition, placing hot foods or hot liquids in the refrigerator imposes an excessive load on the unit. Foods and liquids should be cooled to room temperature before they are placed in the refrigerator. Doing this will reduce the operating time of the unit.

Warm-weather operation. During the summer months, the air usually contains a considerable amount of moisture. When the warm, moisture-laden air comes in contact with the cold surfaces of the refrigerator, the moisture condenses. The result is a condition called *sweating*, which affects the inside surfaces of the refrigerator cabinet. This condition is completely normal during warm, humid weather. The extent to which sweating will take place is contingent upon the number of door openings, the length of time during which the door is left open, the temperature of the air outside the refrigerator, and the relative humidity.

A loose door seal will allow warm air to leak into the cabinet and cause excessive sweating during warm weather. If the door seal is good, the customer must be informed that a certain amount of sweating is normal and also be told that the condition can be kept to a minimum by following these recommended procedures.

1. Keep all liquids and moist foods covered. This will prevent the moisture from evaporating and settling on the interior surfaces of the cabinet.
2. Keep the number of door openings to a minimum—it is extremely important.
3. Defrost often. On standard models which do not have the automatic defrosting mechanism, it might be necessary to defrost manually as often as twice a week during warm, humid weather.

During very hot, humid weather, the refrigerator will run longer. If the control knob is set too high under these prevailing conditions, the ON cycle will be too long, and on models having a cooling coil in the fresh-food section, the entire cooling coil will become heavily frosted. Heat cannot penetrate the thick frost surface, so the result of this condition would be high temperatures in the fresh-food compartment. The solution to this problem is to turn the thermostat to a lower setting until the cooling coil has defrosted completely. Then the control knob should be kept at a low setting to allow an OFF cycle. Each ON cycle will begin with a cooling coil free of frost, and better fresh-food temperatures will be obtained.

As a final word on the subject of customer education, it must be pointed out again that it is the responsibility of service technicians to establish and maintain cordial relations with their customers. If they are able to establish pleasant relationships, their associations with their customers will not only be more agreeable, but immeasurably more effective.

GENERAL BREAKDOWN OF REFRIGERATOR SERVICING

Most home refrigeration units manufactured today contain two separate compartments, the refrigerator and the freezer sections. Each of these compartments is heavily insulated from the other, and each operates at a different temperature. The provision or refrigerator section, for instance, is usually maintained at a user-selected temperature which may range from 34 to 42°F, while the temperature in the freezer section is normally kept at approximately 0°F. The freezer compartment may be either above or below the refrigerator or provision section; some larger home refrigeration units have a side-by-side arrangement.

The problem areas that a service technician must be familiar with to repair a home refrigeration unit are as follows:

1. Cabinet design and installation
2. Air-handling system
3. Electric and control system
4. Performance
5. Refrigeration system

CABINET DESIGN AND INSTALLATION

The cabinet is the housing for the space to be refrigerated. It also houses the self-contained refrigerating system. That part of the cabinet surrounding the food compartment is heavily insulated to retard the flow of heat into the compartment. The door to the food compartment is well insulated and is made to close tightly.

Cabinet Doors

Refrigerator and freezer doors consist primarily of an outer door panel, an inner door panel, insulation, and a gasket. The inner door panel is usually held to the outer door panel with screws located under the gasket flange. Retainer strips, when used, hold the door gasket and add stiffness to the edges of the inner door panel. While fiberglass is still the most popular door insulation, some models use a combination of fiberglass and foam slabs. Many larger models now use foam insulation poured into the outer door panel.

The door swings on heavy hinges and is held closed by a latch mechanism. If the door swings to the left, it is called a *left-hand-door refrigerator*. If it swings to the right, it is called a *right-*

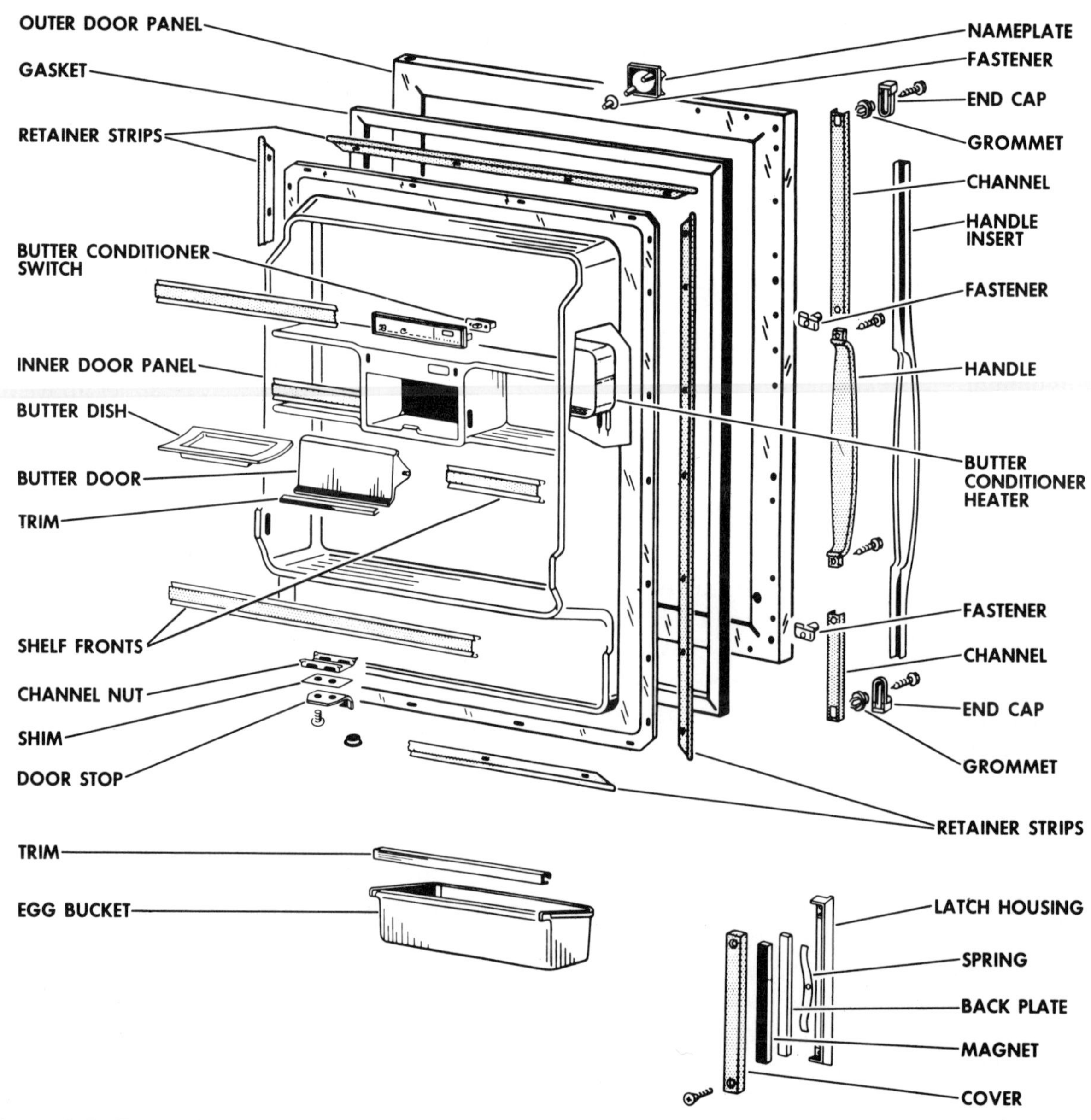

Figure 2-1. Typical food door assembly.

hand-door refrigerator. A door gasket, fitted around the periphery of the door's inner panel, seals against the cabinet when the door is closed, and prevents the infiltration of air into the food compartment. Actually, maintenance of the proper storage temperature is dependent upon the efficiency of the gasket seal. Air leakage will also cause a rapid buildup of frost, condensation, and longer operating cycles.

When servicing the outer door of any refrigerator or freezer, if you have removed any trim, handles, hinges, etc., be sure in reassembly to use permagum or similar material as a sealer around any screws or fasteners which pass through holes in the outer door panel. This will ensure minimum air leakage into the insulation. Here are some general tips on door service problems. For more specific information on this subject, check the service manual for the unit you are servicing.

Door adjustment. The refrigerator or provision compartment and freezer compartment doors can be readily adjusted in the field. A door out of plane with the cabinet can generally be adjusted as follows.

1. Make sure the cabinet is level side-to-side and tilted slightly to the rear.
2. Loosen the screws holding the inner door panel (and the screws at the lower ends of the diagonal cross braces, if used) to allow the inner door panel to "float." *Note*: Failure to loosen all the screws may result in a cracked inner door panel, necessitating a replacement.
3. Pull in the corner or section of the door in the area where the door gasket does not contact the cabinet. While holding the door rigid, tighten the screws at the corners of the inner door panel. Repeat as necessary until the door is in proper plane with the cabinet.
4. Retighten all screws holding the inner door panel. But never overtighten. Overtightening will usually cause the door gasket to dimple.

If the doors are parallel with each other and with the cabinet body of a top freezer model,

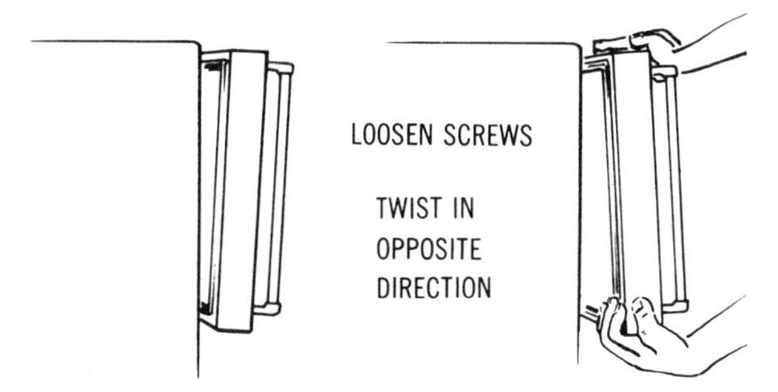

Figure 2-2. Method of adjusting a door out of alignment.

loosen the screws in the top and the center hinges, and shift the hinges to the right or left as necessary to obtain proper alignment. Also if one corner of the door stands away from the face of the cabinet more than the others, the door may be racked or twisted. Loosen the gasket retainer screws and twist the door in the opposite direction to get uniform spacing between the door and the face of the cabinet, then retighten the screws until snug (Fig. 2-2).

Hinge adjustments. In the event a door does not have proper fit, first and foremost, level the cabinet, then check for hinge bind. Hinge bind will cause excessive compression of the door gasket at the hinge side, which prevents the gasket from sealing properly against the pilaster at the handle side. To relieve hinge bind, add shims under the hinges. Also, hinges can usually be adjusted as follows to align the doors with the cabinet.

1. *Vertical adjustment*. The door can be adjusted up or down on most models by adding or removing washers at the bottom hinge pin. *Caution*: Bronze washers must be used (to ensure ground continuity) when an electric circuit is in the door—except on later models where a ground wire is used to ground the door.
2. *Side-to-side adjustment*. Elongated or oversize holes in most hinges permit shifting the door to the left or right after loosening the door hinge bracket screws.
3. *In-and-out adjustment*. Removing and adding shims at the hinge bracket permit in-or-

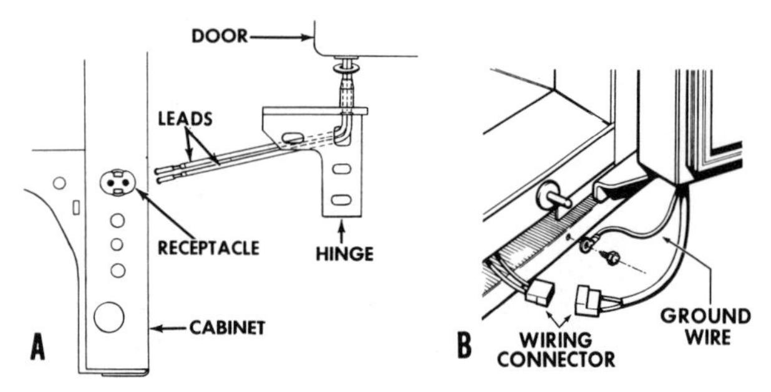

Figure 2-3. Two types of butter conditioner heater disconnects.

out adjustment for most hinges. Some hinges have elongated or oversize holes which allow in-or-out adjustment.

Door removal. When it is necessary to remove the cabinet door from a refrigerator model featuring a heated butter conditioner or some similar feature, disconnect the power cord before disconnecting heater leads. On models having the disconnect receptacle shown in Fig. 2-3A, remove the hinge cover and pull the individual leads from the female receptacle. When reassembling, be sure to push each lead fully into the receptacle.

On models using the wiring connector shown in Fig. 2-3B, pull the wires from under the cabinet until the connector is exposed, then separate the two parts of the connector. If the door has any switch connectors, separate them in the same manner. Once you have disconnected the door switches and heater wires, you can remove the door, in most cases by supporting it and removing the top hinge. Then raise the door to disengage from the bottom hinge.

To reinstall the door, reverse the above procedure.

Changing door swing. Some models are designed so that the door hinging can be reversed to swing either right or left to suit the user's needs. Most refrigerators are shipped with the door hung for a right-hand swing. If they are reversible, pretapped hinge screw holes will be found on the left side of the cabinet. They are sealed with either sealing screws or plastic plugs.

When relocating the doors, the upper one is removed first. Use a knife blade to pry off the plastic cover over the top hinge. Remove the screws attaching the hinge to the top of the cabinet and lift the door off the center hinges (Fig. 2-4A). Be sure to save the nylon spacers.

The lower door is removed after the screws holding the center hinge and shim are unscrewed and the hinges removed. The door is then lifted off the lower hinge pin. Again, save the nylon spacers. They are used to act as bearings as well as to adjust door height (Fig. 2-4B).

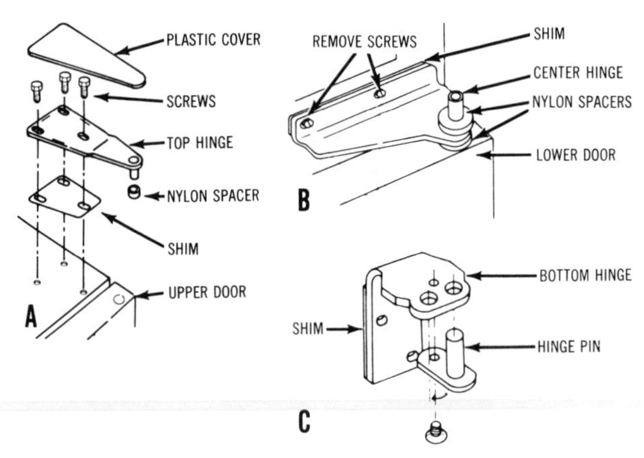

Figure 2-4. Typical (A) left, (B) top right, and (C) center hinges.

While the doors are off, reverse the handles. The screws are usually hidden under covers which must be pried off (Fig. 2-5). Remove the sealing screws or plugs from the right side of the

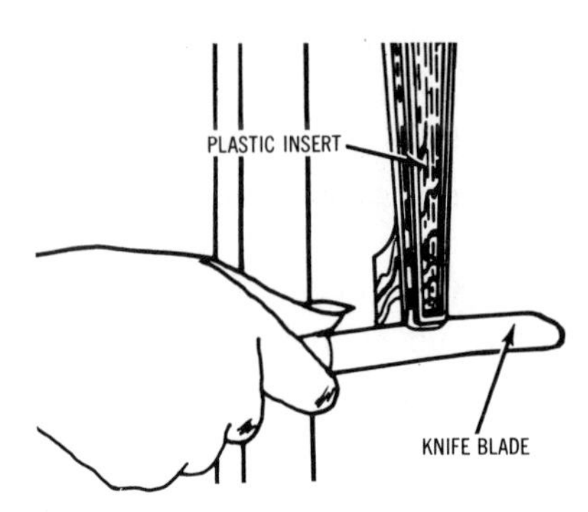

Figure 2-5. Typical method of removing a handle insert.

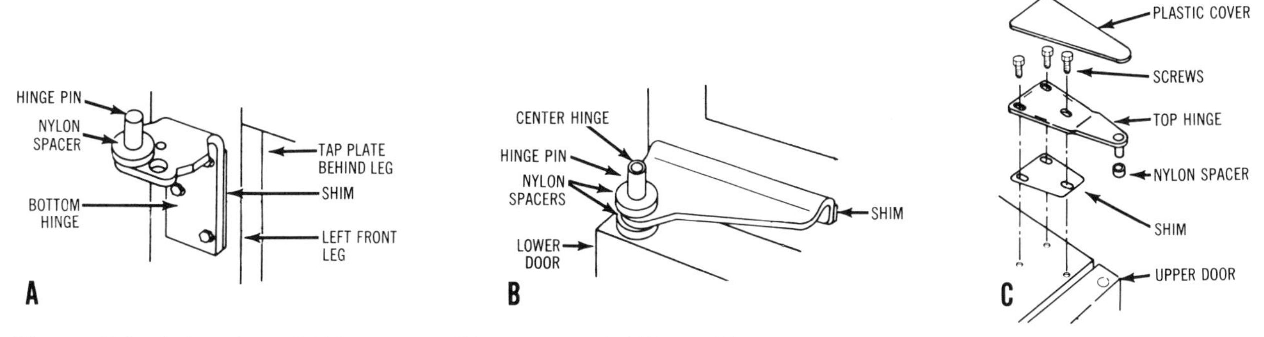

Figure 2-6. Relocation of (A) bottom, (B) center, and (C) top hinges.

door and reinstall them into the old screw holes on the left side. Plugs are also used to conceal the unused hinge pin holes in the doors. When the hinges are reversed, the plugs should be pried out and reinstalled in the holes on the opposite side of the door. All plugs or sealing screws in the cabinet must be removed and reinstalled in the screw holes on the opposite side of the cabinet.

The bottom hinge is generally fastened to the front leg of the cabinet by three screws. These screws are screwed into a tap plate behind the leg. After the toe plate grille is removed, the screws, hinge, and tap plate can be dismounted ready for remounting on the other leg. There are generally locating holes in the hinge for the movable hinge pin (Fig. 2-4C).

The hinge pin, usually held in place by a screw, must be relocated to the opposite hole for the door to hang and swing correctly. Remove the grille filler block from the left front leg and reinstall it on the right leg. It is usually made of plastic and only needs to be pulled straight out for removal and pushed in to install. It fills the opening in the opposite end of the grille not filled by the hinge. Install the bottom hinge to the left leg by holding the tap plate behind the leg and running the screws through the hinge, the shim, the leg, and into the tap (Fig. 2-6A). Use a punch or awl to line up the holes.

With the bottom hinge in place, the lower door can be installed. A nylon spacer should first be placed on the hinge pin and then the door set into position on the pin. The center hinge with its shim and nylon spacer is installed on the cabinet at the top of the lower door and secured by screws (Fig. 2-6B). When the center hinge is reversed, it is rotated 180°.

The top door may be set in position on the

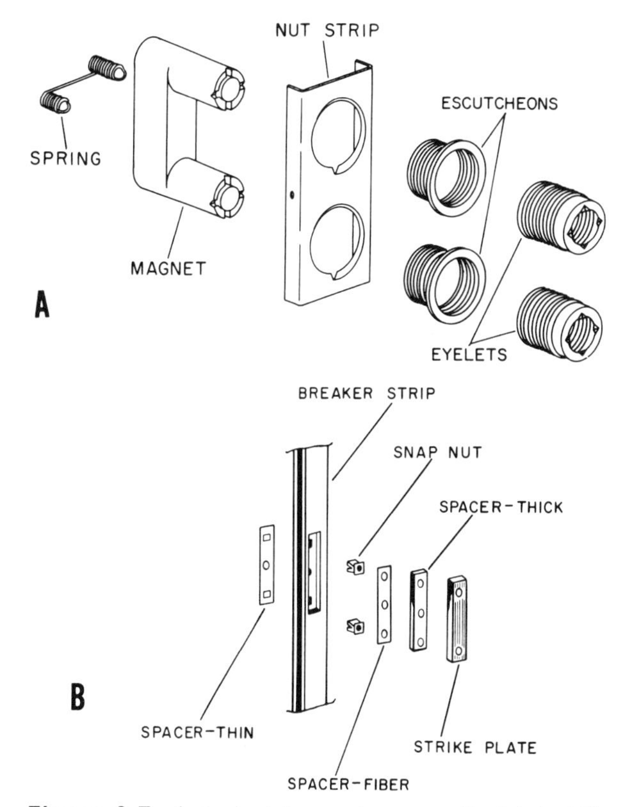

Figure 2-7. A typical two-pole magnetic latch (A). To remove the latch, it is usually necessary to remove the inner door panel. The strike (B) for the two-pole magnetic latch assembly is generally fastened to a side breaker strip.

upper pin and spacer of the center hinge and the top hinge screwed to the cabinet (Fig. 2-6C). When the doors are hung they must be aligned and the door seal adjusted.

Latches and strikes. A federal law, effective in 1958, requires that the force necessary to open the door must not exceed 15 lb along the latch edge of the inside of the closed door. The pull at the door handle should not exceed 14 lb as measured by an accurate scale. The strike and latch of each cabinet are matched to ensure compliance with the legal limit.

The major problem with latches—either magnetic or mechanical—is misalignment. Most latches and strikes are held by screws, and the hole arrangement will usually permit adjustment. Shims can be removed or added under the strike for in-or-out adjustment. Remember that you must ensure a good door seal along the entire length of the door.

Door gasket. Two general types of door gaskets, as shown in Fig. 2-8, are used on refrigerators and upright freezers. The compression gasket (Fig. 2-8B) is used in conjunction with a magnetic or a mechanical door latch. The magnetic gasket is self-sealing against the steel outer case. Some models, however, which use the magnetic gasket also have a mechanical door latch. Most magnetic gaskets have magnetic strips on all four sides. Where the gasket has magnetic strips on only three sides, the side with no magnetic strip is always installed to the hinge side of the door.

On most models, the door gasket is mounted to the door with metal retainer strips. The retainer, secured by screws to the outer door, also holds the inner door panel. To remove a gasket, loosen the screws under the flange of the gasket and pull the gasket out of the retainer strip. When replacing a gasket, note the configuration of the original gasket with respect to the door and install the replacement gasket accordingly. *Note*: The magnetic strip on the hinge side of some door gaskets has less magnetic force than those on the other three sides. The hinge side of the gasket can be identified by ribs on the back of the gasket flange.

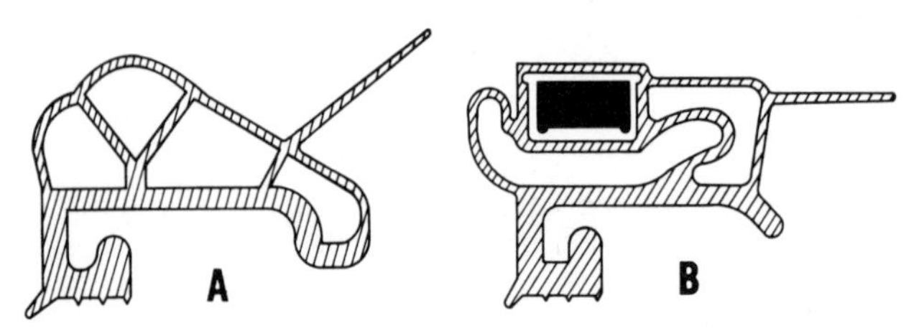

Figure 2-8. Door gasket cross sections.

To check the door seal, proceed as follows.

1. Cut or tear a strip of newspaper about $1\frac{1}{4}$ in wide by 6 in long.
2. Place the strip of paper against the cabinet so the door gasket can be closed against it.
3. Close the door against the strip of paper and pull on the end of the strip of paper. A distinct drag should result.
4. Repeat step 3 at 2- to 3-in intervals around the perimeter of the door.

If a door gasket does not properly seal against the cabinet, check for food packages preventing the door from closing fully, or interference between the inner door panel and breaker strips. Actually, the possibility of an improper door seal with the magnetic-type door gasket is very remote. However, the necessity of a tight door seal cannot be overemphasized. Before any adjustments are made on the cabinet, first and foremost, level the cabinet. A cabinet that is not level can have poor sealing of the door gasket. Poor door seal is, in most cases, caused by hinge bind or improper alignment of the door. If there is insufficient compression of the gasket at the hinged side of the door, or if there is too much compression, adjust the position of the hinge as outlined earlier in this chapter.

There may be instances when the door gaskets will not form a seal on the hinge side of doors immediately after reversing door swing, particularly when gasket temperature is below 75°F, or when the refrigerator is in operation. That portion of the gasket now on the hinge side of the door (formerly on the handle side) has a slightly compressed accordian-fold section and has taken a set. The accordian folds must be relaxed in order for the magnet to pull the gasket over to the cabinet and form a seal. Given a sufficient time, the gasket will relax (expand) and seal. To accelerate expansion of the accordian-fold section, open the doors and pull outward on the gasket progressively from top to bottom of door. Close doors and check for seal. Repeat procedure if necessary. If the gasket does not magnetically form a complete seal within several minutes, warm it with a 150-W lamp or suitable heating device. With doors closed, direct heat on entire length of gasket until seal is formed.

With a good airtight cabinet and gasket seal, and a cool cabinet, it may be observed that, when two door openings are made in quick succession, about 25 to 30 s apart, the second door opening will require a stronger pull on the door handle. This is due to a difference in air pressure between the cabinet interior and the room atmosphere. Opening the door the first time results in spillage of cold air from the cabinet which is replaced by warm air. When the door is closed, this warm air is cooled, reducing its specific volume, thus creating a partial vacuum. A stronger pull on the door handle is required to overcome this pressure difference. Leaving the door closed for 1 to $1\frac{1}{2}$ min will generally allow the air pressures to equalize.

It is important to remember that certain paints and lacquers are not compatible with vinyl gaskets. When refinishing or touching up the finish of a refrigerator, use only a factory-specified paint purchased through franchised parts sources. Using the wrong finish will cause the gasket to deteriorate. It will become sticky and will disintegrate. If this has occurred, the only way to keep replacement gaskets from deteriorating is to remove the improper paint and refinish the cabinet with the factory specified material.

Cabinet Construction

Cabinets of refrigerators and upright freezers consist primarily of the outer case, inner liner, and insulation. The outer case is steel, finished with baked enamel. Outer cases as a rule are not available for replacement.

The liner, on most models, is steel with welded or crimped seams and finished with porcelain or baked enamel. Some models, however, have molded plastic liners. Many combination refrigerators have a separate liner for each compartment, but some later combination models use a single liner with a dividing partition between the freezer and fresh-food compartments. Many models have fiberglass and foam slab insulation, and the liner is fastened to the outer case with screws through nylon spacers and plastic or ceramic supports at the corners. Some models use foam insulation poured between the outer case and liner that bonds the two together. On these models, the liner is usually not available for replacement.

Breaker strips and breaker frames are used around the front edge of the liner to break direct heat transfer from the outer case to the inner liner. Various supports inside the liner are of the 180° twist-locking type or are secured by screws.

Complete disassembly of the cabinet is required if the liner or insulation is replaced. Remove doors, drawers, baskets, shelves, bins, and all other lift-out parts. When replacing the liner, transfer any brackets, braces, nut strips, or supports not provided with the replacement part. Be sure to transfer the rating nameplate when it is affixed to the liner. When servicing the outer case of any refrigerator or freezer, be sure in reassembly to use permagum or similar material as a sealer around any screws or fasteners which pass through holes in the outer case. This will ensure minimum possible leakage of air into the insulation.

Why is the sealing of the cabinet so critical? The refrigerator door is opened and closed several times every day without adversely affecting the cooling function, so why worry about a small air leak through the back of the cabinet? Any air leak is a problem. We must remember that in the event of a door or cabinet air leak, the infiltrating air will travel through the insulation in its effort to reach a cooler area (Fig. 2-10). The air and the moisture vapor it con-

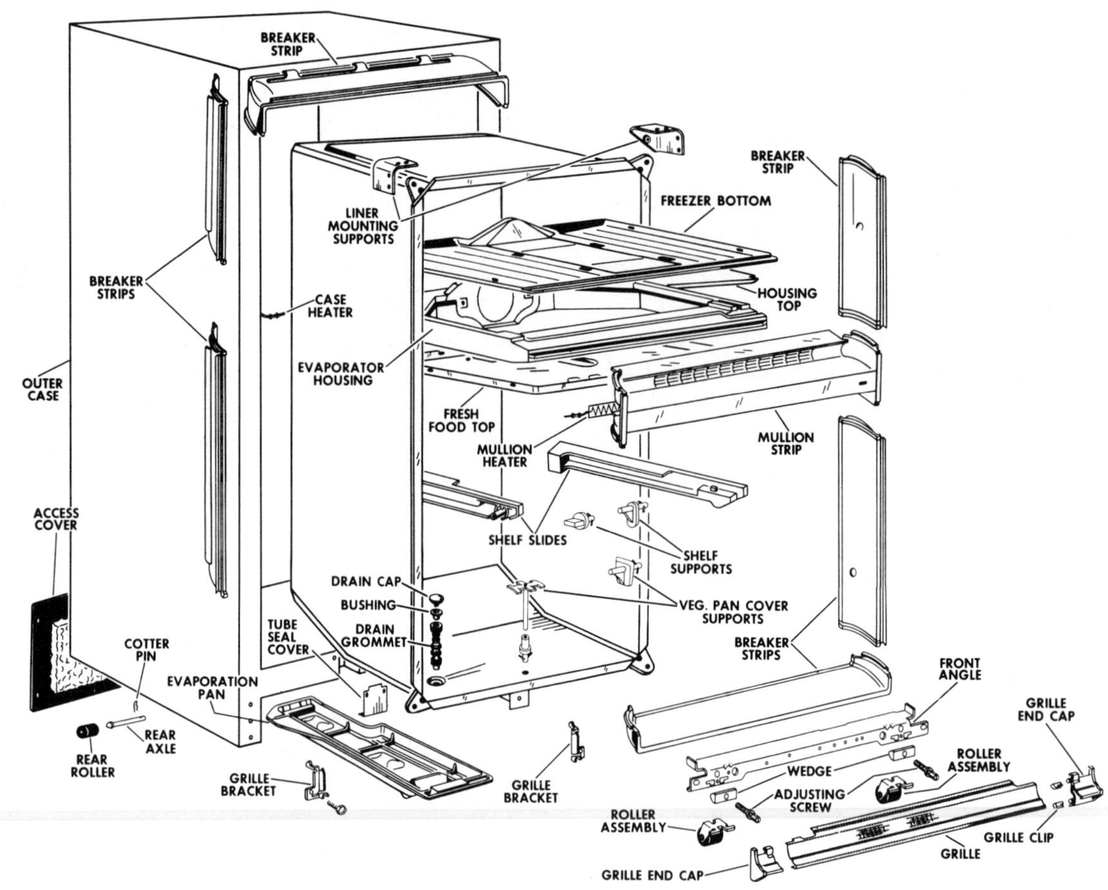

Figure 2-9. Typical refrigerator cabinet assembly.

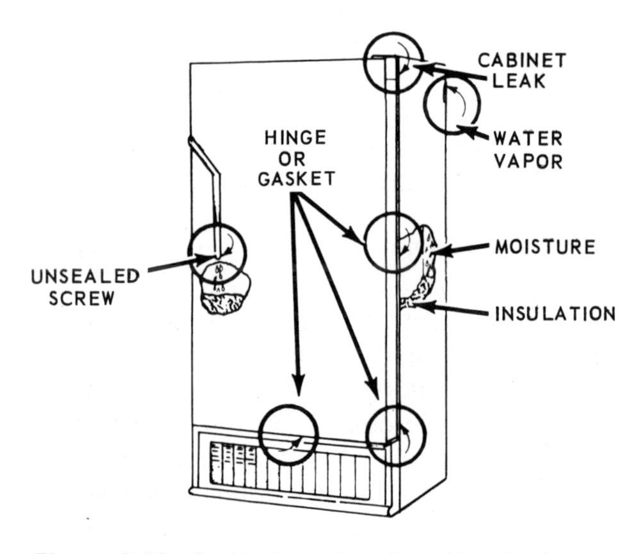

Figure 2-10. Seal leakage that should be checked.

tains are drawn in through the leak by the difference in temperature and humidity. Heat always flows to cold, so the warm air will flow into the refrigerator by any means available. Water vapor in the air will condense onto the coldest surface with which it comes in contact. Water vapor will also migrate to any area which has a lower relative humidity. In summary, both heat and water vapor try to seek their own level, so they will travel toward anything or any place with a lower temperature or humidity.

The air with its water vapor passes through the cabinet insulation to reach a point where it is cooled to its dew point. It condenses into droplets and in a short time the insulation is soaked with condensation. Wet insulation has the same effect as no insulation. Heat can travel readily through water. An immediate indication of wet or frozen insulation is sweating or condensation forming on the outside of the cabinet. To illustrate further, a nail-size hole in a freezer cabinet will allow the migration of over a pint of water a day from the room air into the insulation. Within a few days, the insulation will be a block of ice.

External sweating. Under conditions of high humidity, external sweating on an object is an indication that the surface is colder than the surrounding air. You notice this sweating every day during the hot, humid summer months on the outside of cold drink glasses, cold water pipes in the basement, and the flush tank in the bathroom. Any surface will sweat if it is 5 to 8°F colder than the room temperature, when the relative humidity is high. And yet, if the same thing occurs on the outside of a refrigerator or freezer cabinet, it is considered a defect by most users when it may be perfectly normal. Since refrigerator and freezer cabinets also have slightly cold surfaces, they will naturally sweat as the humidity becomes high. This will usually occur around the door strike, hinges, and breaker strips, which are the coldest surfaces, but condensation may also appear on other portions of the outer case. Reassure your customers that nothing is wrong in these cases when you find nothing wrong mechanically.

Location will have a bearing on external sweating. Cabinets in cellars, on back porches, in garages (a much more common location for freezers than for refrigerators) are much more subject to sweating than those in other locations. Refrigerators installed in kitchens next to gas ranges or dryers may sweat at certain times because of the increased humidity from operation of these units. Naturally, moving the refrigerator or freezer to a better location will help correct complaints on these installations.

Ordinarily, high temperature and humidity in an area will bring an abnormal number of sweating complaints. If exterior sweating occurs on only one or two cabinets at a time, insufficient insulation should be suspected. This type of exterior sweating is generally confined to a small area, and a gap or a thin area in the insulation will usually be found opposite the spot. Fill these gaps with *loose* fiberglass insulation, for, contrary to popular opinion, solidly packed insulation will transmit heat better than loose material. A filler strip of superfine insulation can be used for filling in these voids.

Sweating on the outer case along the door gasket seal can generally be corrected by improving the door seal on all models of freezers and refrigerators. If the door seal is perfect, there probably is a void in the insulation immediately behind the breaker or jamb strip. Pieces cut from fiberglass can be used here. In addition, foam insulation pieces have been used under the breaker strips around the freezer compartment on combination refrigerator models to decrease frost buildup. This foam material can also be used successfully on other models of refrigerators and freezers on which external sweating complaints are encountered.

Speaking of foam, it is important to remember that this type of insulation consists of a great number of tiny individual cells, each one of which is totally enclosed and separate from the others. If the foam is exposed to food such as spilled milk which seeps into the insulation, the only part of the foam insulation which can absorb any odor from the food is the edge just behind the breaker strip. Since this edge is accessible, any small amount of absorption caused by accidental spoilage can be dissipated quickly by cleaning.

Cabinets with case heaters or mullion heaters should be checked to see if the heaters are operating properly and are positioned correctly. Either a defective or improperly located heater could cause sweating on the outer case. Sometimes chest-type freezers (see Chap. 3) sweat only along the top rim of the outer case and a short distance down the sides. If the lid gasket seal is good, then the cabinet breaker strip nearest the sweating area should be removed to check for a gap in the insulation directly under the breaker strip. The insulation should touch both the top flange of the outer case and the breaker strip. If you find a gap, fill it loosely with superfine fiberglass insulation.

A vapor leak in the door or cabinet should be suspected if customer complaints of "excessive ice buildup," "high cost of operation," or "cabinet not cold enough" are received, and inspection reveals that the door gasket seal is good. To check for vapor leaks, remove the interior door panel and check for wet insulation. If insulation is found to be completely dry, check the cabinet insulation. If for any reason a portion of the cabinet's insulation should become water-saturated, the cause must be determined and corrected before permanent remedy can be accomplished. Follow a systematic sequence of checking all drain connections and usage. The most frequent causes are the ones listed below.

1. User failure to adhere to defrost procedures described in the owner's manual or to prevent clogging of drain by food particles or other material.
2. Rough handling of product prior to installation and rupture of drain line connections or cabinet seams.
3. Failure to seal cabinet openings after completion of service operation.
4. Under certain climatic conditions, algae may develop in the drain line. Periodic cleaning will prevent such growth from causing an obstruction.

How to correct wet insulation. Wet insulation will result in a very high heat transfer and will cause long running times. Rusting of the hot tube can occur if the cabinet insulation becomes wet. (The *hot tube* is a functional part of the condenser which provides additional subcooling.)

To inspect cabinet insulation, remove the lower throat molding. If insulation is only damp and not wet up the sides of the cabinet, it may be possible to remove all damp insulation through the lower throat molding and replace with new insulation. Always find and correct the source of moisture.

When cabinet insulation is soaked, it will be necessary to remove the interior liner to correct properly. To change the insulation, proceed as follows.

1. Remove all the wet insulation.
2. Inspect the inside of the cabinet shell to determine the source of the water. Likely sources are (a) drain tubes misaligned or broken (align or replace); (b) evaporator mounting screw holes (put permagum around screws); and (c) ruptures, cracks, or screw holes in the cabinet shell (seal with permagum or similar material).
3. Inspect the entire cabinet and hot tube thoroughly for any rust or corrosion. Put special emphasis on inspection of the hot tube in contact with the wet insulation.
4. If an odor is present, clean the entire cabinet with an odor-removing solution. Sand all rusted areas down to bare metal, wipe clean, and allow to dry. (A heat lamp will reduce drying time.)
5. Remove all rusted or pitted sections of the hot tube and replace with new copper tubing. When it becomes necessary to replace tubing along the cabinet sides, be certain replacement tubing lies flat against the cabinet sides. Use all available clips. Good thermal contact between the cabinet and hot tube is essential. The replacement copper tubing in the bottom of the cabinet need be formed only to the contour of the cabinet bottom.
6. Paint all rusted areas with an acrylic lacquer. Allow the paint to dry thoroughly (allow a minimum of 3 to 4 h drying time before reinsulating the cabinet). A heat lamp will shorten the drying time.
7. Use new high-density insulation to replace the wet insulation. (Usually install it on the bottom and 2 ft up the sides and backs.) Most manufacturers have this insulation available in replacement kit form, already cut to size.
8. Be sure all the openings by which the water entered are sealed to prevent a recurrence of wet insulation.

9. Reassemble liner and components. Be sure the defrost water drains are assembled correctly.
10. Evacuate and recharge the system (see Evacuating and Recharging the System, under Refrigeration System below), using a new service replacement drier.

Sealing is also important between the refrigerated fresh food section and the freezer section of a refrigerator. When you remove the breaker trim you will find the area behind the cross stile (also called the mullion or the cross brace) plugged with permagum or another nondrying puttylike material which has sealing, insulating properties. This sealant will also fill in voids in the two front corners. If it is left out or not installed correctly, heat will migrate and the temperature difference between the fresh food compartment and the freezer section will be affected. Cabinet sweating in the refrigerator and on the outside of the cabinet may result. The warmer air of one compartment will migrate to the colder, and the water vapor it contains will condense on a colder surface.

Breaker trims. Breaker trims are usually used to cover the opening between the cabinet liner and the outer cabinet shell. They serve as a thermal or heat break. We know that heat will travel to cold and metal will provide a very convenient path. It follows, therefore, that the inner liner cannot have a direct metal-to-metal contact with the outer cabinet. If there were, the cabinet surface temperature would be lowered enough to cause sweating. The heat leakage into the food-cooling sections would reduce the efficiency of the refrigeration system.

The liner is usually suspended in the insulated interior of the cabinet and supported by corner gussets and side spacers to the cabinet opening. Rubber or plastic spacers are generally used at the attaching points to reduce heat transfer through the metal brackets. To cover the opening and conceal the brackets and insulation, we use decorative plastic breaker trim strips. Being plastic, they provide a thermal break as they conduct very little heat. They are shaped to snap into place without the need for screws.

The side breaker trims are retained at the rear by a flange which grips the front edge of the liner, and at the front by spring clips behind the cabinet flange. Top and bottom trims are retained by spring clips on the lip of the liner and behind the cabinet flange.

When removing breaker trims from a cold refrigerator it is advisable to warm them to reduce the danger of breakage. A towel wrung out in hot water or a hair dryer are suggested for this purpose. The side breakers must be removed first. Starting at the bottom cuff, apply pressure

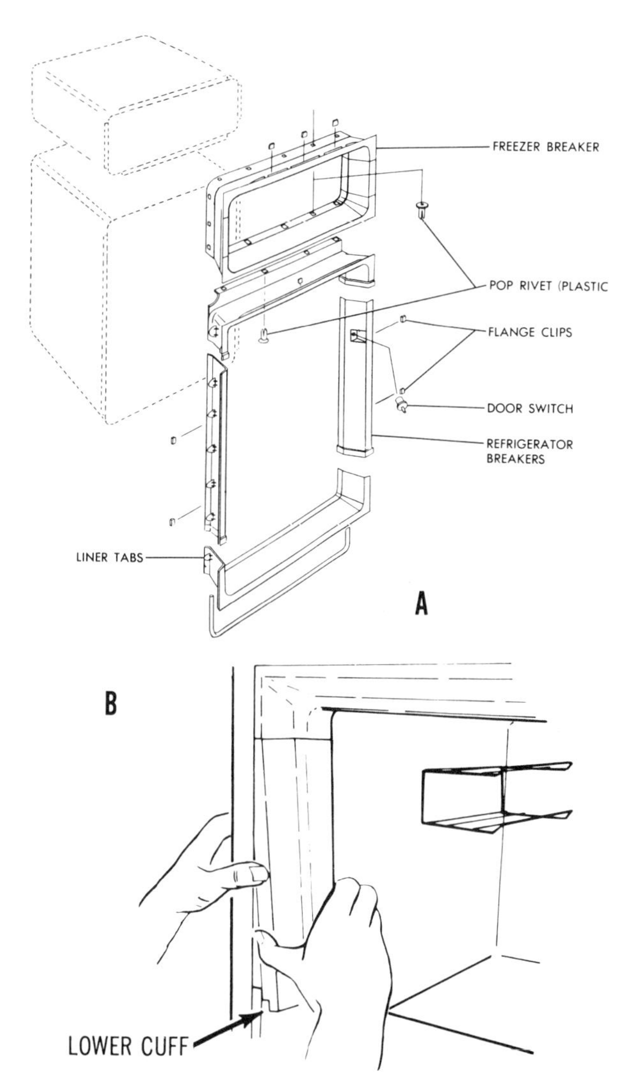

Figure 2-11. (A) Typical breaker strip trim location and (B) how it is usually removed.

on the front surface and work the front edge of the trim out from behind the cabinet flange (Fig. 2-11), then pull the trim forward to disengage the back flange from the front edge of the liner. To replace, start the back flange of the trim behind the front lip of liner, and then press the trim toward the liner until the front flange of the trim can be snapped behind the cabinet front flange.

The top and bottom breakers can be removed after the side breakers are removed. Beginning at one end, pull up on the bottom breaker or pull down on the top breaker to disengage it from the spring retainers. To reinstall, position the trim with both ends behind the cabinet flange, then press it into place between the edge of the liner and the cabinet flange.

The top trim in the lower food compartment of some top freezer models may have screws securing it to the underside of the cross stile. Due to its flat shape, it lacks the rigidity of the other trims. The screws keep it from sagging in the middle. The breaker trims of side-by-side model refrigerators are removed in the same manner as the other models. Care must be taken to reduce breakage when removing and replacing the trim. Expertise comes with experience.

Liner removal. Removal of a refrigerator or freezer liner is a very rare occurrence. It would only be necessary in the event of wet insulation, extreme odor conditions, replacement of hidden wiring, or in certain models, the replacement of the evaporator. Replacement refrigerator liners are not available. If there is porcelain damage, a porcelain repair epoxy kit is available from your local parts source which will, if used as directed, produce an invisible repair lasting as long as the porcelain.

Certain models have foamed-in-place polyurethane insulation. The liner cannot be removed; all wiring connections, however, are accessible without removing liner or foamed insulation. (Check the wiring diagram on the back of the refrigerator for location of connections.) Other models use fiberglass batts and block foam for insulation. These liners are removable if the need arises. When reinstalling the liner, lay the refrigerator on its back (if space permits) and ease the liner down into the cabinet, being careful not to damage the insulation.

Usually the liner does not have to be removed to service the sealed refrigeration system. The one instance where this would be necessary would be the evaporator removal on certain cycle defrost models. These models will be immediately apparent by the height of the front toe plate grille. As shown in Fig. 2-12, some models have a high grille while others have a low profile high grille (approximately 5 in). The liner of the 5-in grille models does not have to be removed. The evaporators and connecting tubing are inside the liner.

The models with the high grille usually have a porcelain partial liner for the fresh food section. The aluminum roll bond evaporator fills the freezer area of the cabinet and is mounted in the same manner as a porcelain liner. The interconnecting refrigerant tubing passes down through the insulation and enters the lower liner through an opening in its back wall. The liner and the evaporator assembly must be removed almost simultaneously after the cross stile is unfastened and out of the way. The lower evaporator plate is dismounted from the liner and carefully swung around so that it can pass

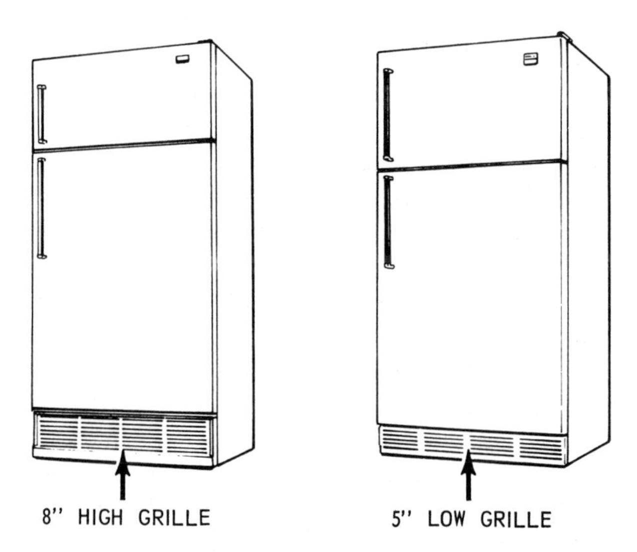

Figure 2-12. High- and low-profile grille arrangements.

through the opening in the back of the liner. When reinstalling such an evaporator and liner, be sure the styrofoam block between the two is in place and the drain hole is in line with the drain hole on the top of the lower liner. The styrofoam block functions as a drain pan as well as an insulator. If it is out of place, defrost water from the freezer will miss the pan and drip down into the lower insulation to collect in the bottom of the cabinet.

Fiberglass insulation adjacent to the aluminum evaporator is encased in a plastic casing like a pillow. Check that the plastic covering is not torn. All tears should be taped up so that the fiberglass cannot contact the aluminum. Wherever the fiberglass and the aluminum come into contact, the aluminum will become pitted as though by electrolysis. If the pitting occurs in the evaporator containing refrigerant, a sealed system leak will result. The pitting problem may be spread over too large an area for patching making a complete evaporator replacement necessary.

Installation of the refrigerator. The refrigerator should always be uncrated before it is delivered to the home of the customer. A careful check should be made for any visual defects, and the refrigerator should be operated and checked for efficient function. If a refrigerator is operated prior to delivery it must be left in an upright position long enough after being turned off to permit the system to balance out prior to being tipped over. Failure to do so before placing the refrigerator in a horizontal position can result in oil displacement in the system, which may cause compressor failure. It is recommended that a refrigerator that has been operated for any length of time be turned off and left standing overnight prior to transporting unless it is to be transported in an upright position.

As previously mentioned, the cabinet must be accurately leveled. Those units not having rollers are furnished with leveling gliders in both front corners behind the grille. If the cabinet cannot be properly leveled at the rear, waterproof shims may be added. Those units having rollers are leveled by adjusting the rollers. To do this, remove the front grille and proceed as follows.

1. Check back of cabinet for level. Add shims below back rollers if necessary.
2. Level cabinet by adjusting the front adjustment screws.
3. Adjust front slightly higher for positive door closing.
4. Tighten brake against the floor.
5. Replace grille.

The location of the refrigerator is, of course, a matter to be decided by the user. The service technician should only be sure that the installation site chosen by the user is proper for the most efficient operation of the unit. For instance, the refrigerator should not be located where it will be subjected to the direct rays of the sun or to heat from a range or radiator. Also it should not be installed where the room temperature will fall below 60°F, as there will not be enough heat leakage into the cabinet to cycle it often enough to maintain satisfactory temperatures. If the refrigerator must be so installed, advise the customer of probable improper operation at times.

An adequate electric supply is required for satisfactory operation. Refer to the specifications in the service manual for electrical information. Also check the data shown on the serial number plate. (Most refrigerators are made to operate on 120 V alternating current.) A separate circuit with a 15-A delayed-action fuse should be provided. Do not use an extension cord connection because of the possibility of a voltage drop and resultant unsatisfactory operation.

Before the refrigerator is set against the wall, and after the interior packing has been removed, turn the control to the ON position. Connect the power cord to a wall outlet and check the following:

1. Interior lights operate when door is opened
2. Compressor operates
3. Fan is operating on applicable models

Then, place the refrigerator in position and adjust leveling legs or rollers, depending on model, so that doors swing closed by themselves from a 45°opening.

In order to familiarize customers with their new refrigerator and to minimize service calls due to installation deficiencies, the following checklist can serve as a reminder during installation.

1. Are the wall spacers mounted on the condenser?
2. Is the adjustable foot attached with its locknut?
3. Is the appliance resting firmly on all four feet?
4. Has the inside packing been removed?
5. Are shelves, drip pan, crisper, and other accessories in their places?
6. Is there proper clearance around the top and sides of the refrigerator? Most manufacturers recommend an air space or clearance of 3 to 4 in at the top and 1 to 3 in on each side except on forced-air condenser models. Leave an inch or two at the back. Models having a fan-cooled condenser may be installed flush against the rear wall since it is not necessary for air circulation behind and above the refrigerator. But always keep in mind that restricted condenser air circulation due to improper installation or dust accumulation of either a fan-cooled or a static condenser model will cause high cabinet temperatures, compressor overheating, compressor kicking out on overload, and high running costs.
7. Does the refrigerator door seal tightly?
8. Has the refrigerator been properly grounded?
9. Is the customer familiar with the instruction manual supplied with the refrigerator?

AIR-HANDLING SYSTEM

Air circulation is important both inside and outside the refrigerator. Outside, the heat produced by the compressor operation sets up a "hot" zone around the lower part of the refrigerator cabinet. Therefore air circulation is created in the form of an escape current rising up the rear part of the refrigerator and an intake draft passing from the lower front part. It is essential that this draft not be interfered with, and for this reason it is necessary for the refrigerator to be spaced away from the wall. For this purpose spacers are attached to the condenser to ensure that the necessary space is provided. This fundamental point should be remembered when installation is being planned.

Inside the refrigerator, natural airflow in a non-frost-free or a non-fan-cooled conventional system is by convection. As we know, there are two zones within the refrigerator: the evaporator (freezer area) and the provision or general-storage compartment, each having distinctly different temperatures. There are, in fact, air currents passing between the top and the bottom of the storage compartment on account of this temperature, and these must be taken into consideration to get the best distribution of the cold in the storage compartment. The relatively warm air of the lower part of the liner rises to the upper zone via the front space, between the

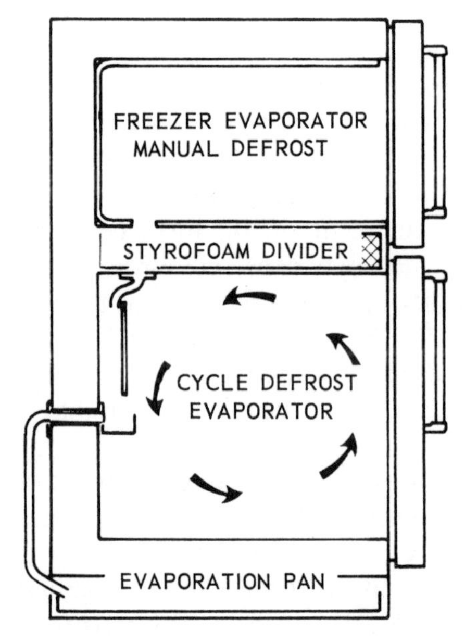

Figure 2-13. Gravity air circulation.

door compartments and the shelves. In return, a current of cold air descends from the evaporator to the lower zone of the liner between the shelves and the liner.

A refrigerator using gravity circulation has its evaporator or cooling coil located near the top of the compartment. Warm air rises and cool air drops. As the air is cooled by the evaporator, it drops to the lower section of the refrigerator. The warmer air rises up to the evaporator where it is cooled and again drops. This sets up a circulation of air in the refrigerator. As it circulates, it picks up heat from the food and deposits it on the evaporator (Fig. 2-13). If anything blocks the natural airflow, the food-cooling action of the refrigerator is reduced. This is the reason open wire shelves are used instead of solid shelves. An overly fussy housewife who places paper on the shelves will soon find the food in the lower area of the refrigerator spoiling while food in the upper area near the evaporator will be too cold or freezing. Packing the food too closely together so that air cannot circulate will also cause the same problem.

The two-door cycle defrost refrigerator has the freezer evaporator in the upper section and an off-cycle defrosting cooling plate evaporator in the lower section. The only air control in these models is the thermostat. It cycles on and off according to plate temperature in the food section.

In ordinary refrigerators, a front-to-back air-

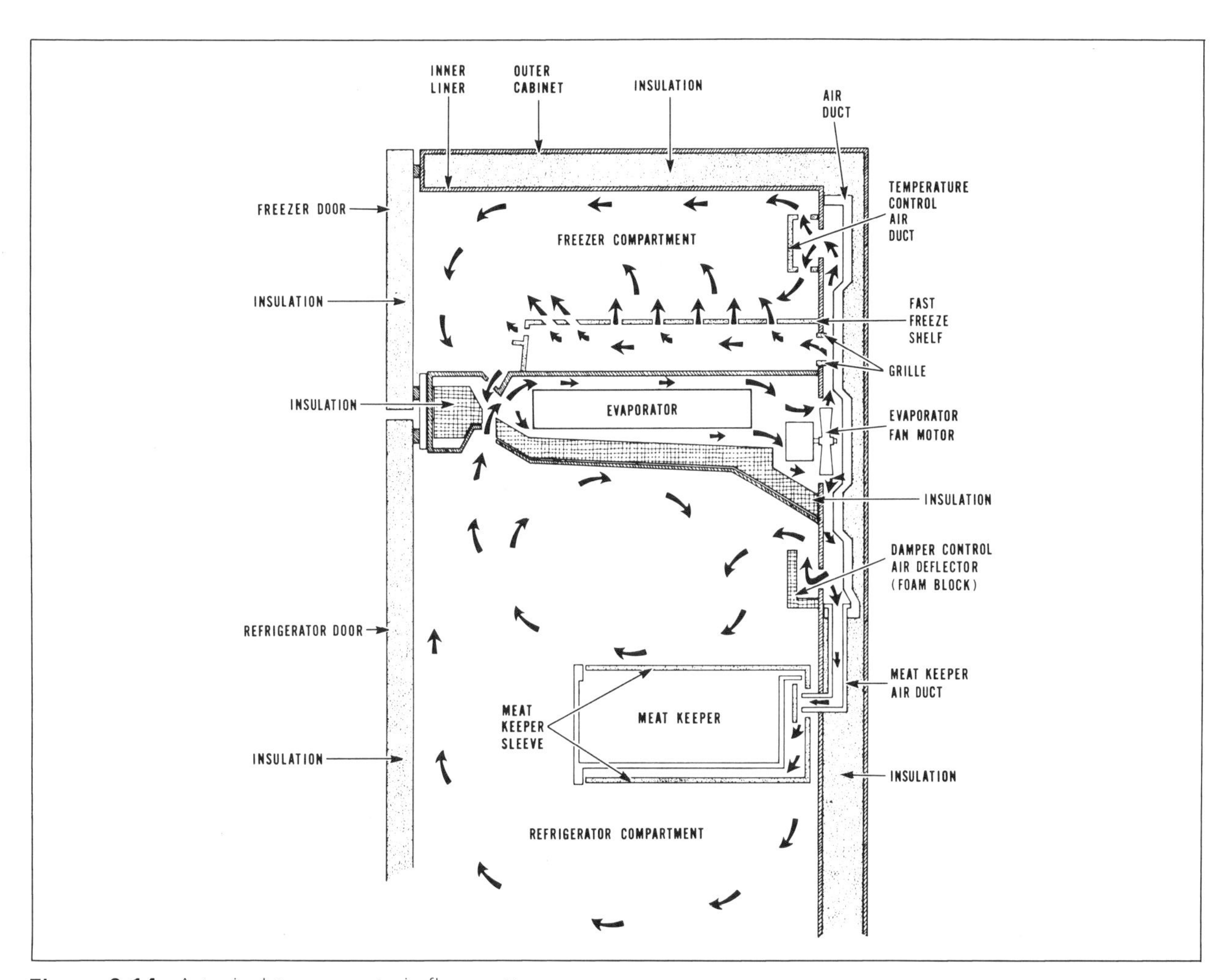

Figure 2-14. A typical top-mount air-flow pattern.

flow is created between the upper breaker frame and the evaporator door, which allows a uniform cold distribution. In addition to the primary front-to-back circulation, there are other secondary currents which develop in the center of the liner and cause air circulation between the drip pan and the evaporator. These models are equipped with a device which has the function of regulating the airflow (by means of a deflector) in the refrigerator, and, as explained in the user instructions booklet, the temperature in the lower part of the liner may be increased with the thermostat unchanged. It must be kept in mind that the main function of this device is to make the inside temperature more uniform with respect to the outside temperature. In places where the room temperature is 85°F or more, it should be kept permanently open.

The airflow in a typical forced-draft top-mounted refrigerator-freezer unit is illustrated in Fig. 2-14. In it, the air in the freezer section is pulled into the front grille and over the entire width of the cooling coil. Heat and moisture are given up to the coil with the result that cold, dry air is discharged out the air duct into the freezer compartment. An air deflector is used to divert air directly over the ice service area. The fan or blower housing is designed so that some of the air is diverted to the food compartment. The position of the modulating air regulator or manual damper will determine the amount of air which flows into the food compartment as indicated by the open arrowheads. The air from the food compartment returns to the freezer section through the slots in the divider located on each side of the control housing. In the freezer it mixes with the freezer return air and passes through the evaporator and back to the fan or blower motor housing. The defrost water is removed from the freezer compartment through the drain tube, then down the rear wall of the food compartment to the drain opening in the bottom of the food compartment liner. Defrost water then goes through a drain trap into a disposal pan in the machine compartment, where it evaporates. Periodic cleaning of the drain trap is necessary.

The forced-air system in the bottom-mount refrigerator-freezer (Fig. 2-15) is usually the same as in a top-mount model.

Airflow in the food compartment of the vertical freezer or side-by-side model (Fig. 2-16) is determined by a modulating air regulator or a manual damper. The temperature control thermostat senses the temperature of the food compartment top liner. An opening in the bottom of the liner divider between the freezer compartment and food compartment provides cooling for the meat keeper. The typical side-by-side forced-draft unit shown here utilizes a single-coil evaporator to cool both compartments. Only one fan is used for air distribution. Start-

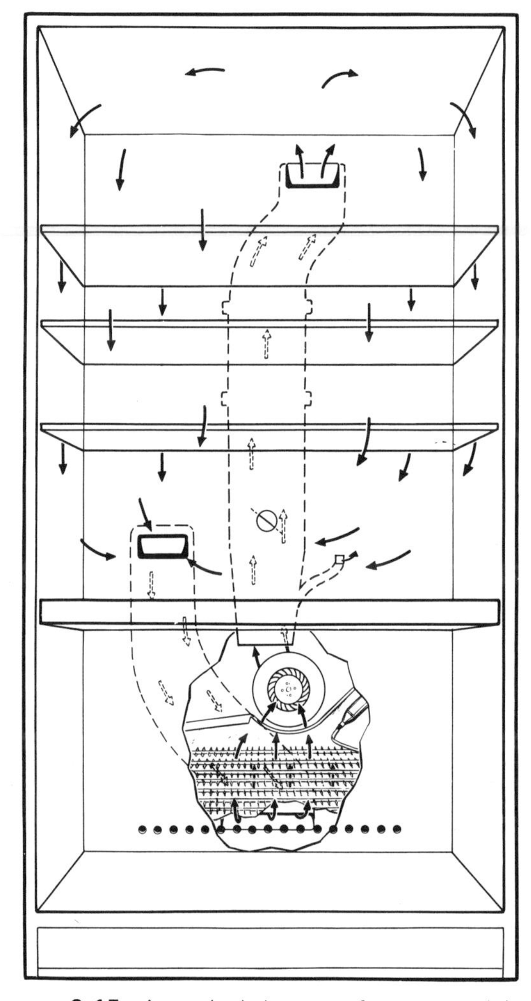

Figure 2-15. A typical bottom freezer model air-distribution pattern.

ing in the freezer compartment, air is drawn into the grille at the bottom of the compartment and through the coil where the heat and moisture are removed. The design of the fan or blower motor housing and insulation ducts provides direct cooling in this area. This air flows downward through the lower section, cooling this section on its way to the return-air grille. Part of the cold, dry air entering the fan housing is diverted upward, through ducts to the top of the freezer compartment. A deflector, mounted on the top freezer panel, directs air over the ice service area. From here, the air filters down through the shelves on its way to the return-air grille. The air duct from the blower housing is so designed that some of the cold, dry air is diverted to the top of the food compartment. The amount of air entering the food compartment is determined by the manual damper. As the air enters the food compartment, the cold air flows to the bottom of the compartment. In the bottom left rear corner of the food compartment, a return-air opening is provided in the foamed divider for air return to the cooling coil, and the cycle is repeated.

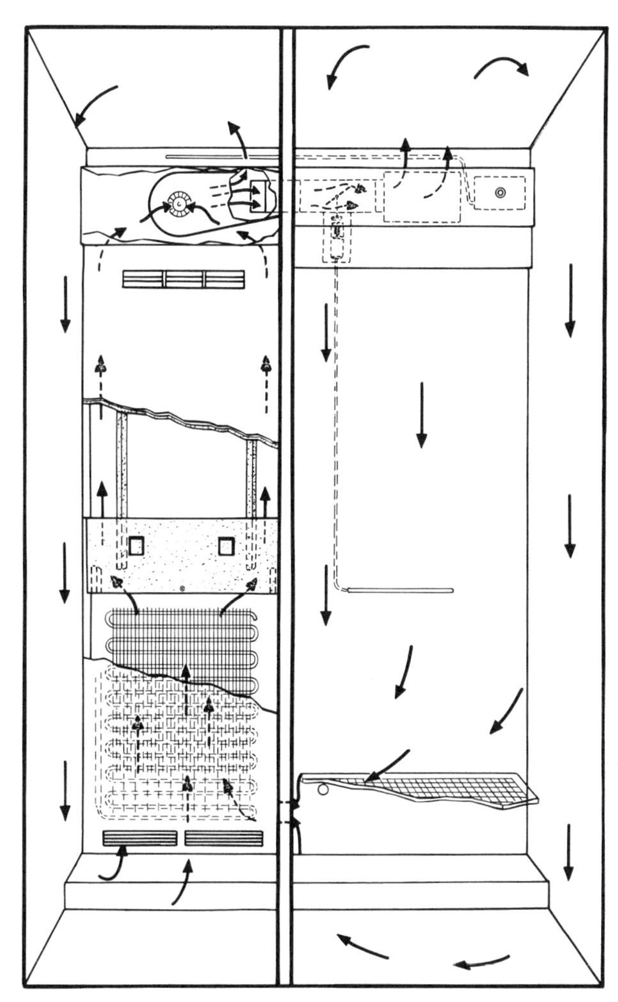

Figure 2-16. A typical side-by-side model air-distribution pattern.

Thus, in any refrigerator, airflow over the condenser must be carefully maintained. The following are causes of poor air distribution.

1. Fan blade in or out too far on fan shaft.
2. Fan cover loose.
3. Evaporator blocked with ice.
4. Control tube not located in its proper position.
5. Permagum or similar sealing compound out of position or missing.
6. Spacers and air ducts broken or out of position.
7. Evaporator not secured tight to liner, or evaporator cover spaced away from evaporator.
8. Fan housing blocked by lint or other foreign objects.
9. Insulation or other debris blocking airflow through unit compartment. Be sure condenser is clean. To clean it, use a bottle brush and/or a vacuum cleaner and attachments.
10. If there is an air divider, check its position. In most cases, it should be touching the floor between the condenser (inlet side) and the drain pan (outlet side).

The fan blade must be properly placed on the motor shaft and the shaft centered in the evaporator cover orifice. As a rule, the mounting-bracket screws of the fan motor can be adjusted to center the shaft in the orifice. If the fan is in too far, not enough air will be moved into the refrigerator section, and warm temperatures will result because the coil will not be loaded and the percentage of running time will decrease. If the fan is out too far, too much air will be

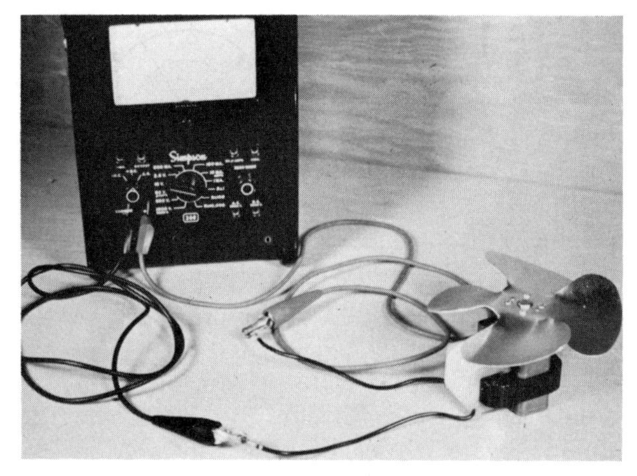

Figure 2-17. Method of checking a condenser fan.

moved into the refrigerator section, resulting in a low cabinet temperature, high-percentage running time, and a tendency toward a warmer-than-normal freezer. If the fan shaft is not positioned horizontally, the fan blades will either be too far inside the evaporator cover at the bottom and outside at the top or vice versa.

Any of these conditions will produce a different flow of air than normal. If you find these conditions, or if after you make repairs it is necessary to change the adjustments, exercise extreme care to ensure that the correct conditions are met. Also, remember that if the fan cover is not installed properly, the air distribution will be adversely affected, resulting in warm cabinet temperatures.

To check the condenser fan (Fig. 2-17), remove the fan motor lead wires and make a continuity check across the leads; a resistance should show on the ohmmeter. No movement or full deflection of the needle indicates an open or grounded fan motor winding; in either case replace the motor.

On models incorporating an evaporator fan switch (see p. 56), check to see that the fan operates when the door is closed. This can be done by removing the fan cover and closing and opening the door; observe fan operation. The fan operates only when the compressor operates. If the fan does not operate, check the switch and motor with a low-voltage tester or ohmmeter. Remove the wires from the fan switch; continuity should exist between the fan switch terminals when the plunger is fully depressed.

Some models use an additional fan to circulate air in the fresh-food area. This fan is usually located in the bottom and rear of the fresh-food compartment. It is operated by the fresh-food control and moves the air from the bottom of the fresh-food compartment through the evaporator and up a duct to the top rear of the fresh-food liner.

If the preceding steps indicate the differential is incorrect or if freezing in the refrigerator section exists, the quantity of air to the refrigerator section can be controlled by adjusting the manual damper or modulating automatic air regulator. To check modulating air regulator, proceed as follows.

1. Remove control.
2. Warm the sensing element by holding it in your hand. The damper should open all the way at the midpoint setting. If it does not open, replace the control.
3. Insert control sensing element into a glass of crushed ice and water, stabilized at 32°F. Damper should close or not be open more than $\frac{1}{16}$ in at the midpoint setting. If the control does not respond in this manner, replace it.

Ice Cube Shrinkage

Ice cubes may shrink when stored in the ice bucket or container in forced-air models. The shrinking of the cubes is caused by the very dry evaporator air passing over the cubes and absorbing some of the cube moisture, which is carried back and deposited on the evaporator tubes. This is the forced-air system at work, also picking up frost and moisture from the surface of the frozen-food packages. The process by which the ice cubes give up moisture is called *sublimation*, the changing of a solid (ice) directly to a gas (water vapor).

The ice cube shrinkage is more noticeable in the wintertime than in the summertime, as the usage of ice cubes is not as great. If objections to the cube shrinkage are raised, suggest that cubes be ejected one tray at a time as they are needed. Frozen-food packages should be wrapped and sealed to preserve the moisture inside. This is especially necessary with the forced-air system.

Odor Complaints

Refrigerator odor is one of the basic nuisance problems of the refrigeration industry. Past experience indicates that as the season for hot weather approaches, a number of odor complaints will undoubtedly arise. During the cooler months of the year an odor complaint is a rarity. The manufacturing process uses the same type of material during the cooler months as it uses during the warmer months. The conclusion can be drawn that the cause of odor complaints is not odorous materials used in construction of the refrigerator. The same conclusion is further borne out by the fact that service technicians all over the country replace various parts of the cabinet in an effort to help the customer, but that no emphasis is laid on any particular part. It seems logical to assume that none of these operations actually affects the odor of the cabinet but merely changes the customer's attitude.

Test results show that as soon as food is stored in a refrigerator, the food storage compartment and adjacent parts absorb a "natural" food odor. It would be difficult to try to describe to the uninitiated what is meant by this term, but every service technician after a few observations should be able to distinguish between cabinets which have "natural" food odors and those which have excessive or unusual odors.

Excessive or unusual odors can result from various conditions such as too much uncovered food, highly seasoned food, and citrus fruits. In some instances, it has been found that unnatural odors originated from food that was spilled in the cabinets and has decayed. Another condition which may cause a complaint develops when the customer stores a dish of cooked and uncovered cauliflower or some such food in the refrigerator for possibly only one day. This type of food gives off a very intense and easily recognizable odor, and the customer is not too much concerned at the time even if other foods taste of cauliflower, recognizing the distinctive odor and knowing the cause. Naturally, this will raise the odor level of the cabinet for a considerable period of time after the offending cauliflower has been removed. In another few days, the odor of the cauliflower loses identity as it becomes mixed with other food odors, but the intensity or amount of odor in the cabinet will be increased to the point where the customer, having forgotten about the cauliflower and not recognizing it, now assumes that there is something wrong with the refrigerator. At this point the service technician is called and is probably confronted with a cabinet with a very high odor level of unrecognizable origin. The customer, of course, has forgotten that several days before an uncovered dish of some such food as cauliflower, cantaloupes, or the like was stored in the food compartment.

Much of the success of handling odor complaints depends upon the initial contact by the service technician. Foods of one kind or another are responsible for most odor complaints, and the service technician must recognize this. Explain to the customer that nothing in the refrigerator itself is causing the odor, and that very rarely will replacement of any part actually affect the odor level of the cabinet.

Some things, of course, can be done to check for outside causes of odor. First, it is a good idea to remove the bottom breaker strip and check the insulation underneath. It will usually be found dry, and the strip is replaceable, but this is a precaution taken to be sure that no liquids have been spilled beneath the strip that might cause an odor. If the odor has not been allowed to build up too much, a complete cleaning of the interior by washing with a solution of water and baking soda will be quite effective.

If washing with a solution of baking soda does not rid the refrigerator of a food odor, it is advisable to use an activated-carbon-filled motor-driven deodorizer for a few days. In fact, some refrigerators have a built-in charcoal air purifier. Located in the freezer compartment, this purifier is usually placed in the return airstream of a forced-draft-type unit. But when a complaint is received concerning a refrigerator that has a purifier, remember that activated charcoal will hold only a certain amount of odor. After its limit is reached, it will start returning the odors into the airstream. For this reason, the charcoal must be replaced periodically. How often will depend on the usage of the refrigerator, the way foods are covered, and the types of food stored. When odor saturation has been reached, the customer must recharge the purifier. This is done by removing the retaining screw and cover, discarding the old charcoal, and replacing it with a fresh supply.

Some models of refrigerators contain a filter whose task is to remove food odors. But it must be remembered that the effective life of a refrigerator filter is about one year. If the filter has been in use for more than one year, it should be replaced.

In rare cases, it may be necessary to replace the insulation (see How to Correct Wet Insulation, p. 50), or to bake and air it for 3 to 4 days to rid it of an odor it has absorbed. When this is done, it is also advisable to heat the no-oxide grease on the inside walls so that it flows slightly and gives off any odor it might have absorbed.

Automatic-defrost refrigerators have their own peculiar problem because a drain pan is provided in the compartment to catch and evaporate the defrost water. Of course, this will also catch any foods which are spilled in the compartment and which may be carried down by the water. This pan should be cleaned out with soap and water periodically. If this is not done regularly, it can be a source of odors and eventually a nest of insect life.

To summarize, lack of air handling or circulation in a refrigerator may be due to any of these causes.

1. Failure of automatic defrost arrangement
2. Excess moisture in air
3. Poor door gasket seals
4. Refrigerator or freezer door left ajar
5. Air-circulating fan motor malfunction
6. Light switch(es) malfunction
7. Motor and compressor off because of defrost cycle or temperature control satisfied
8. Damper control malfunction
9. Capillary tube has lost charge
10. Improper adjustment of damper control knob
11. Overcrowding of shelves or stacking of packages in front of air outlet grilles
12. Improper seals: blower scroll to liner, evaporator cover to header plate, or control to liner

ELECTRICAL AND CONTROL SYSTEM

There are two basic electric circuits in a refrigerator: the circuit that operates and controls the cooling cycle and the one that does the same for the defrost cycle. As described in Chap. 1, the cooling circuit includes the cold control (thermostat), current relay, overload, and motor-compressor. The function of the cold controls is to sense and control temperatures by starting and stopping the motor-compressor. The current relay energizes and de-energizes the compressor start winding. The overload protects the compressor against high amperage and high temperatures. The motor-compressor compresses and circulates the refrigerant. Diagnosing and checking these parts will be covered later in this chapter. Here we shall concern ourselves with the defrost system and the various other controls. But, before doing this, let us take a look at power supply.

Power Supply

If the house wiring is inadequate, causing low voltage during the period of peak electrical use, the refrigerator may run and perform well for a part of the day (normal voltage), and stall at another part of the day (low voltage). Low

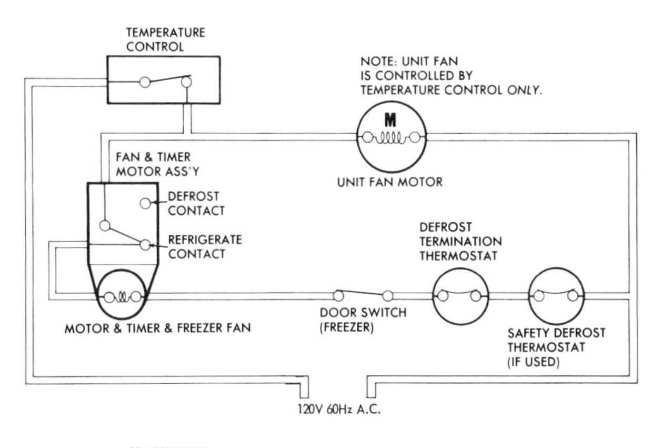

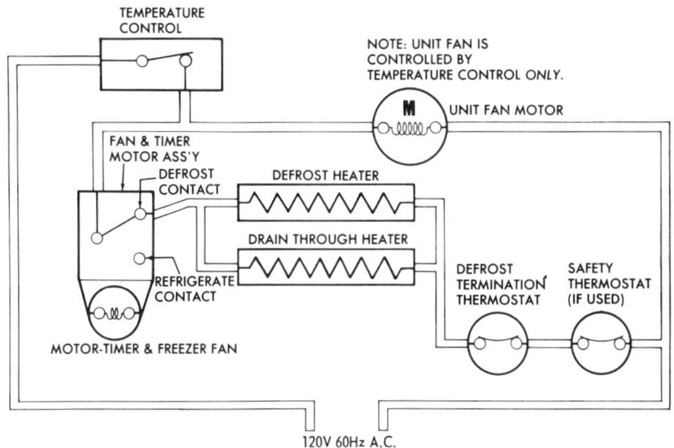

Figure 2-18. (Top) Basic wiring diagram for cooling cycle. (Bottom) Complete diagram including defrost system.

voltage or voltage fluctuation will cause the motor-compressor to stall. With the use of a volt-watt meter, check for the *correct voltage*. The voltage must be at least 105 V at the compressor at the instant of start. If the voltage drops below 105 V while the compressor is starting or at any time during the run cycle, damage to the compressor winding could result. Where possible, the refrigerator-freezer should be on a separate circuit.

To check power supply and polarity. With the refrigerator or freezer plugged in, make certain the interior light bulb operates. Remove the light bulb. With a line voltage tester, check from a liner screwhead or ground to the metal screw shell of the light socket. No voltage should be indicated. If voltage is present (reversed polarity is indicated), check the outlet, service cord, and appliance wiring for possible cause.

To check a three-wire 120 V ac outlet using a line voltage tester, proceed as follows.

1. Check the outlet for correct voltage.
2. Voltage should be present between the small slot and ground receptacle; if voltage is present between these points, outlet is grounded and has correct polarity.
3. If voltage is not present between the small slot and ground receptacle, check between the large slot and ground receptacle. If voltage is present, outlet has reversed polarity. If voltage is not indicated between either slot and ground receptacle, a no-ground condition is indicated. Indications of reversed polarity or no-ground conditions must be corrected; contact a qualified electrician.

All modern refrigerators are equipped with a power supply cord incorporating a three-prong (grounding) plug and a ground wire which is attached to the refrigerator and outer-wrapped for protection against shock hazard. Each electric component is connected through a ground wire to the outer wrapper to complete the ground. Be sure the electrical wall receptacle is of the three-prong type and is properly grounded in accordance with the National Electrical Code or local codes. (See *Basics of Electric Appliance Servicing* for more details.)

When only a two-prong electrical wall receptacle is available, we strongly recommend it be replaced with a three-wire grounded and polarized outlet. If the customer insists on using the two-wire outlet and a ground adapter, the outlet must pass the same test as for a three-wire 120 V grounded and polarized outlet. Once the outlet (with ground adapter installed) is determined to be grounded and polarized, the ground adapter must be marked to indicate the position in which it must be installed for the correct polarity. But, do not under any circumstances cut or remove the round grounding prong from an appliance plug. It is a good idea to mention to the customer that many appliance manufacturers do not recommend ground adapters.

The ground wires on all components are green or green with a yellow stripe and are not to be used as electric conductors. When a grounded component is replaced, the replacement *must* always have a ground wire and should under no circumstances be used if it does not. Whenever a ground wire is disconnected for replacement of a component, it *must always* be reconnected. Never consider a service call completed until you are satisfied that all ground wires are connected. Leaving a ground wire off could cause a hazard to the users or other service technicians.

When replacing the compressor in these models, it is imperative that the replacement be grounded in the same manner as the original. All replacement compressors now have provisions for a grounding screw in the relay housing. If you should encounter a replacement without a hole, it will be necessary for you to drill a $\frac{1}{8}$-in diameter hole in the relay housing to attach the ground wire.

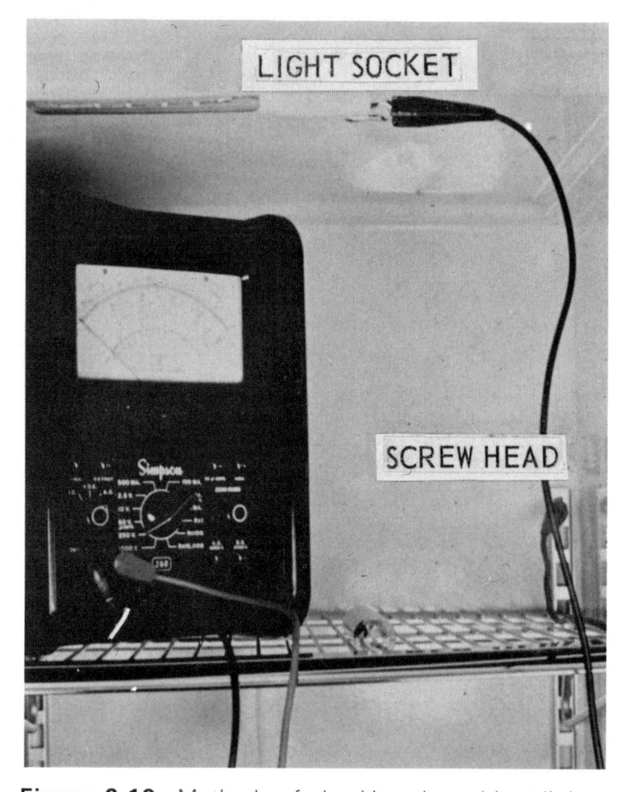

Figure 2-19. Methods of checking the cabinet light.

Basic Electric Circuit

The basic electric components of any home refrigerator include the cabinet lights, the unit temperature control, and various heaters and fans.

Lights. A refrigerator may have one to four lights working off a switch. This switch must turn the cabinet interior light(s) off when the door is closed. Open the door, insert a small, thin object (pencil) between the cabinet and door gasket (Fig. 2-19), then close the door and observe the light. It should go out. Do *not* actuate the plunger with your finger. Actuating the plunger with your finger does not ensure proper operation of the switch because the door may not contact or depress the plunger properly.

Unit temperature control. The operation of the temperature control on the thermostat is fully described under Thermostats in Chap. 1. While designs vary slightly, most refrigerator thermostats have just two adjustments, range (which varies both cut-off and cut-on) and differential (which varies cut-on only). The constant differential usually has two methods of control, a knob which the user sets and a differential screw adjustment which is adjusted by the service technician. The range adjustment is also a screw arrangement, but it should be changed only by the technician.

The constant differential adjustment is one that makes for the proper coldness of the unit. As a rule, this is accomplished by turning a numbered dial. While numbers usually have no direct relationship to any actual temperature setting, they do give the user a point of reference as to whether the temperature in the refrigerator will get colder. The differential setting of the thermostat control will raise or lower only the cut-out temperature. The effect of changing the setting of the differential knob is to cause the compressor to run for a shorter period of time before cut-out if it is set higher and for longer periods if it is set lower.

To check the calibration of the control

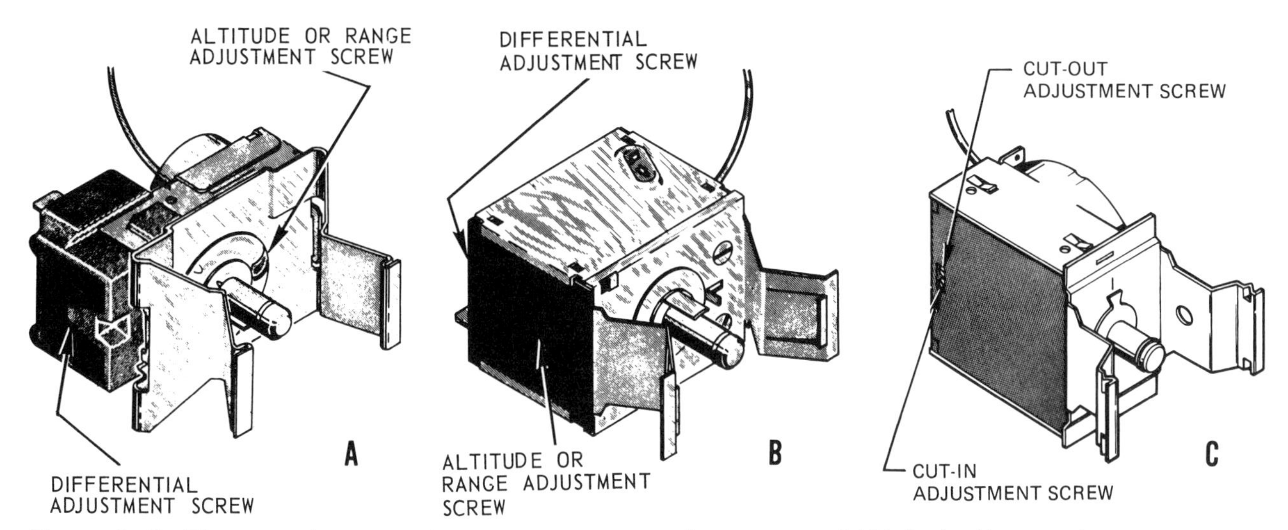

Figure 2-20. Three popular types of thermostats and their adjustments: (A) G.E. thermostat; (B) Ranco thermostat; and (C) Cutler-Hammer thermostat.

thermostat, a general procedure such as this may be followed.

1. Remove the control bulb of the thermostat from its well and place a thermistor in the control bulb well. Then place the thermostat's control bulb back into the well and label this thermistor No. 1.
2. Tape a second thermistor to the control bulb so that the thermistor is about $1\frac{1}{2}$ in from the point where the control bulb enters the well. Label this thermistor No. 2.
3. Do not place tape on the thermistor; tape the lead to the control bulb.
4. Allow the unit to cycle off and on twice and record the temperatures of both thermistors.
5. Take the second reading of each thermistor. Use thermistor No. 1 to read the cut-on temperature and thermistor No. 2 to read the cut-off. Use the second set of readings and discard the first.

If the control is found to be out of calibration, you can usually adjust the control through the range adjustment screw, the differential adjustment screw, or the control knob. For example, if the control is out of calibration, either too high or too low, an equal amount on both the cut-on and cut-off, the logical way to adjust the control would be to turn the control knob to a colder or warmer setting until the approximate desired temperatures are reached. However, if the customer objects for some reason, the same thing can be accomplished by turning the range adjustment screw.

If you find that only the cut-off temperature is too high or too low, you will have to use the range adjustment screw to bring the cut-off temperature back into line. But if you use the range adjustment screw to vary the cut-off temperature, it will usually be necessary to use the differential screw to bring the cut-on temperature back into line. If the cut-on only is out of calibration, you can use the differential screw to bring it back into calibration. Remember that this procedure may vary slightly with the different thermostats; always check the service manual.

Frequently, the range adjustment is called the *altitude control* and for good reason. The lower barometric pressure at higher altitudes causes temperature controls to cycle at colder temperatures than they do at sea level. Thus it may be necessary to reset the range control temperature limits warmer for high-altitude installations. The table below gives typical adjustments necessary for different altitudes above sea level.

Altitude, ft	Screw adjustment*
Sea level to 1,400	No adjustment usually necessary
1,400–3,200	$\frac{1}{8}$ turn clockwise
3,200–5,000	$\frac{1}{4}$ turn clockwise
5,000–7,000	$\frac{3}{8}$ turn clockwise
Over 7,000	$\frac{1}{2}$ turn clockwise

* Some unit control thermostats are built so that the screw adjustment must be made in a counterclockwise direction; check manufacturer's service manual for the exact procedure.

To check if a control thermostat is operational, disconnect the wires from the terminals and with a low-voltage test light or ohmmeter check across the terminals. With the control tube warmed above the normal cut-on temperature and the control turned to the ON position, continuity should exist; if not, replace the control.

Heaters. In most modern refrigerators, there is a series of heaters to perform various functions. These include, for example, a dewpoint heater (to prevent sweating on the cabinet in the freezer area), a mullion heater (to prevent sweating on the mullion bar), a butter-keeper heater (to keep butter spreadable), an air duct heater (to prevent ice buildup around the air duct opening), a meat-keeper heater (to warm meat up to 28 to 32°F in a 0°F area), and a divider or stile heater (to prevent sweating on top of the provision compartment panel). Operation of these and other heaters can be checked in the same way: Remove the heater leads and connect an ohmmeter or a low-voltage test light across the terminals. If continuity does not exist, replace the heater unit. You should also take a ground check by placing one lead on the terminal and the other on an unpainted portion of the refrigerant tube. If continuity exists, replace the heater unit.

Still other electrical control devices are found in some refrigerators. For instance, one of the more popular is the so-called quick-chill compartment which is designed to chill beverages, salads, pies, etc., from room temperature or warmer to 45°F in half the time required to chill the same item(s) when normally placed on a refrigerator shelf. The two electrical devices in this compartment that may concern the service technician are these:

1. A mechanical timer (mounted on the side

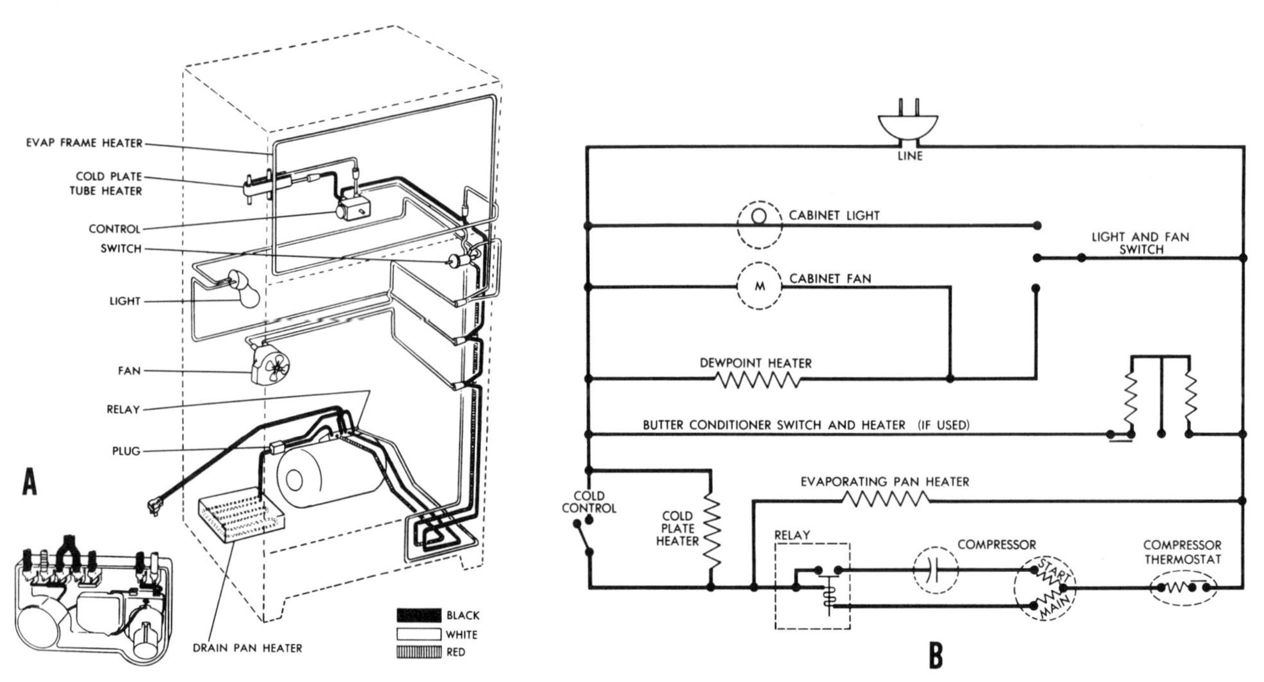

Figure 2-21. (A) Typical pictorial and (B) schematic diagrams of the electrical system.

of the food compartment) which operates an ON-OFF switch. The ON-OFF switch controls the quick-chill fan.

2. The quick-chill fan which circulates air within the quick-chill compartment. The quick-chill fan usually mounts on the rear of the food compartment liner near the refrigerator damper control.

To check a typical quick-chill compartment for proper operation, proceed as follows:

1. Make sure the unit is running and the food compartment is approximately 38° to 40°.
2. Turn the timer clockwise and check the quick-chill fan for operation.
3. Place a bottle filled with 90° tap water in the quick-chill compartment, in front of the fan air outlet.
4. Place a test thermistor or thermometer in the 90° tap water (Fig. 2-22).
5. Set the quick-chill timer to 20 min and close the quick-chill compartment door and then the refrigerator door.
6. The water temperature should drop 5° for every 5 min of elapsed time.

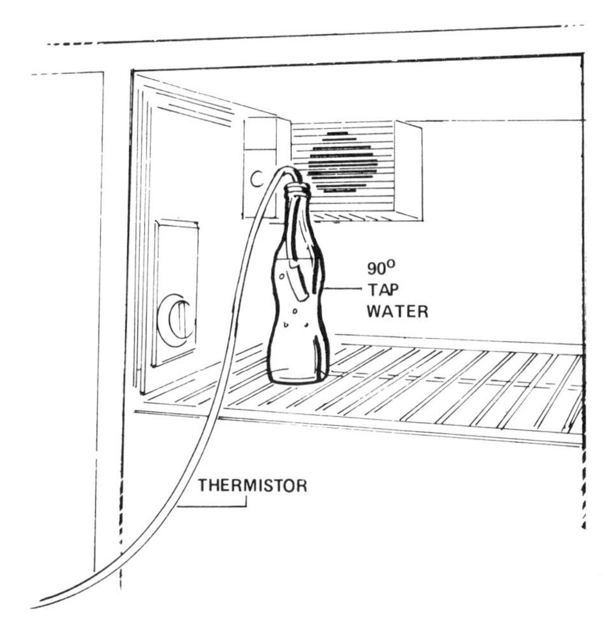

Figure 2-22. Typical quick-chill compartment test.

To check these devices, test the timer for continuity in the same manner as the unit thermostat, and test the quick-chill fan in the same way as the condenser fan. If you do not obtain continuity, replace the part.

Electric Defrost System

To properly diagnose a defrost problem, the service technician must thoroughly understand the functions of the electric defrost system. As described in Chap. 1, during the cooling cycle, the evaporator collects moisture in the form of frost. This accumulation of moisture (frost) must be removed periodically to maintain proper cooling temperature. If frost is allowed to build up, it will block air moving over the fins of the evaporator, no heat transfer will take place, and the fresh-food and freezer temperatures will rise. To remove this frost, the refrigerator must be defrosted.

The defrosting procedure may be accomplished automatically or manually. In the latter case, the defrost procedure should be carried out as follows.

1. Remove all frozen foods from the freezer section and cold-storage tray. Wrap the frozen foods in several thicknesses of newspaper to keep them from thawing.
2. Position the damper so it will direct the defrost water into the cold-storage tray
3. Turn the temperature control dial to the OFF position.
4. Place deep pans of *hot* water into the freezer section and close the refrigerator door. *Note*: If the complete refrigerator is to be cleaned at the time of defrosting, the foods and interior equipment can be removed and the interior of the refrigerator can be cleaned.
5. After the frost has melted, wipe the freezer section clean. *Caution*: Never use a sharp instrument such as a kitchen knife or ice pick to scrape frost. Always use a blunt instrument made of plastic or rubber.
6. Replace the cold-storage tray and put back the damper in its original position.

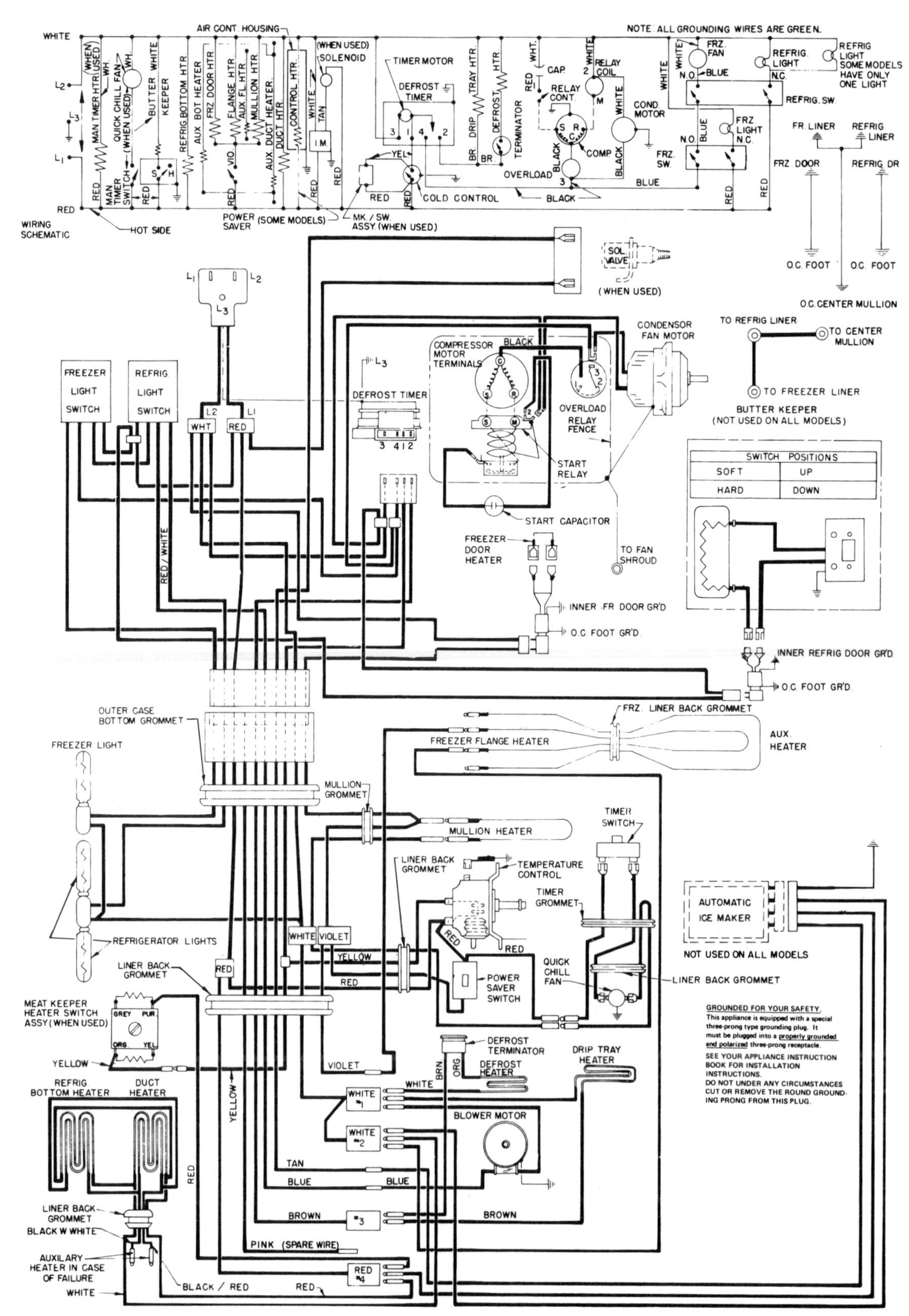

Figure 2-23. Typical electrical diagram as it appears in a service manual.

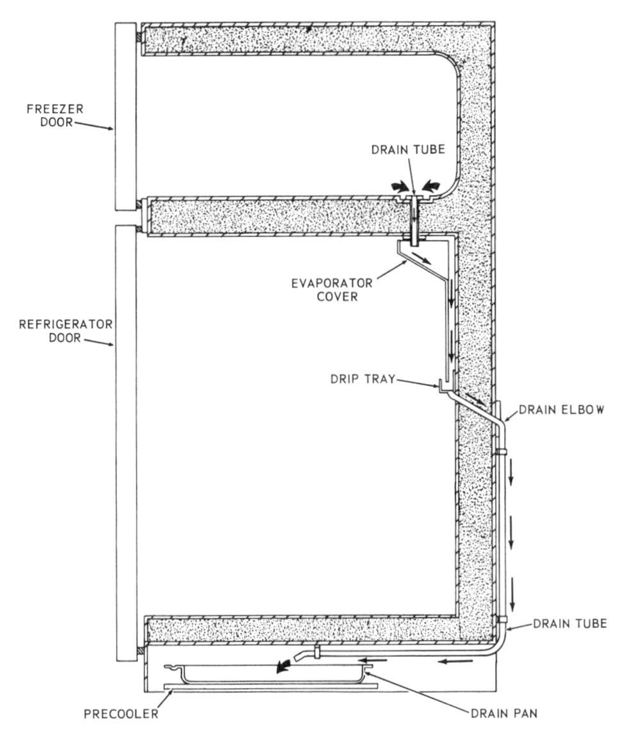

Figure 2-24. The defrost drain system.

7. Set the temperature control dial to the original position.
8. Replace all frozen food, interior equipment, and foods. As part of customer instruction, recommend that the freezer section be defrosted when the frost accumulation is about $\frac{1}{4}$ in thick.

As the name implies, automatic defrosting, which is called by a variety of names, requires no work on the part of the user. As described in Chap. 1, while there are several ways of achieving automatic defrost, the two most commonly used methods of initiating the defrost cycle are the cycle defrost and the timed defrost. Also, the actual process of defrosting can be accomplished in several ways, but the two most popular methods are with heating coils and with hot gas.

Timed defrost with heating coils. By far the most popular way of automatically defrosting a refrigerator is the timed-defrost method employing heating coils. This type of defrost system is usually made up of the following four basic items.

1. *Defrost timer.* This automatic timer determines or initiates when the refrigerator goes into its defrost cycle. As a rule, the defrost timer consists of a timer motor, a gear train, a cam, a set of contacts, a terminal board, a solenoid, a mounting plate, and a case. In a few models, the defrost cycle is initiation by a door switch arrangement in which a cam on the bottom of the refrigerator door trips a counter-mechanism. When the door has been opened and closed a predetermined number of times—for example, 20 times—the counter cam closes a defrost circuit and the defrosting operation begins. Otherwise the timed-defrost system works in the same manner as with an automatic timer. But since the timer is much more widely employed, it is used in our discussions here.
2. *Sheath or defrost heater.* This is the electric heating element which defrosts the evaporator. Most defrost heaters are either a rod type which is embedded in the fins of the evaporator or the radiant type which is located in close proximity to the evaporator. Either type is easy to replace.
3. *Defrost thermostat or limiter.* This thermostat, usually the bimetal type, controls the length of time heat is applied to the evaporator. It is mounted in a definite location to sense the evaporator temperature, thus indicating when all the frost is removed. Most defrost thermostats turn off when the evaporator temperature reaches approximately 55 to 70°F. The defrost heater cannot be energized again until the bimetal limit switch has been cooled down to its resetting temperature at about 30°F. For this reason, if the defrost limit switch is accidentally moved too close to the defrost heat, it could attain its cutoff temperature too soon and before the whole evaporator is defrosted. The unit would return to the freezing cycle with an incompletely defrosted

coil. After a few cycles, the evaporator will be blocked with ice. No air can flow through the blocked coil to cool the freezer and refrigerator. Food spoilage and "warm refrigerator" complaints by the customer will surely follow.

4. *Drip-trough heater.* This heater is generally the blanket type sealed to aluminum foil and wrapped around the underside of the drip trough. The trough heater prevents defrost water from freezing in the drip trough and possibly blocking the drain and airflow to the evaporator. Again, the length of time it is energized is determined by the defrost thermostat. In models that employ a radiant defrost heating element, a drip-trough heater is not necessary. The radiant heating element is below the evaporator and serves as both the evaporator and drip-trough heater.

To accomplish the defrost operation, the timer must stop the compressor and condenser fan motor. The point in stopping the compressor is obvious, in that we certainly would not want to continue to refrigerate while trying to remove frost with a heater. We stop the fan motor in order to keep from blowing warm air throughout the freezer area. The automatic timer is often called the "heart" of the defrost system, and to properly diagnose a defrost problem, the service technician must thoroughly understand its function, as well as those of the defrost thermostat and heaters. In Fig. 2-25 below we have pictured an internal view of a typical electric defrost timer. The middle leaf (which connects to terminal No. 1) is always electrically energized. In the run cycle, continuity exists only between terminals Nos. 1 and 4. As the cam turns, depending on the frequency of the defrost cycles, contact No. 4 falls from the ledge of the cam, opening contacts between terminals Nos. 1 and 4. At this time contacts Nos. 1 and 2 close and energize the defrost circuit.

At no time should continuity exist between terminals Nos. 1 and 2 and 1 and 4 simultaneously. If this condition should exist, the timer must be replaced. (*Note*: To check this timer for continuity, you must remove *all* wires on terminals Nos. 1, 2, and 4 to eliminate any possibility of a false continuity reading.) Actually, this schematic illustrates the simplicity of most timer circuits. The incoming current connects directly to terminal No. 1 on the timer; the neutral side of the line connects to terminal No. 3 on the timer. Terminal No. 1 is connected to the center leaf which has two sets of contacts. In the defrost cycle, contacts on leaves Nos. 1 and 2 are closed, and current flows to the defrost thermostat. Most defrost thermostats close at 20°F $\pm 5°$ and open at 55°F $\pm 5°$. (Refer to the service manual for correct temperature of various models.) If the evaporator temperature is 15°F or lower, the thermostat contacts should be closed and current would flow to the defrost heater (embedded in the evaporator) and the drip-trough heater directly below the evaporator. As an example, when the defrost thermostat senses 55°F $\pm 5°$, it should open and remain open for the remainder of the defrost cycle. As the timer cam continues to turn, leaf No. 1 will fall from the ledge on the cam. At this time contacts Nos. 1 and 2 will open and contacts Nos 1 and 4 will close and initiate the run or cooling cycle. When the defrost thermostat senses an evaporator temperature of 20°F $\pm 5°$, its contact will close and await the next defrost cycle. The next defrost cycle will be initiated when leaf No. 4 falls from the ledge on the timer cam. As was stated earlier, the number of defrost periods varies from one to four in 24 h, depending on the particular timer employed. The number

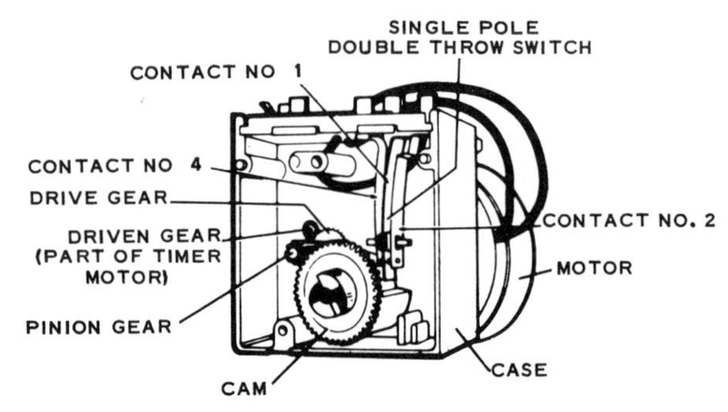

Figure 2-25. Cutaway view of a typical defrost timer.

of minutes the compressor is off varies from 15 to 45, depending on the design of the defrost system.

Checking defrost timer. When checking for a defrost timer malfunction, it is important to check the timer shaft for movement before advancing the shaft manually. If the timer shaft is advanced manually, it could momentarily correct a malfunction such as a stalled timer motor or broken teeth on the timer gears. If the timer contacts stick together, the defrost heaters would be on while the compressor is running. When this occurs, condenser temperatures are much higher than normal. A high-wattage draw would also indicate this condition.

Prior to checking the defrost timer, the unit temperature control must be "closed" and the compressor running. The evaporator must be approximately 20°F or colder. Then remove the defrost timer and observe the window on the back of the drive motor. If the drive motor does not rotate, manually advance the defrost timer until you hear a click, indicating that the center contact has moved from the refrigeration contact to the defrost contact. (The timer may be manually advanced by inserting a screwdriver in the slot provided. Under no circumstances should the timer manual advancing screw be turned backwards.) If the drive motor rotor starts to rotate, you have either an open defrost thermostat or open defrost heater. However, if the drive motor still does not rotate, you have a defective timer. Replace the defrost timer.

The defrost timer can also be checked for continuity as described earlier in this section.

Checking defrost heater. Remove the defrost heater leads. Then connect an ohmmeter or a test light across the two defrost heater terminals; continuity should exist. If not, replace the defrost heater. Also ground-check the defrost heater by checking each terminal to an unpainted refrigerant tube; if continuity exists, replace the defrost heater.

Checking defrost thermostat. To check the defrost termination thermostat, disconnect the wires coming from the evaporator. Using an ohmmeter or test light, make a continuity test.

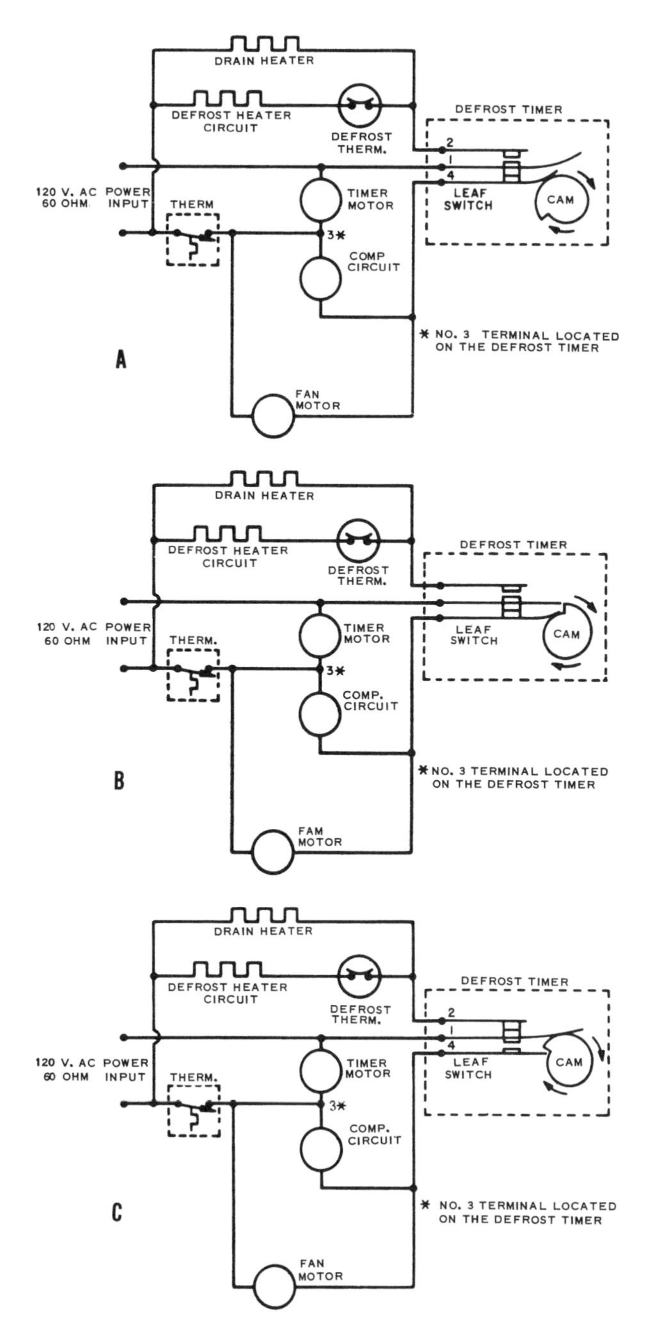

Figure 2-26. Timer connections: (A) normal cycle; (B) start of defrost cycle; and (C) end of defrost cycle.

The thermostat should be closed if the evaporator temperature is 20°F or below. When the evaporator warms up to approximately 55°F, the defrost thermostat should open, and no continuity

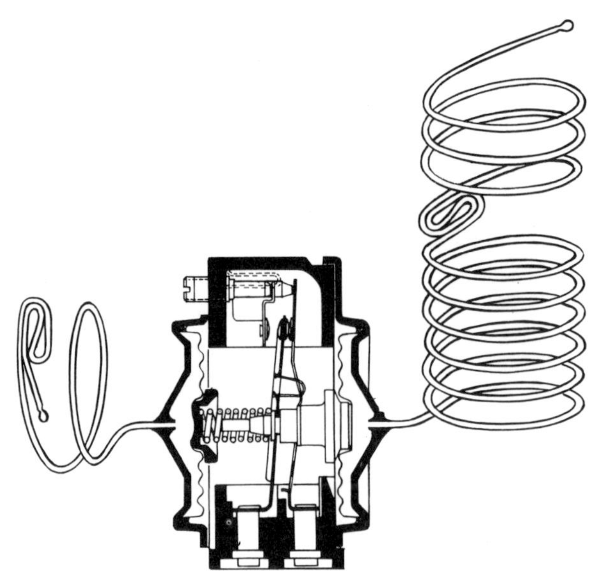

Figure 2-27. Cutaway view of a typical automatic defrost thermostat.

should exist. If either tests fail, replace the defrost thermostat.

The lower barometric pressure at higher altitudes can cause the defrost thermostat to trip out of defrost at lower temperatures than at sea level. Thus it may be necessary to reset the defrost thermostat warmer for higher-altitude installations. Only a defrost thermostat with a bellows tube can or needs to be adjusted for different altitudes. Owing to the number of turns required for higher-altitude adjustments, it is suggested that a piece of tape be placed on the shank of the adjusting screwdriver and an index mark added to the screwdriver tape and to the control face. Turn the screwdriver one-half turn each twist of the hand until the last fraction of a turn is reached. Once the screwdriver is inserted into the screw slot, do not remove it until the adjustment is complete. Move your hand around the screwdriver to make the next twist. Except for altitude adjustments, no adjustments in the defrost thermostat limits should be made.

Checking drip-trough heater. Remove the the drip-trough heater leads. Then connect an ohmmeter or a test light across the two drip-trough heater terminals; continuity should exist. If not, replace the drip-trough heater. Also ground-check the drip trough by checking each terminal to an unpainted refrigerant tube; if continuity exists, replace the drip-trough heater.

Timed defrost with hot-gas method. In order to use a hot-gas defrost, a solenoid valve must be used to direct the refrigerant through the condenser during cooling, and through the hot-gas tube during the defrost cycle. A counter or timer must also be used to initiate the defrost cycle. To defrost the freezer evaporator, the following conditions must exist.

1. An automatic timer control or door cam switch must initiate the defrost cycle.
2. The unit must be running to pump the hot gas through the evaporator.
3. The solenoid must be energized so the gas will bypass the condenser and heat exchanger. *Note*: The freezer fan does not run during a defrost cycle, when the compressor is not running, or when the freezer door is open.

To accomplish the conditions mentioned above, the defrost control, unit control, freezer fan, solenoid, and defrost heater are all interlocked electrically. The defrost control is fed through the unit control (the unit control must be closed to the unit in order to feed the defrost control). We said the defrost control is a single-pole, double-throw switch. One side feeds the freezer fan, and the other side feeds the solenoid and defrost heater. When the solenoid is energized, the plunger in the valve is pulled up, which then opens the port to the hot-gas line. It does not, however, close off the discharge line to the condenser.

As soon as the valve is opened, the gas is pumped through the hot-gas line because this is the path of least resistance, owing to the higher pressure in the condenser and capillary tube as compared with the low pressure in the hot-gas tube. After a period of time, the pressure in the condenser and capillary tube will tend to drop; however, the gas still passes through the hot-gas tube in that it is still the path of least resistance because of the restriction presented by the capillary tube. Thus, we can say that there is gas in the condenser and capillary tube, but it is inactive gas and has no bearing on the actual defrost cycle.

The refrigerant is pumped through the hot-gas line as a hot gas into the evaporator and cold plate where the refrigerant turns to cold gas and some liquid. This change occurs because the evaporator and cold plate are now functioning like a condenser (reduces temperature of gas while pressure stays constant). The cool gas and liquid then passes through the suction line back to the compressor where the cycle is repeated again and again, until the evaporator (where the defrost control bulb is attached) reaches a temperature of 45°F, at which time the defrost control switches to the other contact, which de-energizes the solenoid and defrost heater and starts the freezer fan. The defrost cycle will last anywhere from 8 to 30 min, depending upon the ice buildup on the evaporator.

Methods of checking the various timing devices, defrost heaters, and fans have already been covered fully. To replace the defrost valve, proceed as follows.

1. Discharge the system by cutting the suction line in a straight section of tube near the compressor.
2. Using diagonal-cutting pliers, cut all three tubes near the solenoid valve.
3. Using a tube cutter, cut the sealed ends off the tube on the replacement solenoid valve, and cap the tubes.
4. Cut the connector tubes where they will conveniently connect to the tubes on the replacement solenoid valve; allow enough extra tube for swedging the end of the tubes.
5. Weld all tubes; then follow the procedure for evacuating and charging (see Evacuating and Recharging the System, under Refrigeration System below).

Cycle defrost. As was stated in Chap. 1, the cycle defrost system makes use of a wide-differential thermostat which has a cut-in point (contacts closed, compressor running) above freezing. The above-freezing cut-in point allows the evaporator to completely defrost before the next running cycle. The defrost heater (which is in direct contact with the evaporator) is wired across (in parallel with) the thermostat contacts so that when the contacts are open, the defrost heater operates. The heater operates on every OFF cycle of the compressor.

In some cycle-defrost models an ambient compensator is located between the evaporator plate inlet and the outlet of the refrigerant tubes. Wired in parallel with the defrost thermostat, the ambient compensator is energized (heats) during the compressor OFF cycle (thermostat contacts open) or when the thermostat is turned to the OFF position. To check the ambient compensator, as well as the defrost thermostat and the defrost heat, use a simple continuity test. If proper continuity is not established, replace the part.

With a cycle defrost system it is wise to inform the customer that the freezer evaporator must be defrosted manually at least twice a year. Frost should not be allowed to build up to a thickness of more than $\frac{1}{4}$ in. Some excess frost can be removed without complete defrosting by scraping it away with a plastic spatula or scraper (never use a metal scraper). To completely defrost the freezer evaporator, turn the thermostat to OFF position. Open both doors and allow the frost to melt off as described earlier in this chapter.

Throughout the electrical checks we must assume that only one part has gone bad at a time. While it is remotely conceivable that you

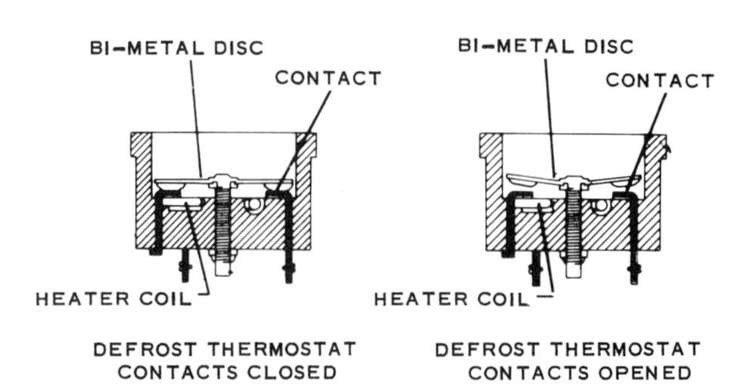

Figure 2-28. Defrost thermostat parts: open and closed.

will find a multiple-part failure, it is highly improbable and would not pay us here to go into a lengthy discussion of how to diagnose a multiple-part failure. Also, under electrical checks (and this holds true for those of the refrigerant system given later in the chapter), we shall make no effort to relate the failure of the electrical part to the particular complaint, as this will be done in the diagnosis section. The purpose of this portion of the chapter is merely to acquaint you with the correct place to attach a continuity tester or ohmmeter in order to determine whether the part in question is functioning properly electrically. Remember to break the source of supply before attempting to service an electrical component of the refrigerator. Always follow color coding when rewiring or when connecting leads to electric components. The wiring diagrams show resistance values to assist in diagnosis. Unless otherwise stated, assembly operations are the reverse of disassembly procedures. For rewiring and connecting of leads, refer to the wiring diagram in the service manual, or to the wiring diagram which many manufacturers paste someplace on the back of the cabinets.

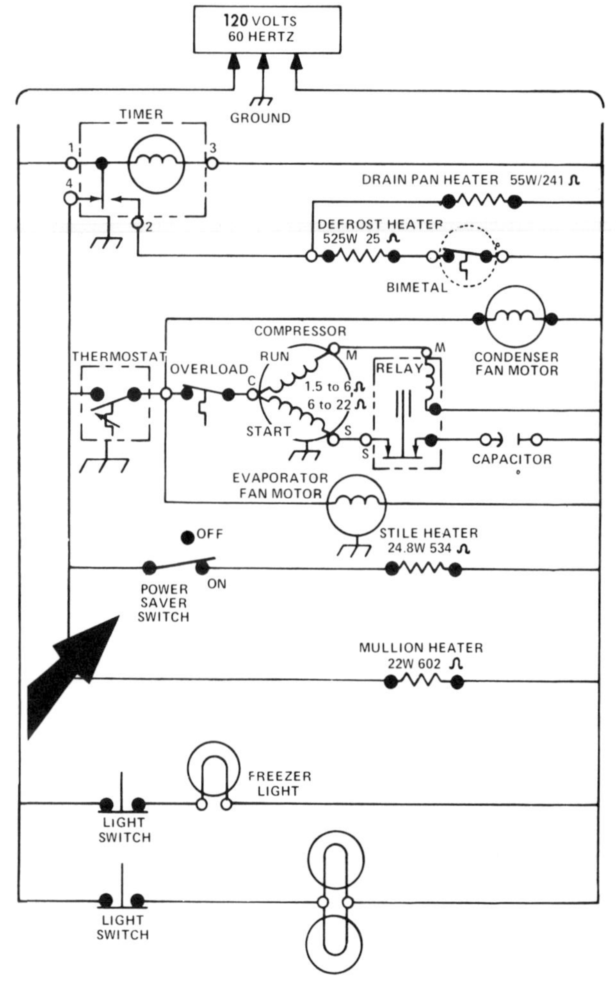

Figure 2-29. Power-saver switch in series with the stile heater.

Power-saver switch. The power-saver switch, also known as the "power miser" or the "economy switch," is found in some refrigerator models. As a rule, this switch will be found on the side of the liner in the fresh-food section. It is an ON and OFF switch which is in series with the stile heater on some models and with both the stile and mullion heaters on others. It will help reduce operating costs if the user lives in an area where moisture and humidity is not a problem. The user can snap it to LOW or OFF (Fig. 2-29). This would (in this example) remove a 24-W heater, normally energized continuously, from the circuit. The savings on different models would depend on the wattage rating of the heaters. If the humidity is high enough to cause condensation or sweating on front edges of the cabinet around the freezer section, the switch must be switched to HIGH or ON.

Test adapter. Some manufacturers have available for their models a *r*apid *e*lectrical *d*iagnosis (or *red*) which is a quick and accurate method of diagnosing electrical faults in refrigerators and freezers. From the front of the cabinet, without disconnecting the power cord, without disassembling any parts, and without unloading any food packages, the electrical system can be evaluated through a multicircuit connector. The multicircuit connector, when separated, isolates parallel circuits in the wiring harness to permit testing of all electrical components and related wiring within the main harness.

To perform necessary electrical tests, the special adapter or red (Fig. 2-30) is connected to the wiring harness through the multicircuit connector. The wiring harness is thus conveniently extended to a row of jacks, serving as test points, on each side of the test adapter panel. The jacks are identified by number, corresponding to the actual wiring terminal location in the multicircuit connector. Above the two rows of jacks, the electrical symbols for the terminals in the multicircuit connector are indicated. On the left-hand side, female terminals are indicated by the arrow tail, and on the right-hand side, male terminals are indicated by the arrow point. The upper push-button switch completes the circuit through the test adapter to jack No. 3 on the left-hand side ("3 female"). This terminal in the multicircuit connector is always the line ("hot") side of the power cord. The lower push-button switch completes the circuits through the brown and orange jumpers, extending from the bottom of the test adapter (Fig. 2-31).

The test adapter is used in conjunction with an ordinary ohmmeter for resistance testing of component circuits. As long as the ohmmeter is in good working condition (fully charged batteries, leads not broken, etc.), the reading should be satisfactory since the resistance values to be measured are relatively low.

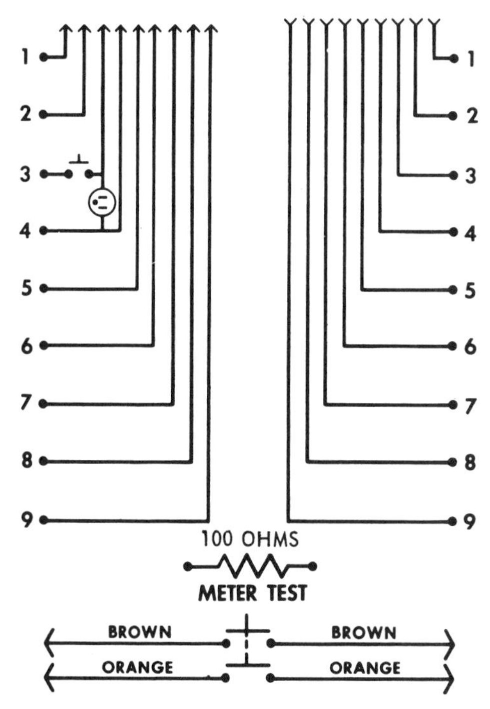

Figure 2-31. Schematic of a typical test adapter.

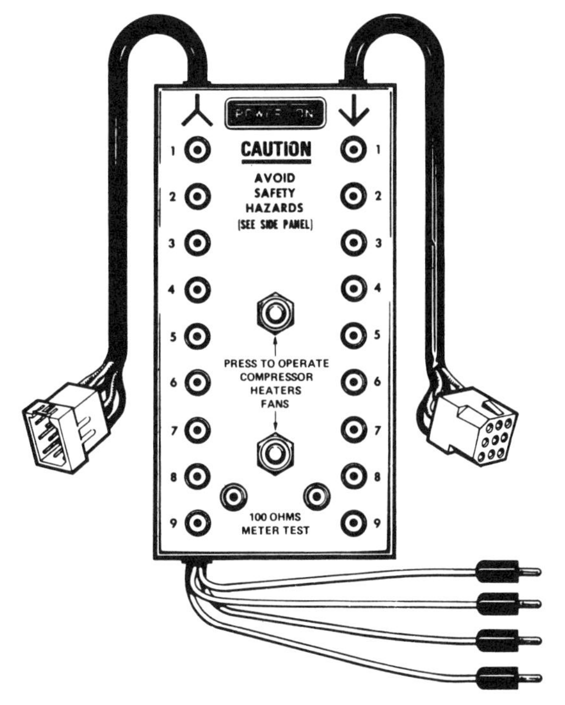

Figure 2-30. Typical rapid electrical diagnosis or test adapter.

PERFORMANCE

One of the best indications of performance of a refrigerator is the amount of power consumption. Of course, the electric energy required to operate a refrigerator per day is determined by many variables. Some of the more important variables are the following.

1. Volume of storage compartments
2. Storage temperatures
3. Control setting
4. Room air temperature
5. Room air humidity
6. Frequency of door openings
7. Food cooling and freezing load
8. Ice freezing load
9. Cleanliness of condenser

Because of these many variables, it is impossible to state the exact number of kilowatt hours per day required to operate a particular model in an individual home. However, numerous factory tests have been made under no-load conditions (refrigerator doors kept closed) to obtain data for the No-Load Performance Range table. Power consumption for an actual installation will always be more than for any of these no-load conditions. Actual power consumption can be estimated by multiplying one of the figures in the no-load table by a usage factor. The usage factor is obtained from the following table after estimating the climate and usage condition (obtain no-load performance data from the table in the section Performance Data and Their Use, p. 123).

No-load performance range.

Climate and usage condition	Usage factor
No usage (no-load)	1.0
Minimum usage	1.2
Moderate usage	1.35
Heavy usage	1.5
Extremely heavy usage	1.7

Example:

The following shows an application of material in this table to calculate power consumption.

Refrigerator model	
Average room temperature	80°F
No-load kWh/day, avg.	1.5 kWh/day
Estimated usage	Moderate
Usage factor (from table)	1.35
Estimated power consumption	
	1.5 × 1.35 = 2.02 kWh/day

Most temperature complaints can be resolved on the first call by taking instantaneous temperature readings with a service technician's pencil-type thermometer. In checking fresh-food temperatures, it is best to take a sample of a beverage which has been in the refrigerator for 24 h. Freezer-compartment temperatures are best checked by taking ice cream temperature. To accurately check ice cream temperature, insert the thermometer into the ice cream for a period of 1 min. Then insert the thermometer in another section of the ice cream for 2 min. Read the thermometer with the bulb remaining in the ice cream. Concurrently with checking temperatures, check the possible causes of high temperatures that are listed in the service diagnosis section of this chapter.

Many complaints of long running time or high power consumption can be resolved on the first call by questioning the customer as to the exact details of the power consumption complaint. Does the refrigerator cycle at all? How often? A lint-covered or dirty grille, or condenser or a piece of sound insulation or improperly positioned drain pan blocking the airflow can cause long running time. A loose control bulb clamp or control on COLD position or out of calibration will cause long running time with too-cold temperature.

If the power consumption and/or temperature complaint cannot be quickly resolved, a temperature-cycle recorder should be placed in the refrigerator. The recordings of this instrument between the hours of midnight and 6 A.M. should be compared with the no-load performance data given for that model in the data and diagrams section of a service manual. Since there is usually very little loading or usage during the 3 h preceding midnight and then no usage after midnight, the hours between midnight and breakfast time can be considered the period when the refrigerator is operating under nearly no-load conditions.

For more information on refrigerator performance data and ways to use them, see Chap. 3.

REFRIGERATION SYSTEM

As thoroughly covered in Chap. 1, the refrigerant cycle consists of heat-laden refrigerant being pumped from the compressor to the condenser on the cabinet back. In the condenser the heat is passed off into the surrounding room, and the refrigerant condensed into a liquid. The liquid refrigerant then passes through the capillary tube to the evaporator. The refrigerant enters the larger evaporator tubing, which releases the pressure on it. This allows the refrigerant to boil and absorb the heat from the cabinet. The heat-laden vapor returns from the evaporator via the suction tube.

To be able to successfully diagnose a problem in a defective refrigeration system, it is important to understand the characteristics of a normally operating system. A correct charge of refrigerant is indicated by a normal wattage draw and temperature differential as specified in the performance chart for the particular model. (See Performance Data and Their Use, p. 123.)

When the refrigerant charge is normal, most of the refrigerant will be in the evaporator during the cooling cycle, and only the final few passes of the condenser will contain liquid refrigerant. Approximately 90 percent of the condenser will be high-pressure gas in the process of heat dissipation and condensation.

Approximately 3 min of the OFF cycle is required for the high- and low-side pressure to equalize and thus permit the compressor to start at the beginning of the next run cycle. The temperature of the suction line rises and the temperature of the discharge line lowers during the OFF cycle. At the beginning of the run cycles there is a noticeable lowering of the suction line temperature and a raising of the discharge line temperature.

These, then, are the characteristics of a normally operating system and are again listed.

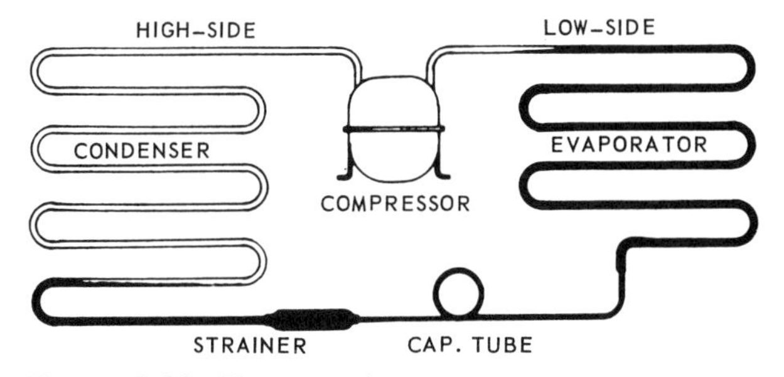

Figure 2-32. The normal system.

1. The evaporator will contain the greater part of the refrigerant.
2. The condenser will contain very little refrigerant. (A simple check is to use a match and heat the various passes of tubing. Feel with your hand to locate where the liquid level is, as indicated by the cooler temperature at that point.)
3. No temperature drop at the drier inlet and outlet.
4. Evaporator temperature within 5°F at the inlet and outlet at shutoff.
5. Pressures will equalize after cut-off.
6. First 30 to 60 s of the ON cycle, the suction line is cooler at the compressor.

Causes of System Failures

The sequence of steps necessary to properly diagnose a complaint of no cooling or not enough cooling is as follows.

First, be sure that the compressor is running. If it is not running, make sure that the unit is plugged in and that all circuit breakers (or fuses) are all right. Use the test cord as shown here to determine whether the compressor is good. Depress the momentary switch for 2 or 3 s. If the compressor does not start, replace it. If the compressor does start and continues to run, trouble elsewhere in the system is indicated. Check all electric components in the compressor circuit, such as the starting relay, thermostat, start capacitor, defrost timer, and overload protector, and look for broken wires or loose connections.

There are three main causes for system failure. In the order of their occurrence, they are:

1. Restrictions
   a. Partial
   b. Complete
2. Improper refrigerant charge
   a. Undercharge
   b. Overcharge
3. Compressor failure
   a. Stuck
   b. Defective motor
   c. Low-capacity compressor

Assuming that the compressor is operating, the first check of the sealed system should be to look for a frost pattern on the evaporator. If no frost pattern is seen, this will indicate that no liquid refrigerant is entering the evaporator. This could be caused by either a complete restriction or loss of refrigerant.

Undercharged system. When the system is short of charge, the wattage and temperature differential will be lower than normal in a performance check. An undercharge will affect the evaporator in the same way as will a partial restriction. The extent of the undercharge will determine the amount of liquid refrigerant in the refrigeration or cooling produced. When an evaporator is undercharged, a portion of it may frost because of the reduced low-side pressure. The more extensive the undercharge, the warmer the suction line will be during the running cycle.

With an extensive undercharge there will be more high-pressure hot gas and less liquid in the condenser and consequently higher temperature near the condenser outlet. A quick temperature rise in the condenser tubing when heat is applied indicates gas and not liquid in the tubing at that point.

While examining the evaporator for a frost pattern, also listen for a definite hissing sound near the point where the capillary tube enters the evaporator (evaporator inlet). A hissing sound will indicate that the capillary tube is not restricted (plugged) and, therefore, the system is undercharged because of a leak.

Other indications which will verify an undercharged condition are a cool, quietly running compressor, with low wattage being drawn because it is doing little work. The condenser will be at or near room temperature from top to bottom because it contains little or no liquid refrigerant. Having reached this conclusion, the service technician's next step is to install access-valve assemblies, as described in Chap. 1, and connect the manifold and gauges to the system. The gauge readings should verify the conclusion by showing very low pressure on the high-side gauge and a deep vacuum on the low-side gauge. Also, the pressures will equalize when the compressor is shut off.

Overcharged system. Under certain conditions it is possible to encounter an overcharged system. When this occurs, the wattage will be higher than normal and the temperature differential will be lower than normal in a performance check. With an overcharge, liquid refrigerant from the evaporator will flood over into the suction line and possibly even into the compressor, depending on the extent of overcharge. In most cases this will cause frosting of the suction line and possibly even the compressor dome. Both temperature and humidity will affect the extent of frost deposit.

With an extensive overcharge the liquid level in the condenser will be higher than normal with a marked temperature difference between the top and bottom passes of condenser tubing. Application of heat to the return bends of the condenser will indicate the liquid level.

If the system is considerably overcharged, the compressor will more than likely be damaged if permitted to operate, as it will be attempting to pump liquid refrigerant (slugging).

Restricted system. With a partial restriction

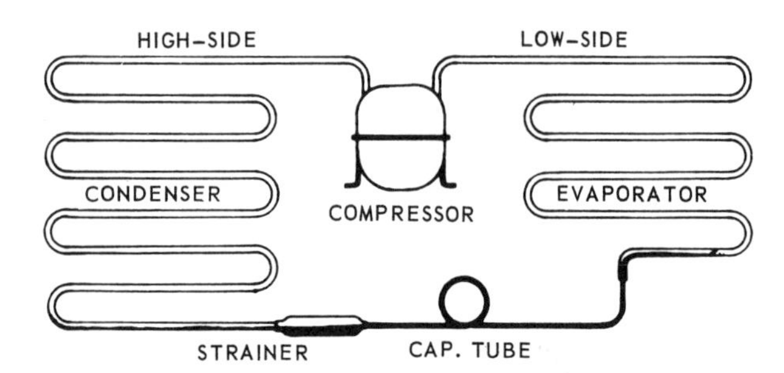

Figure 2-33. The undercharged system.

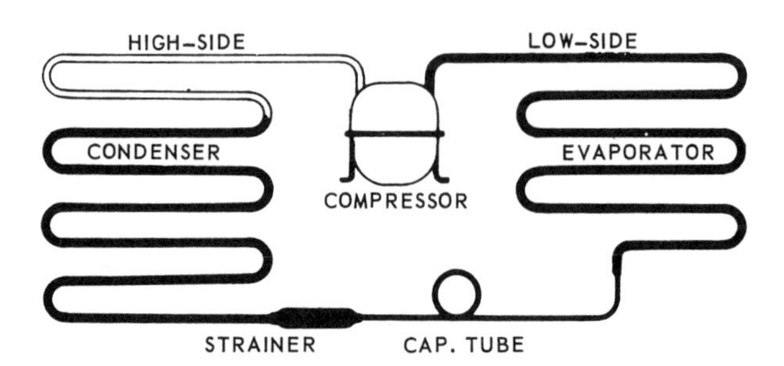

Figure 2-34. The overcharged system.

the evaporator will be starved for liquid refrigerant and will react much the same as with an undercharge. The suction will be warmer than normal, but the evaporator will likely frost in the area where liquid is present. If the restriction in the strainer or capillary tube is nearly complete, there will be a marked temperature difference before and behind the point of restriction.

With a partial restriction the condenser will contain an above-normal amount of liquid refrigerant, and there will be an above-normal temperature difference between the top and bottom passes of condenser tubing. The discharge line will be very hot, but the wattage will be lower than normal. It will take a longer-than-normal time for the system to equalize when the compressor is stopped.

When a complete restriction is present, both the evaporator and condenser will be at room temperature and the wattage will be below normal. With a complete restriction, once the compressor has stopped, it will rarely start again.

Table 2-1 provides a summary of the various refrigerant troubles and how to differentiate between them.

Table 2-1. **Refrigeration system troubleshooting chart.**

When unit has . . .	Compressor suction line will be . . .	Compressor discharge will be . . .	Capillary tube will be . . .	Evaporator will be . . .	Condenser will be . . .	Wattage will be . . .
Correct charge of refrigerant	Cold, but slightly warmer than evaporator; compressor not sweating	Very hot	Warm	Cold	Very hot	Normal
Undercharge of refrigerant	Warm, near room temperature	Hot	Warm	Partially warm near outlet; very cold, possible frost near inlet	Hot	Lower than normal
Overcharge of refrigerant	Very cold sweating of suction line and compressor; frost under low-evaporator-load conditions	Slightly warm to hot	Cold	Cool to cold	Slightly warm to hot	Higher than normal
High-side restriction (partial)	Warm, near room temperature	Very hot	Cool	Partially warm near outlet; very cold possibly frost, near inlet	Low passes cool compared to top	Lower than normal
High-side restriction (complete)	Room temperature	Hot and then room temperature	Room temperature	Cool and then room temperature	Warm and then cool to room temperature	High and then low

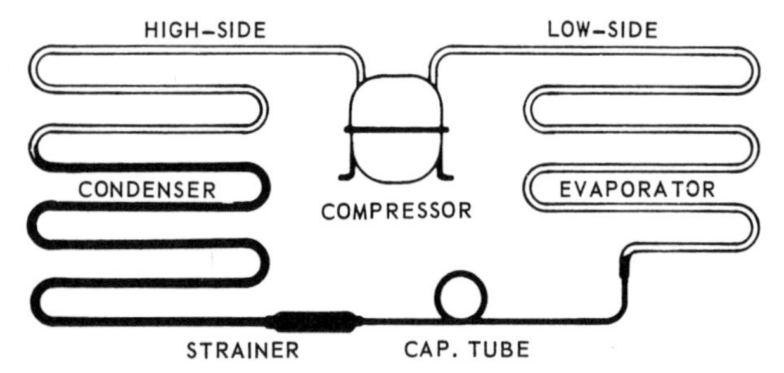

Figure 2-35. The restricted system.

Compressor troubles. If the compressor will not run, the following procedures should be carried out in order to establish which of the compressor units is at fault, the compressor itself, the starting relay, or the overload.

To check the voltage, use a conventional voltmeter. The voltage may vary 10 percent from that stated on the nameplate value. If it is more or less than this tolerance, check with the local utility company. Remember that an appreciable difference usually means that the wire size is too small or the run is too long. Either of these problems usually requires an electrician to correct.

If the voltage is within the nameplate tolerance, remove the terminal box cover which is usually mounted on the compressor. As a rule, this will expose the compressor terminals, the starting relay, and the overload protector. To check for open compressor motor windings, see if continuity is registered on an ohmmeter or test lamp between the COMMON and START terminals and the COMMON and RUN terminals. If there is no continuity, there is an open circuit. If continuity is registered between any of the terminals and the compressor housing, this indicates a grounded winding. Also if the amperage is over 15 percent of the value printed on the compressor nameplate, there is probably a short circuit in the electric windings of the compressor. Either an open or grounded winding makes replacement of the compressor necessary.

As was stated in Chap. 1, the starting relay has the task of aiding the compressor in starting. When the compressor starts, a high starting current flows into the relay, energizing the solenoid which causes the closing of the movable contacts and the subsequent closing of the circuit. As the compressor reaches its normal rotating speed, the current requirement decreases to the point at which the solenoid is no longer able to attract the movable contacts. When relay contacts open, the starting winding is no longer energized and the compressor continues to operate with the running winding only. To test the starting relay, proceed in the following order.

1. Disconnect the feeding cable plug.
2. Set the thermostat knob to STOP.
3. Insert the circuit tester plug in the socket.
4. Touch the relay terminal, in series with the line, with the circuit lamp tester and the relay terminal corresponding to the wire originating in the overload cut-out with the circuit lamp tester. If the lamp goes on, try to join the two relay terminals with a wire junction, with the refrigerator plugged in and the thermostat knob toward an operating position. If the compressor starts, this is evidence that the movable part of the relay is blocked with the contacts open, which requires the replacement of the relay.

If, on the other hand, the compressor operates with excessively short cooling cycles and you are sure that the thermostat is normal, you can check the relay as follows.

1. Insert the refrigerator power cord plug in the wall socket.
2. Set the thermostat knob to the STOP setting.
3. Connect the circuit tester ends to the relay terminal to which the current arrives and the end where the conducting wire from the thermostat is located.
4. Turn the thermostat knob to an operating position and watch the test lamp. If it remains unlit for a few seconds and then lights up and does not go off any longer, it means that the relay is blocked with the contacts closed and must be replaced.

It is important to keep in mind that a hot-wire relay can be faulty in any one of the following three ways:

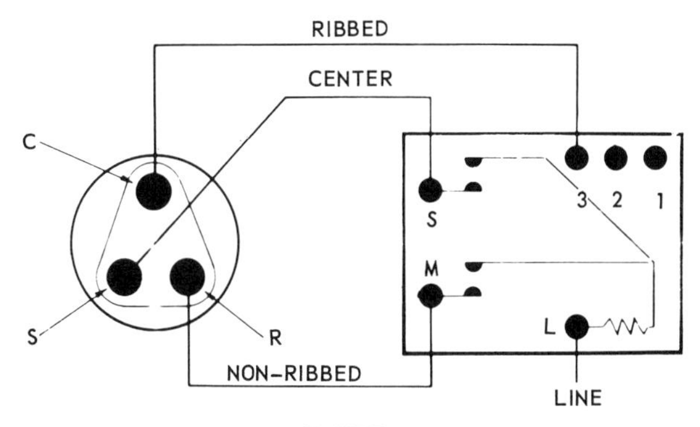

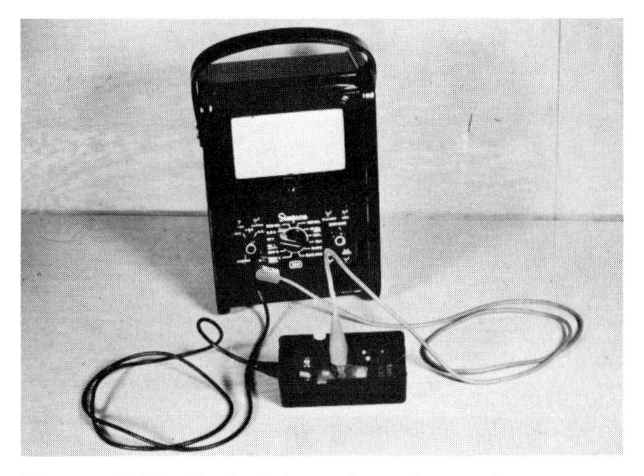

Figure 2-36. Typical hot-wire relay and motor to relay cable socket (top). Continuity check of a hot-wire relay (bottom).

1. *Contacts stuck open.* If the contacts stick in the open position the compressor will not run unless wired direct. With the service cord unplugged, remove motor to relay socket from compressor. With a low-voltage test light or ohmmeter check for continuity between terminal L to M and S on the relay (Fig. 2-36). Continuity should exist (compressor should start); if not, replace relay.
2. *Contacts stuck closed.* If the contacts stick in the closed position, the compressor would run for 5 s or less. If correct voltage is present at the instant of start, replace relay.
3. *Calibration.* If the compressor attempts to run, check for correct voltage at the instant of start. Voltage must be at least 105 V. If voltage is correct, wire compressor direct; if the compressor runs, replace relay.

On most refrigerator compressors, the overload cut-out consists of a bimetallic disk or bar which is connected to a resistance; it protects the compressor against any electrical overloads. The electric resistance is connected in series to the starting and running winding of the compressor, and its dimensions allow the passage of the current required for the normal operating of the compressor. In fact, the bimetallic arrangement is so sensitive that when the current intensity rises above normal, the heat generated by the resistance causes a deformation of the disk or bar which is sufficient to open the circuit and stop the compressor. The same thing happens when the compressor heats excessively. Once open, the bimetallic disk or bar remains open until the temperature of the resistance of the compressor has become normal. When this takes place, the overload contacts close, inserting the circuit again.

The test of the overload cut-out may be effected by inserting a junction between the two ends of the overload cut-out so that it is excluded from the circuit. If the compressor starts, it means that the overload cut-out is defective and must be replaced.

The overload protector and the starting, to be mounted on the compressor, are chosen with specific characteristics. Should replacement be required, it is not sufficient to make sure that the new components are suited for the same power of the compressor, it is necessary that the new parts have the same characteristics as those printed on the replaced components.

The motor-compressor can be wired direct by using a starting test set such as the one shown in Fig. 1-30. This device, as shown in Fig. 2-37, bypasses all control circuits and will operate virtually on any hermetic compressor. If the compressor runs when wired direct, it is an indication that the trouble is in the cold control (thermostat) timer, relay, overload, or connecting wiring and *not* in the compressor

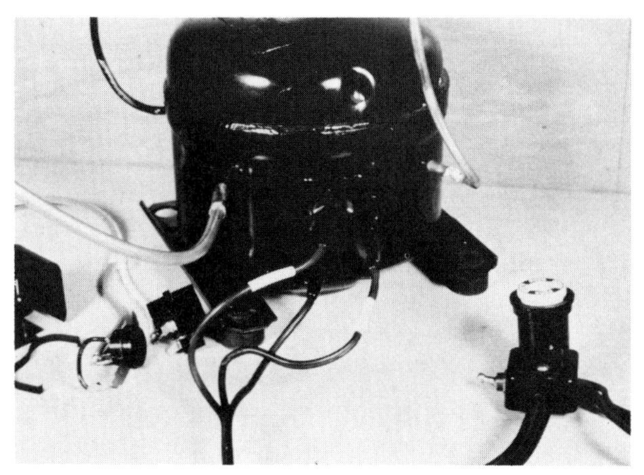

Figure 2-37. The use of a starting test set to wire a compressor direct.

itself. If compressor does not run when wired direct, before condemning the compressor, discharge the system by scoring the cap tube 1 in from the strainer or drier with a three-corner file. After system is discharged, wire direct again. If compressor runs after discharging, a restriction is evident. If compressor does not run after discharging, first check your wire direct cord. If wire direct cord checks okay, replace compressor. If the compressor has a burnt odor, inspect the compressor oil. Place a clean white cloth over suction line and tip compressor. If oil is dark and/or sweet smelling, the entire system must be flushed.

Typical compressor terminal configurations are shown in Fig. 2-38. To successfully wire the compressor direct, the service technician must be familiar with the terminal arrangements on the various compressors. When a terminal arrangement is in question, the common, start, and run terminals can be determined as follows:

Using an ohmmeter set on the R × 1 scale, find two terminals having the highest resistance reading between them. These are the start and run terminals. The remaining terminal is the common terminal. Connect one lead of the ohmmeter to the common terminal. Then touch the other lead to the remaining terminals (one at a time). The highest resistance reading shows the start terminal. The remaining terminal is the run terminal.

Check the motor-compressor for ground before wiring it direct. Continuity check from each compressor terminal to an unpainted refrigerant tube coming from the compressor. No continuity should exist; if continuity exists, replace the compressor. Its windings are internally grounded to the compressor shell.

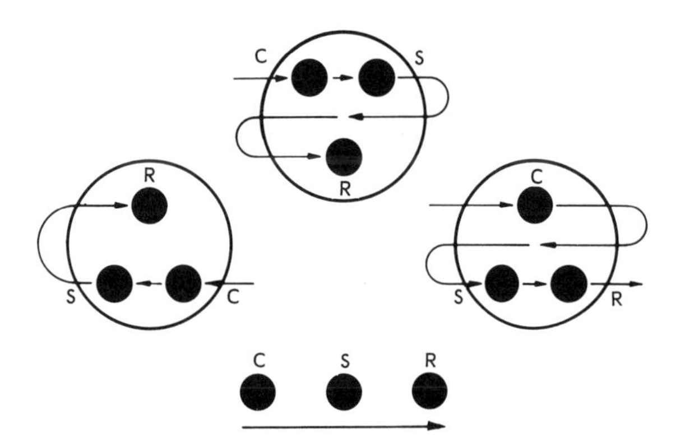

Figure 2-38. Typical compressor terminal configurations.

Compressor stuck. If the compressor hums when the momentary switch on the starting cord is depressed, but does not start, the compressor is probably stuck and should be replaced. *Caution*: Never attempt to free a stuck compressor by applying voltage higher than the nameplate rating except in extreme emergencies, and then only for a very short period of time. High voltage will cause the winding insulation to burn and thus contaminate the system.

If, after you have replaced a defective compressor, it runs for a short period but trips the overload protector, this condition may be caused by having incorrectly connected the precooler to the compressor, causing the compressor to overheat. Correct this condition by referring to the assembly information for compressor replacement and follow the instructions for connecting the precooler.

Little or no load on compressor. If the compressor coasts to a stop when it shuts off, either little or no refrigerant, or a defective compressor, is indicated. Either condition requires entering the system. However, a defective compressor is extremely rare.

If the suspected compressor is pumping at all, the overall temperature of the evaporator will be higher than normal, and the inlet and outlet temperatures of the evaporator will be the same. Connecting a manifold and gauges into the system will show a lower-than-normal high-side pressure and a higher-than-normal suction pressure. Wattage draw of the compressor will be lower than normal, and the unit will rarely or never cycle (will run continuously). A troubleshooting chart is included under Restricted System, p. 76, which can be used as a quick reference in the diagnosis of inoperative sealed refrigeration systems.

Repair Procedures for Refrigeration Sealed Systems

Ninety percent of sealed-system entry is for repairs of the high side of the system. This is known as *dry-system work* because some refrigerant still remains and it has not yet been contaminated with moisture. Dry-system repairs include high-side leaks, restrictions, and compressor replacement.

The remaining 10 percent of entries for repairs are due to either moisture in the system or a low-side leak. A leak in the low side is often subjected to a vacuum in the sealed system; air and moisture are drawn in. It therefore must be classified as a wet system.

High-side leak repair. High-side repairs can easily be done in the home. The floor, of course, must be protected during the operation.

To start the repair, remove the unit sufficiently to permit access to the suction line, second discharge line, and drier. Avoid excessive kinking or bending the tubing which might cause restriction or breakage. Locate the leak using one of the methods listed under Leak Detectors in Chap. 1.

Install access valves on both the second discharge line (high side) and the suction line (low side). *Note*: The high-side valve is installed on the second discharge line instead of the first to prevent oil being removed from the system. *Caution:* When installing access valves on a charged system, locate the saddle nuts a safe

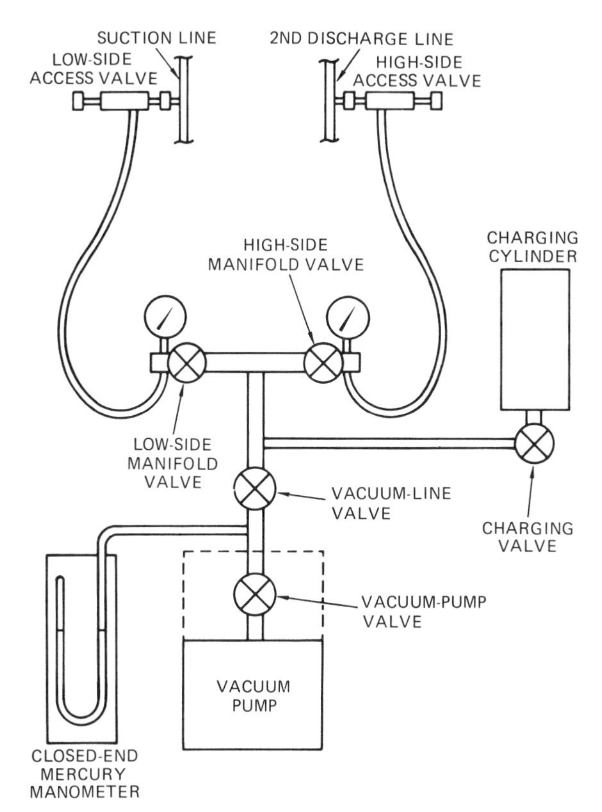

Figure 2-39. One method of setting up an evacuating and charging system.

distance from any brazed joints and apply only enough heat to flow the silver solder when brazing the nuts to the tubing. Excessive heat may rupture the tubing or a nearby joint, causing oil and refrigerant to spew forth. Actually it is a good idea to install a core-removal tool (see Valve Core Removal Tool, under Service Tools in Chap. 1) on both the high- and low-side access valves.

After the source of the leak has been found by the leak detector, the unit must be purged. To do this, connect a purge hose and two charging hoses to a tee. Attach the core depressor ends of the charging hoses to the high- and low-side access valves through the core tools. Run the purging hose outside the working area, preferably outdoors. Place its end in a container and cover it with a cloth to catch any oil which may be discharged with the refrigerant and to protect floors, walls, or shrubbery from oil spray. It is also a good idea to remove all food from the unit and close the door.

After the pressure has been purged, attach a vacuum pump in the system and evacuate for about 5 min to remove any refrigerant left in the oil or the unit. Otherwise it will continue to vaporize and could interfere with the brazing operation. In any purging operation, always cut out the strainer or drier with diagonal cutters. Do not unbraze a drier. Applying heat will drive any trapped moisture back into the sealed system. It will just return to the new drier, and you will have gained nothing.

Next repair the leak, replace the defective part, or correct the restriction as necessary. Most restrictions generally occur in the capillary tube because of its small inner diameter. If the high-side repair was due to an inoperative compressor, unsolder the four tubing connections.

When installing a new compressor, additional tubing may be needed to make connections to the process stubs of replacement compressors. This is because one model of service compressor is used as the replacement for many different refrigerator models. Suppliers keep their inventories of service compressors at a workable minimum. On smaller units, the replacement compressor may be of a slightly higher horsepower than the original. The installation instructions will call for a special suction line extension. This piece of tubing will contain a restrictor or orifice and will be recognized by the scored marks on the tubing. The restrictor is engineered to balance the sealed system with the new compressor. If it is left out, the system will be out of balance and will not operate correctly. Restrictors are made in more than one size. Use the correct part number as specified in the installation instructions.

Any tube you cut that will eventually be welded must be reamed and faced (see Chap. 1). Use a fine file to file the face of the tube smooth. Hold the tube toward the floor so filings do not enter the tube. Then use the end of a file, or pointed tool, to ream (deburr) the inside of the tube. To cut the tube (all except capillary and suction line), use a tubing cutter wherever possible. If there is not enough room to cut with a tubing cutter with the part in position, use a pair of diagonal pliers to cut tubes first, and then pull the component out so you can cut again with a tube cutter. Methods of cutting capillary and suction line are given in Suction and Capillary Tube Assembly Repair, p. 88.

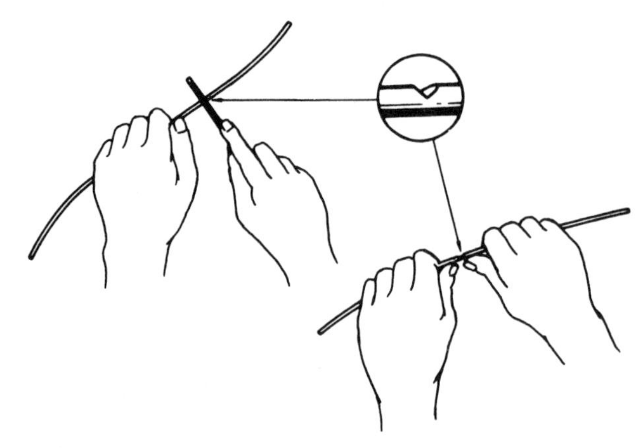

Figure 2-40. Scoring and breaking the capillary tube in order to obtain a clean, unrestricted hole.

The last component to be installed during any sealed-system entry is the *new* service drier or strainer. The sealed ends of the drier ends are not removed until you are ready to use the drier. You do not want air or moisture to enter and contaminate the drying agent. Actually, before inserting the capillary tube into the drier, score the tube with a file, as shown here, and break off the end to be sure of a clean, unrestricted hole. Do not bottom the capillary tube in the drier. To do so may puncture the screen. *Hint:* After cleaning the capillary tube, and before brazing, rub the last $\frac{1}{2}$ in or so of the tube with your fingers. The natural oils in your skin will prevent the silver solder from flowing into the tube and causing a new restriction.

Evacuating and recharging the system. The evacuation and recharging operations may be easily followed on the diagram shown here. Be sure the mercury keeper is removed from the manometer. Check to see that the charging valve on the charging cylinder is closed. Open both high- and low-side access valves, both high- and low-side manifold valves, the vacuum-line valve, and the vacuum-pump valve. Remember that the efficiency of the vacuum pump can be checked by closing the added valve. The service technician can determine whether a leak is in the equipment or in the sealed system. All the hoses and evacuation and charging equipment can be vacuum-checked by manipulation of the hand valves.

Before evacuating the sealed system, check your vacuum pump alone for efficiency. The pump and its connections must be able to pull a 29.8-in vacuum and hold it. Now start the vacuum pump and watch the manometer. If a vacuum reading of at least 28 in is not reached within 3 min, a large leak is present, and the system and evacuation equipment must be test-charged with refrigerant and leak-tested. After the leak is repaired, or if no leak was indicated by the manometer, continue to pump the system down until the mercury reads 29.8 in or more. Then close the valve on the evacuation pump and watch the mercury. It should hold steady for 30 s or more. If it does not hold, a small leak is still present, and the system must be test-charged with refrigerant and leak-tested. Repair the leak and repeat the pump-down procedure.

Once the mercury will hold steady for 30 s, the system is tight and evacuated. Apply heat to the evaporator, condenser, and compressor to drive out moisture by using the heat gun discussed in Chap. 1.

When the evacuation is completed, the system is ready to be recharged with refrigerant. Determine the correct number of ounces specified for that model. Also check to make sure the charging cylinder contains enough refrigerant to completely charge the system. Then align the scale on the charging cylinder to the cylinder pressure gauge reading. Shut off the low-side valve on the manifold. Close the valve leading to the manometer and vacuum pump. Leave only the high-side manifold valve open.

We are going to charge into the high side of the system. When the valve at the cylinder base is opened quickly, the refrigerant will rush through the manifold and the access fitting into the system. The flow will be fast because the sealed system was under vacuum. Watch the charging cylinder scale. When the specified number of ounces has been reached, close the cylinder valve. Sometimes the flow will stop before the full charge has entered the system. The pressures in the unit and in the charging cylinder have equalized. The last few ounces must be pulled into the low side by the refrigerator compressor. Attach a test cord to the compressor for manual operation. Close the cylinder valve and the high-side manifold valve. Start the compressor and watch the low-side gauge. Crack open the low-side manifold valve just enough to keep the gauge at about 10 or 15 lb pressure. This will remove liquid from the hose slowly. It will not slug into the compressing chamber and damage the compressor. After the hoses are cleaned, the cylinder valve can be cracked open. The balance of the charge can be metered slowly into the unit and the valve closed.

Any refrigerant left in the hoses or manifold must now be pulled into the low side. Replace the high-side valve core, or close the hand valve if used. Crack open the high-side manifold valve and observe the pressure gauges. When pressures are down to 5 or 10 lb, turn off the unit. Replace the low-side valve core and remove both charging hoses.

To complete the job, remove the core-removal tools and seal the access valves immediately with caps. As a rule, these caps should only be hand-tight; that is, do not tighten with pliers, or the seal will be damaged. But be sure to check the caps for leaks.

Clean all flux from tubing and components with a cloth and hot water. If flux is allowed to remain, it will cause corrosion. Reinstall the unit in the cabinet. Be careful not to kink the tubing. If a tube is too close to the cabinet or another tube, adjust it to eliminate the possibility of noise or rattle. Paint solder joints.

Low-side leak. A low-side leak is classed as a wet system because the low-side pressures are often in a vacuum. Air will be drawn in, and the air contains moisture. Water or moisture in a refrigeration system will cause freeze-up at the cap tube outlet. It will react with the refrigerant and form acids, sludge, and corrosion. Actually, air or water in any form cannot be tolerated in the system. To be certain that none is left, the low-side leak or wet-system repair procedures require the changing of the compressor oil and the applying of heat. Fortunately, only 10 percent of all sealed-system entries fall into this category.

The wet-system repair procedure of finding the leak, discharging or purging the system, and repairing the leak is the same as for the high side. But before the new drier or strainer is installed, the oil in the compressor must be changed to eliminate most of the water in the refrigeration unit. To do this, unsolder all restraining tubing connections to the condensing unit. Tip the unit up and pour all possible oil out the second discharge stub. Catch the oil in a container so that the amount removed can be measured.

Replace only an amount of oil equal to the amount you removed from the unit. Use the type of refrigeration oil recommended by the manufacturer of the refrigerator. A squeeze bottle such as a liquid-soap container makes an ideal oil dispenser. These containers must be cleaned and dried before they are used. Certain brands have a tapered outlet, and some are sized just right to fit in a process stub. The oil may be added to the compressor only through the second discharge or the discharge return stub. If both lines are unsoldered from the compressor, the oil will flow in much faster.

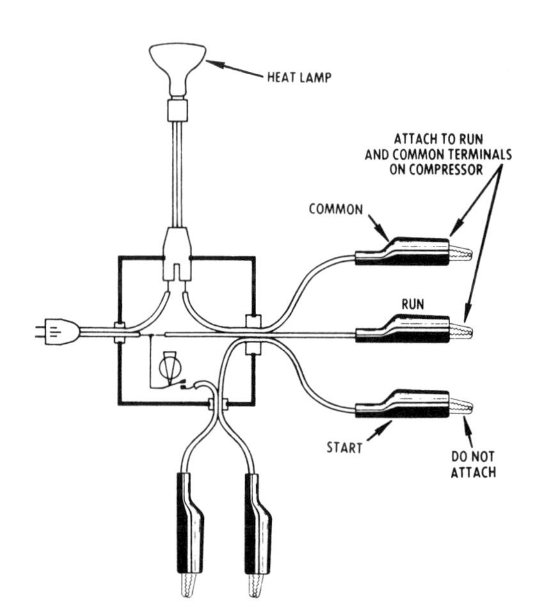

Figure 2-41. In this typical starting test set (Fig. 1-40), a 250-W heat lamp is used in place of the fuse. When the run and common connections are made to the compressor, the lamp is in series with the run winding.

If water was found in the old oil, it is recommended that the condenser and precooler be backflushed with refrigerant vapor. Direct the coil outlet into a cloth to catch oil spray or contaminants. Inspect the cloth after each flushing until signs of debris are no longer found. A Kwik-connect adaptor on the coil outlet makes an easy hose connection. Make it tight.

Once the new refrigeration oil has been added, reassemble the sealed system and install a new drier. The charging board, mercury manometer, and vacuum pump can now be connected and the system evacuated of air and moisture as described earlier in this chapter. However, one new factor is added for a wet-system evacuation: heat. Controlled heat from a heat gun or lamps should be directed on the compressor, precooler, condenser, and evaporator. The run winding of the compressor may be warmed by placing it in series with a 250-W lamp. Replace the test cord fuse with the heat lamp (Fig. 2-41). Attach the compressor terminal adaptor to the compressor and turn the test cord knob to RUN. The 250-W lamp will use about 90 V of current, leaving about 30 V to the compressor. This internal heat will help drive moisture and air from the compressor during evacuation. External heat can be applied simultaneously by the heat gun.

The evacuation of the wet system is accomplished in the same way as for the high-side repairs. If a 28-in vacuum is not reached within 3 min, a leak in the system or evacuation connections is apparent. Pressurize the system, leak-check it, and repeat the evacuation. When a 29.8-in vacuum is attained and holds steady after the vacuum pump valve is closed, the system can be considered evacuated. It is ready to be charged with refrigerant, as described above in Evacuating and Recharging the System, p. 82.

Changing a Compressor

When a compressor must be changed, use the following general procedure.

1. Bleed the refrigerant slowly by cutting the process tube on the compressor with diagonal cutters. If the refrigerator has not been in operation for some time, oil may be discharged with the refrigerant. Use care when bleeding the refrigerant. Place a cloth over the process tube to prevent the oil and refrigerant from splattering the room. Preferably, run the compressor, if operative, until the dome becomes warm, to separate the refrigerant from the oil. *Caution*: Ventilate the room while purging, especially when open-flame cooking or baking is taking place in the kitchen.
2. Using the diagonal pliers, cut the discharge, suction, and oil cooler lines. Crimp the tubes that remain on the compressor dome to prevent oil leakage during shipment.
3. Remove the wire leads from the relay. Remove the mounting retainers. Then, remove the compressor from the machine compartment.
4. Remove the filter drier by cutting the $\frac{1}{4}$-in inlet tube 1 in from the brazed connection. Using a file, score the capillary tube uniformly, approximately 1 in from the brazed joint at the filter drier. Then break off the capillary tube.
5. Transfer the rubber mounts, if used, from the inoperative compressor to the replacement compressor. Set the replacement compressor in place and install. Some replacement compressors require special mounting kits. In such cases, the mountings are installed as directed in the instructions that come with the kit.
6. Since most replacement compressors are shipped with an oil and nitrogen charge, remove the line caps and bleed off the charge of nitrogen. Using a suitable tool, cut suction, discharge, and oil-cooler extension lines to the required lengths. Swedge the lines as required, and join to the lines on the cabinet for brazing.
7. Install the replacement filter drier and braze it to the refrigerant lines and the compressor.
8. Evacuate and recharge the system as previously described.

Replacing a burned compressor. A burned compressor is one in which the windings have been extremely overheated, the wire insulation has been burned, and the circuits usually shorted or grounded. The compressor oil will be cooked. A strong odor is present when the system is opened.

There are four major causes of compressor motor burnout.

1. *Low line voltage.* When the motor winding in a normal or new motor gets too hot, the insulation melts and the winding short circuits. A blackened, burned-out run winding is the result. Low line voltage causes the winding to get very hot because it is forced to carry more current at the same compressor load. When this current gets too high, or is carried for too many hours, the motor run winding fails. A burnout caused by low voltage is generally a slow burnout and contaminates the system.
2. *Loss of refrigerant.* In a hermetically sealed motor-compressor, the refrigerant vapor passes down around the motor windings. The cool refrigerant vapor keeps the motor operating at proper temperature. If there is a refrigerant leak and little or no refrigerant to cool the motor, the windings become too hot, and a burnout results. The overload protector may not always protect against this type of burnout since it requires the transfer of high heat from the motor through the refrigerant vapor to the compressor dome.
3. *High head pressure.* With high head pressure, the motor load is increased and the increased current causes the winding to

overheat and eventually fail. Poor circulation of air over the high-side condenser can cause motor failure for this reason.

4. *Moisture.* It takes very little moisture to cause trouble. In the compressor dome, refrigerant is mixed with lubricating oil, and there is heat from the motor windings and compressor operation. If any air is present, the oxygen can combine chemically with hydrogen in the refrigerant and oil to form water. Just one drop of water, no matter how it gets into the system, can cause trouble. When water comes into contact with the refrigerant and oil, in the presence of heat, hydrochloric or hydrofluoric acid is formed. These acids destroy the insulation on the motor winding. When the winding short-circuits, a momentary temperature of over 3000°F is created. Acids combine chemically with the insulation and oil in the compressor dome to create sludge, which quickly contaminates the refrigerating system. Sludge collects in various places throughout the system, and is very hard to dislodge. A purge of the refrigerant charge, or blowing refrigerant vapor through the system, will not clean the system.

The repair procedures for a burned compressor require one more step than does ordinary compressor replacement; that is, the system must be flushed out before the new compressor and drier are installed. The worst debris will be in the compressor, drier, precooler, and condenser. The compressor and drier will be replaced, but the precooler and condenser can usually be flushed out.

The condenser and precooler should be backflushed with *liquid* refrigerant instead of being blown out with refrigerant vapor. They may be cleaned individually. Solder a piece of copper tubing, with a flare nut, to drier end of the condenser coil, or use a Kwik-connect coupler. Connect a hose to an inverted refrigerant drum. Direct the opposite end of the condenser coil into a clean cloth. Open the drum valve and flush the coil a short time. Inspect the cloth for debris. Repeat the flushing until the cloth remains clean. Proceed to flush the precooler in the same manner. Incidentally, the use of a Kwik-connect coupler for attaching the flushing hoses will speed the operation. Many comparable types of fittings are available from your local refrigeration supply house.

When installing the new compressor, tip the replacement unit to see if oil comes out of the discharge return stub. If the compressor does not contain oil, or if the amount it contains is in doubt, drain all remaining oil from the unit. Refill by injecting the recommended refrigeration oil (see Low-Side Leak, p. 83).

Once the new strainer or drier is installed, the system may be evacuated and recharged in the same manner as outlined earlier. An evacuation pump capable of pulling a 29.8-in vacuum and the mercury manometer to measure it are the successful secrets of any repair procedure of a refrigeration sealed system. Pumps should be checked before each job. Be alert for leaky charging hoses and seals. Old hoses can leak through hose casings, and a leak detector may not sense it. Checking individual hoses with the mercury manometer will isolate the offender.

Checking and Replacing the Evaporator

To replace the evaporator, usually proceed as follows.

1. Disconnect the service cord from the power supply.
2. Loosen the evaporator unit and, using a tubing cutter, cut the capillary tube from the condenser and the suction line from the compressor. Cap or plug all open holes.
3. Remove the condenser from the cabinet, making sure not to kink the high-side tube. Straighten the suction and capillary tubes.
4. Once the inoperative evaporator assembly has been removed, install the replacement unit by reversing steps 1 to 3.
5. Evacuate and charge the system as described in Evacuating and Recharging the System, p. 82, to complete repair.

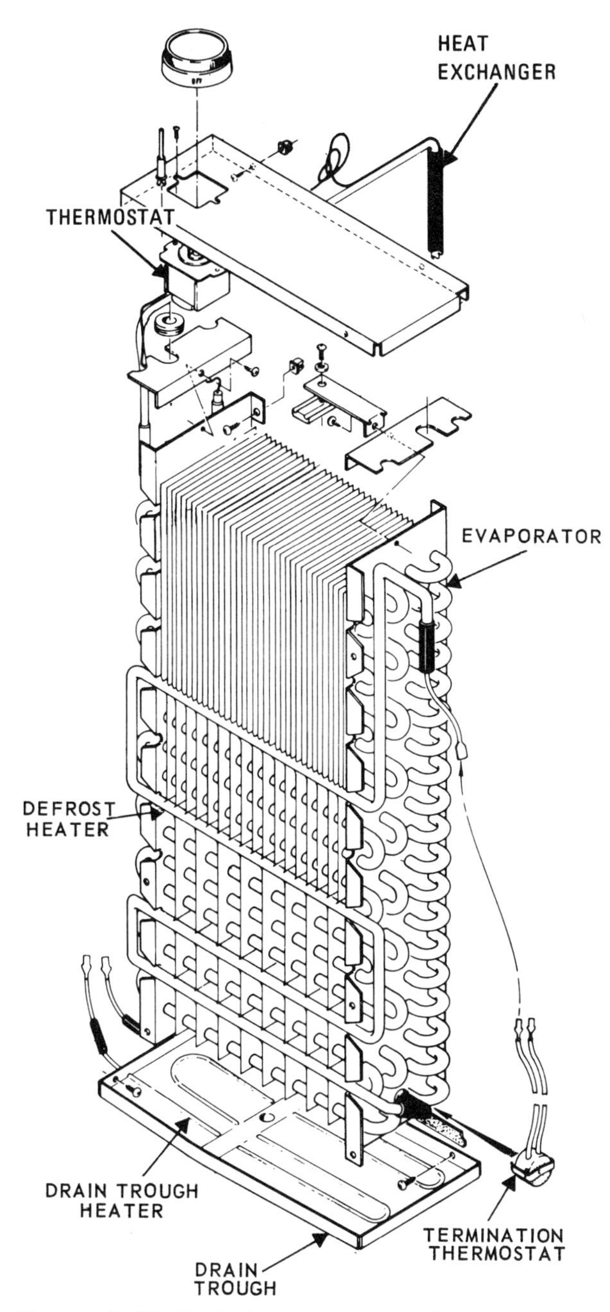

Figure 2-42. Typical arrangement of an evaporator and drain trough.

Checking and Replacing the Condenser

The condenser is usually, as described in Chap. 1, a series of tubes soldered to the plate mounted on the back of the refrigerator and so arranged as to give off the heat from the condensing refrigerant. For most efficient operation, the condenser must be kept free from dirt and lint, and there must be no obstruction to free flow of air over the condenser. To check the condenser for leaks, a procedure such as the one below may usually be followed.

1. Turn the control to the coldest position. The condenser can best be checked while the compressor is running. At that time the highest pressures are reached in the condenser.
2. Leak-check all tube connections and solder joints with a good leak detector.
3. If a small pinhole leak is detected, it can be repaired.
4. Cut the tubing at point of leak and place a sleeve over both ends of tubing and silver solder joints.

If proper tools are not available or in case of large leaks in the tubing, replace the condenser as follows.

1. Disconnect the service cord from the power supply.
2. Discharge the system at the high side.
3. Disconnect the capillary tube from the condenser.
4. Disconnect the high-side tube from the compressor.
5. Remove the condenser from the back of the refrigerator.
6. It is not necessary to replace the filter unless the oil in the compressor is burned or dirty.

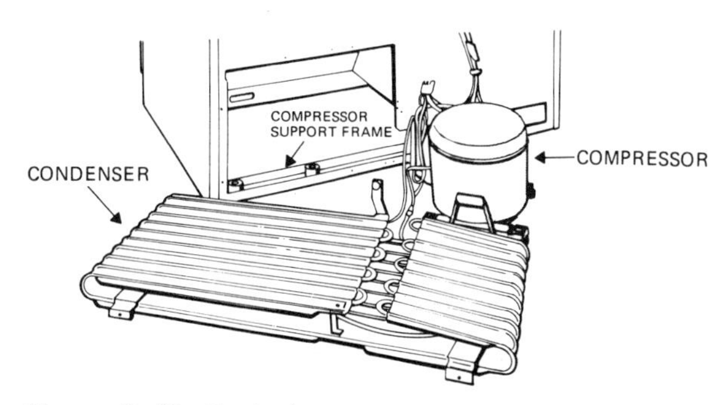

Figure 2-43. Typical compressor-condenser assembly out of the machine compartment.

7. Reverse steps 1 to 5 to install the replacement condenser.
8. Evacuate and recharge the system.

Before attempting to remove the evaporator from any model, be sure you first unplug the service cord. The refrigeration system should be evacuated and at zero pressure. Remove all evaporator trim, doors, trays, etc., and unclamp the thermostat sensing tube. Remove breaker trims and the model serial plate and cover. Disconnect the suction line from the compressor and the cap tube from the drier. Remove the screws holding the evaporator to the liner and swing the evaporator forward and out. Use care when handling the evaporator and heat exchanger assembly. Pull the heat exchanger out from the insulation area behind the right-side breaker trim and from the entry slot at the toe plate. Replace the assembly by reversing the procedure.

The terms "frost-free" and "no-frost" evaporators are misnomers. We know that to attain zero temperatures in the freezer section we must have an evaporator temperature much colder than zero. There has to be a temperature difference for there to be a transference of heat.

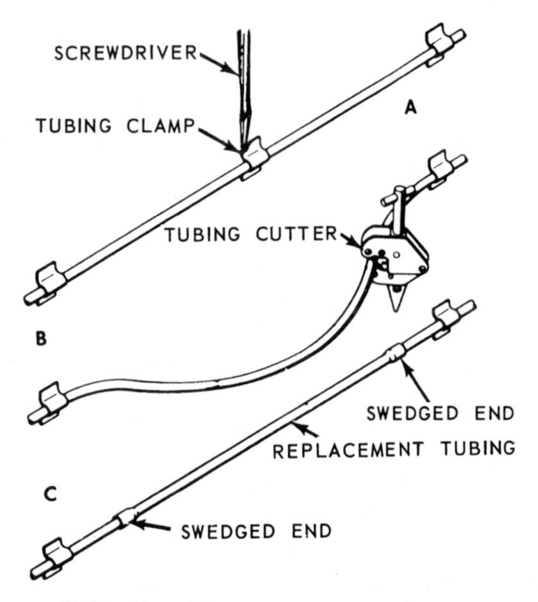

Figure 2-44. Repairing a wrapper condenser.

Heat always travels to cold. The greater the temperature difference, the faster the heat travels. We cannot get this cold without a frost buildup. The evaporators of the frost-free or no-frost refrigerators therefore do accumulate frost but it cannot be seen since the evaporator is concealed by a cover panel. A fan circulates freezer and refrigerator air behind the cover panel and over the finned tube surfaces of the evaporator. Periodically a timer energizes an evaporator heater which defrosts the accumulation of frost, turning it into water which will drain to an evaporation pan in the compressor department. It is called a no-frost refrigerator since the customer never sees frost and never has to defrost the refrigerator.

The evaporator may, determined by the refrigerator model, be located under a false floor in the freezer or behind a cover panel on the back wall of the freezer. The evaporator under the freezer floor is called a mullion evaporator because it lies flat behind the mullion cross stile. To service this type of evaporator, the defrost heater, or the heater limit switch, it is necessary to remove the side and bottom breaker trim, the vertical air duct on the back wall, and the floor attaching screws. The freezer floor can then be lifted up out of the freezer. To completely remove the evaporator, the cross stile and refrigerator section breaker trim must also be removed. After evaporator retaining screws are removed, the evaporator may be lifted up and out of the cabinet. The heat exchanger and the evaporator are usually an assembly and not available separately.

Suction and Capillary Tube Assembly Repair

The capillary tube is a carefully engineered length of small-diameter tubing used to carry the liquid R-12 from the condenser to the evaporator, and is attached to the suction tube for added efficiency for heat exchange. The length, diameter, and position of the capillary tube are very carefully determined and accurately specified to give the necessary restriction to the flow of refrigerant from the condenser to the evaporator.

To clear a restriction from the capillary tube, generally proceed as follows.

1. Discharge the system at the high side.
2. Remove the capillary tube from the filter and cap the filter.
3. Remove the suction tube from the motor compressor.
4. Connect the charging line to the suction tube.
5. Open the charging drum valve and charging line valve allowing pressure on the gauge to reach 150 lb.
6. Hold a white cloth over the end of the capillary tube so it will be possible to determine the cause of the restriction.
7. It may be necessary to warm the charging drum by placing it in a pan of hot water to obtain 150 lb pressure.
8. When a pressure of 150 lb is reached, close the charging drum valve.
9. If the obstruction has not been dislodged from the capillary tube by the application of pressure, heat the capillary tube with a torch, starting at the inlet end and working toward the evaporator. Heating the capillary tube will expand it, thereby increasing its inside diameter so that when the point is reached where the tube is obstructed, the obstruction may be loosened and expelled from the inlet end of the capillary tube.
10. Install a new filter and evacuate and recharge the system.
11. If the obstruction was not removed by the above procedure, replace the filter, and suction and capillary tube assembly.
12. Evacuate and recharge the system.

To replace the suction and capillary tube assembly, proceed as follows.

1. Remove the evaporator (see Checking and Replacing the Evaporator, p. 86).
2. Remove the suction and capillary tubes from the evaporator by unsoldering with a proper torch.
3. Replace the suction and capillary tubes. Be sure to silver-solder and evacuate and recharge the system.

Capillary tube separation from the suction line can cause colder-than-normal refrigerator temperature. The loss of heat transfer (warm liquid in the cap tube being cooled by the vapor in the suction line) would mean that the liquid refrigerant entering the evaporator would not be able to perform its maximum work. Thus the evaporator would not cool to a temperature low enough to cause the cold control to reach cut-out temperature in a normal time lapse. This means that the evaporator fan would be supplying air to the fresh-food section for a longer period of time than originally engineered (at temperatures much lower than 32°F). Good thermal contact between the capillary tube and suction line also ensures that no liquid will reach the compressor.

When cutting a suction tube, always cut it first with a pair of diagonal pliers in order to partially pinch off the tube and allow the gas (if any) to escape slowly. After practically all the gas has escaped, finish by cutting with a tube cutter, then ream and face the tube. When cutting capillary tubing, as previously mentioned, score around the impedance (capillary) tube with the edge of a file until it is scored sufficiently to break in two pieces. Whenever it is necessary to bend or form tubes, exercise care to prevent kinking of the tubes. When bending and forming tubes $\frac{1}{4}$-in OD and smaller, no special tools are required. When bending and forming tubes of $\frac{5}{16}$-in OD or larger, use either inside or outside bending springs whenever possible. This will prevent kinking of the tubes. It is always necessary to clean all tubes before cutting them. Use fine emery cloth to clean the paint off the tubes.

Heat Exchanger Replacement

The number of times you will need to replace a complete heat exchanger on refrigerators that have such a unit will be few, if any; however, for the sake of completeness, the following instructions are included. A restriction at the weld joints at either end of the capillary tube or suction line can be corrected by cutting out the weld joint and rewelding the joint. A leak in either the capillary tube or suction line can

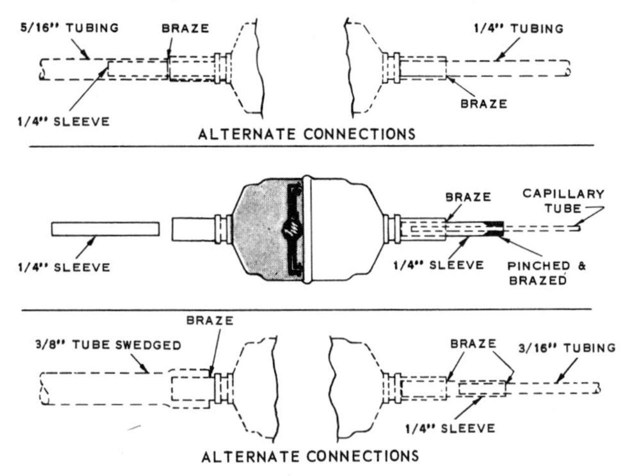

Figure 2-45. Replacement of a filter drier or strainer.

be repaired by using a copper union and cutting out the leak point and welding in the splice. If you should ever find it necessary to replace a heat exchanger, proceed as follows.

1. After gaining access to the heat exchanger, use a pair of side cutters to cut the suction line about 2 in from the compressor, and discharge the system.
2. Using a tubing cutter, cut the condenser outlet tube as near the drier as possible. Then employ a tubing cutter to cut the suction line at the evaporator end a minimum of 6 in from the copper-to-aluminum joint. You must do this in order to braze a copper-to-copper joint on the replacement heat exchanger.
3. "Unbraze" the capillary tube from the evaporator inlet. Then carefully remove the old heat exchanger as you must use it as a guide to form the new heat exchanger.
4. Install the new heat exchanger, brazing it in place.
5. Put in a new drier (Fig. 2-45).
6. To complete the repair, evacuate and recharge the system.

Trouble Diagnosis for Electrical Refrigerators

To properly and successfully service refrigeration equipment, there are several fundamental factors to guide the service technician.

1. A thorough understanding of the theory of refrigeration.
2. A good working knowledge of the purpose, design, and operation of the various mechanical parts of the refrigerator.
3. The ability to diagnose and correct any trouble that may develop.

When you have analyzed the trouble and properly determined the cause, apply the correction in a straightforward manner.

A prime requisite on the initial contact is always to allow the customer to explain the problem. Many times the trouble can be diagnosed quickly on the basis of the customer's full explanation. Most important of all, do not jump to conclusions until you have heard the full story and have evaluated the information obtained from the customer. Then proceed with your diagnosis.

Problem: Compressor will not start.

Possible cause	*Solution*
1. No power at outlet.	1. Check with test lamp or voltmeter. Check fuse circuit breaker. If blown, check for cause of overload and replace fuse or trip circuit breaker.
2. Broken service cord.	2. Check voltage at relay. If there is none but voltage is indicated at outlet, replace service cord or plug on cord.
3. Loose electric connection or broken lead.	3. Check with ohmmeter. Repair as necessary.
4. Air circulation is poor.	4. Clean condenser. On static condenser models, allow 3 in of free air space above cabinet. Check condenser fan (some models); replace if faulty.

5. Inoperative cold control or thermostat (thermostat stuck open or turned off).	5. Place jumper across terminals. If compressor does not start, use ohmmeter to check switch operation. Replace thermostat if necessary.
6. Inoperative relay and overload.	6. Use starting cord to check compressor directly. If compressor starts, use ohmmeter to check relay and overload separately. If compressor does not start, replace compressor.
7. Faulty start capacitor.	7. Check capacitor using test cord and 150-W light bulb. If capacitor is bad or shows signs of leakage, replace it.
8. Motor windings open, shorted, or grounded.	8. Check winding with ohmmeter. See wiring diagram for resistance values. Replace compressor if motor is defective.
9. Overheated compressor.	9. Clean condenser, check fan, or clearance around the cabinet.
10. Stalled unit.	10. Use starter cord and check unit. If compressor fails to start because it is too hot, allow it to cool to room temperature and repeat. If wattage is too high, or unit will not start or blows fuse, replace the compressor.
11. Inoperative defrost timer.	11. Replace timer assembly.

Problem: Compressor motor runs but there is insufficient or no refrigeration.

Possible cause	*Solution*
1. Moisture restriction, characterized by heavy frost around evaporator inlet.	1. Heat frosted area. If frost line moves farther along coil after heating, restriction was probably caused by moisture freeze-up. Discharge unit, evacuate using wet-system procedure, and recharge.
2. Permanent restriction.	2. First check for moisture restriction. Then check for crimped or damaged tubing. Repair or replace restricted component.
3. Low charge or no charge.	3. Check for leak. Add leak charge if necessary to get internal pressure. Repair leak or replace leaking component. Then, evacuate and recharge.
4. No-capacity or low-capacity compressor.	4. Check operating wattage and pressures. See performance chart for wattage and high- and low-side pressures. Do not judge compressor to have low capacity until restrictions and low charge have been ruled out. Replace compressor, if necessary.
5. Air circulation on high side:	5. To fix:
a. Condenser or grille restricted by lint.	a. Clean condenser and air passage with vacuum cleaner.
b. Condenser fan not running or running	b. Disconnect fan motor leads and and check separately. Replace

5. (*cont'd.*)

slow.	motor if defective.
c. Lower air baffle missing.	c. See that air baffle is in place under condenser. Replace if necessary.

Problem: Unit attempts to start but compressor kicks out on overload.

Possible cause	*Solution*
1. High ambient temperature and/or abnormal usage.	1. On initial pulldown in high ambient temperature, the compressor may cut off on overload. Instruct customer.
2. Low or high voltage (outside 10% tolerance).	2. Check voltage with voltmeter. Voltage at outlet should be 110 to 125 V. Low voltage may cause false starts. High voltage may cause compressor to overheat.
3. Defective start capacitor.	3. Check capacitor. Replace if defective.
4. Air circulation impeded on high side:	4. To Fix:
a. Condenser or grille restricted by lint.	a. Clean condenser and air passage with vacuum cleaner.
b. Condenser fan not running or running slow.	b. Disconnect fan motor leads and check separately. Replace motor if defective.
c. Lower air baffle missing.	c. See that air baffle is in place under condenser. Replace if necessary.
5. Inoperative start relay and/or overload protector.	5. Replace with parts known to be good.
6. Motor winding shorted or open.	6. Check winding with ohmmeter. See wiring diagram for resistance values. Replace compressor if motor is defective.
7. Overcharge.	7. Check for high wattage and frosted suction line. Evacuate and recharge with correct charge.
8. Leads broken or connections loose.	8. Check for broken leads and loose terminal connections. Make necessary connections.
9. Stuck compressor.	9. Direct-test compressor. If compressor does not start, replace.

Problem: Unit hums and shuts off.

Possible cause	*Solution*
1. Low voltage.	1. Check electric source with volt-wattmeter. Under a load, voltage should be 120 V ±10 percent. Check for use of extremely long extension cord, or several appliances on same circuit.
2. Faulty capacitor.	2. Replace capacitor.
3. Inoperative relay.	3. Replace relay.
4. Inoperative temperature protector.	4. Replace protector.
5. Condenser fan.	5. If obstruction prevents fan from running or winding is open, compressor will cycle on protector.
6. Stalled compressor.	6. Check with test cord. If compressor will not start, replace.
7. Broken lead.	7. Repair or replace lead.

Problem: High noise level.

Possible cause	*Solution*
1. Blower motor.	1. Check for blower wheel rubbing housing. Readjust. Replace motor if noise is excessive.
2. Compressor noise.	2. See that mounting studs and compressor are centered. See that the unit base is not applying too much pressure on top mount. Check mounting pads for clearance. If noise is internal, replace compressor.
3. Vibration of tubing.	3. Run your hands over various lines. This can often help determine the location of the vibration. Gently re-form tubing or tape it into position to eliminate vibration problem. Check discharge line from compressor to top of condenser to eliminate possible rubbing of line against condenser.
4. Superheat coil or condenser.	4. Tighten tubing clamps. Bend or straighten fins. Check drip-disposal pan to see that it is not touching the superheat coil.
5. Location of cabinet.	5. An out-of-level floor may be a factor in causing noise. Also, certain types of flooring transmit vibration more readily than others. Investigate the floor construction where unit is located.
6. Vibrating water-disposal pan.	6. Check drain pan positioning. Make sure drain pan clamps and screws are tight and in proper position.
7. Loose parts.	7. Tighten loose parts such as relay and overload protector.
8. Abnormal fan noise.	8. Some fan noise is to be expected. Noise caused by loose or unbalanced fan blades can be corrected by tightening or replacing the blade. Motor noise can be reduced by isolating the motor on rubber grommets.
9. Noise may be normal.	9. If noise appears normal, the customer must be instructed that some operating noise is to be expected.

Problem: Unit runs excessively.

Possible cause	*Solution*
1. Normal summertime operation.	1. Customer education.
2. Too-frequent door openings, heavy usage.	2. Customer education.
3. Poor gasket seal.	3. Adjust door hinges and check for tighter seal; if worn, replace gasket.
4. Control set too cold.	4. Instruct customer to return control to normal setting and check temperatures.
5. Light staying on.	5. Check light switch for proper operation and making contact with door.
6. Altitude adjustment not	6. Turn altitude adjustment screw to warmer

Possible cause	Solution
6. *(cont'd.)* properly set on cold control.	or colder position as required.
7. Control clamp loose.	7. Check clamp bracket. If loose, position correctly and tighten.
8. Airflow around condenser restricted or bypassing.	8. Adequate airflow over condenser is necessary. Advise user of this if cabinet is built-in type without proper circulation of air.
9. Placing sudden full load on unit such as all fresh water in ice cube trays or heavy loading after shopping.	9. Explain to customer that this heavy loading will cause long running time until temperatures are maintained. This running period may be several hours after heavy loading of the cabinet. Refer to use and care instructions regarding running time.
10. Thermostat mounting.	10. Be sure bulb is properly located and positioned for air to affect it.
11. Freezer drain trap not in place, allowing warm air to travel to the freezer.	11. Install trap in proper place.
12. Incorrect refrigerant charge: too much or too little gas.	12. Discharge, evacuate, and recharge with proper amount of gas.
13. Restriction or moisture.	13. Replace component where restriction is located. If moisture is suspected, replace drier-filter.
14. Normal operation.	14. Familiarize yourself with what is considered to be a normal cycle for a given model. It may be the fan that customer hears.
15. Manual damper.	15. Set freezer control higher.
16. Frosted coil (on automatic defrost models).	16. If air cannot get through to switch bulb, unit will not cycle.
17. Obstruction in air duct.	17. If inlet or outlet duct has an obstruction or frost accumulation, air cannot actuate switch bulb. Remove obstruction.
18. Condenser fan.	18. Check for fan motor operation and obstruction of fan blade.

Problem: Unit runs continuously with warm cabinet temperatures.

Possible cause	*Solution*
1. Excessive number of door openings. High temperature and humidity.	1. Instruct customer on proper location and use of refrigerator.
2. Poor door seal.	2. See if cabinet is level. Make necessary front and rear adjustments and, if necessary, put shims under back of refrigerator.
3. Gaskets not sealing.	3. Check door gasket. Realign door, and, when indicated, replace door gasket.
4. Interior light burns constantly.	4. Check light switch operation. If light does not go out when door is closed, replace the light switch.
5. Insufficient	5. Check position of

Possible cause	Solution
air circulation.	refrigerator. The rear and sides must be several inches away from the walls. Locate in correct position.

Problem: Unit runs continuously with too-cold temperatures.

Possible cause	*Solution*
1. Loose connection of thermostat bulb.	1. Check connection and make necessary adjustments.
2. Malfunctioning thermostat.	2. Turn thermostat to OFF position. If unit continues to run, replace the thermostat.
3. Control knob set too cold.	3. Reset cold control (thermostat) to correct setting.
4. Abnormal usage.	4. The household refrigerator is designed to be used in the home under normal household temperatures. Location on back porch or other unheated place may cause a complaint of too-cold temperatures in winter.

Problem: Both the food (refrigerator) and freezer compartments are too warm.

Possible cause	*Solution*
1. Inoperative or erratic thermostat.	1. Bypass thermostat. If unit starts and temperature goes down to normal level, replace the thermostat.
2. Liner cycling control open.	2. Check continuity of liner cycling control. If it is inoperative, replace heater.
3. Excessive frost on evaporator.	3. Check defrost timer, defrost thermostat, and defrost heaters for proper operation.
4. Fault in sealed system.	4. Refer to Refrigeration System Troubleshooting Chart, p. 77, under Restricted System.
5. Excessive door openings and food loads.	5. Customer education.
6. Door gasket.	6. Check for proper seal of door gasket. Replace if necessary.
7. Inoperative evaporator fan.	7. Check fan motor connections and fan switch.

Problem: Both the food (refrigerator) and freezer compartments are too cold.

Possible cause	*Solution*
1. Abnormal location.	1. Instruct customer as to new location.
2. Thermostat set too cold, or contact stuck.	2. Check control setting; if too cold reset to warmer position. If thermostat is faulty, replace.
3. Control capillary loose.	3. Check capillary control clamp and if loose, retighten clamp.

Problem: Food (refrigerator) compartment is too warm; freezer temperature is normal.

Possible cause	*Solution*
1. Improperly positioned freezer cold control.	1. Set freezer control to a lower number to permit more air to be transferred to the food section.
2. Loose air seal gaskets.	2. Check that all gaskets in the air transfer from the freezer to the fresh-food section are firmly in position to prevent air leakage.
3. Improperly positioned freezer fan blade.	3. Reposition fan on motor shaft.
4. Fresh-food air return path	4. See that the air return passage is not

4. (*cont'd.*)	
blocked.	blocked by insulation or ice buildup. Check all heaters used to prevent ice buildup in the air duct systems.
5. Air transfer duct blocked.	5. See that the ducts in the air transfer system are not blocked.
6. Intermittent freezer fan motor or fan switch.	6. Check connections to fan motor and fan switch. Make sure fan operates when door depresses fan switch plunger.
7. Malfunctioning defrost system.	7. Induce defrost cycle and check that all components are operating properly and freezer coil is defrosting.
8. Stalled fan.	8. Check for fan blade hitting fan housing and for ice blocking fan blade. Also check that fan is not frozen to fan orifice. Test defrost heater and check position of drain trough heater loop around fan orifice.
9. Evaporator heavily frosted.	9. Needs defrosting more often. Check door gaskets and usage.
10. Chiller tray baffle positioned out of tray.	10. Baffle restricts airflow down rear of food liner. Set control knob to colder position. Instruct customer.
11. Evaporator door improperly adjusted.	11. Check door alignment to see if there is $\frac{1}{8}$ in from liner lip to door.
12. Food compartment thermostat control baffle.	12. Check fresh-food control settings; make sure airflow from freezer section is sufficient. Check fresh-food cold control baffle for proper operation. Instruct customer.
13. Inoperative food compartment control thermostat.	13. Check operation of control thermostat to make sure it opens food compartment duct.
14. Heavy load of warm foods in frozen-food compartment.	14. Instruct customer on proper use.

Problem: Freezer compartment is too warm, food (refrigerator) compartment temperature satisfactory.

Possible cause	*Solution*
1. Inoperative or erratic cold control.	1. Check cold control by bypassing control. If unit starts and freezer temperature returns to normal, replace control.
2. Excessive frost on evaporator coils.	2. Check defrost timer, defrost thermostat, and defrost heaters for proper operation.
3. Inoperative evaporator fan.	3. Check fan motor connections and fan switches. Make sure fan operates when door switches are depressed.
4. Freezer control.	4. Control set too warm; reset. Also the freezer load could be too light, resulting in low running time.

Problem: Outside of cabinet sweats.

Possible cause	*Solution*
1. Unusual	1. Location in damp

location.	areas such as basements will cause sweating. Locating forced-air models in a confined location or too close to a wall may cause sweating on back or sides. If cabinet sweats for the above reasons, and it is not possible to relocate, add antisweat heaters as required.
2. Void in insulation.	2. If insulation void is found, pack the area with insulation.
3. Inoperative heaters.	3. Mullion or stile heaters may be burned out. Check with ohmmeter. Replace defective heaters.
4. Low-side tubing.	4. The suction line or other low-side tubing may be too close to cabinet. Reposition tubing.
5. Wet insulation.	5. Replace wet insulation. Correct the cause, which may be bad cabinet seal or stopped-up drain.
6. Door seal.	6. Adjust door for proper door seal.

Problem: Inside of cabinet sweats.

Possible cause	*Solution*
1. Abnormal usage.	1. Instruct user to cover foods and liquids. On manual defrost models regular defrosting is necessary. In summertime defrosting may be necessary twice a week.
2. Door seal.	2. Check door seal. Make door adjustment if necessary.
3. Cabinet seal.	3. Wet insulation is usually a sign of bad cabinet seal. If a bad cabinet seal is suspected, remove liner, unit, and insulation, and reseal cabinet seams.
4. Inoperative liner bottom heater.	4. Condensation on bottom of refrigerator compartment could be result of inoperative liner bottom heater. Check heater with ohmmeter. Connect reserve heater if necessary.

Problem: Ice in the bottom of the freezer and/or water in the bottom of refrigerator (food) compartment.

Possible cause	*Solution*
1. Stuck float.	1. Remove and discard float. Clean drain fitting and install new service drain trap.
2. Clogged drain.	2. Clean drain. Replace plastic drain cap with metal cap, if available.
3. Drain fitting deformed or out of position.	3. Reposition drain fitting and seal between fitting and Styrofoam plastic foam with permagum.
4. Gasket on back of drain trough.	4. Check for proper positioning of drain-trough gasket.
5. Drain funnel.	5. Check for positioning of drain funnel.
6. Evaporator position over drain trough.	6. See that evaporator is in position over drain trough.
7. Cabinet not level.	7. Level cabinet.
8. Defective defrost timer.	8. Timer may not be energizing defrost circuit or may be terminating defrost

Possible cause	Solution
8. (*cont'd.*)	cycle too soon. Replace timer if defective.
9. Inoperative drain heater.	9. Liner drain heater may be defective or not making proper contact with bottom of liner. Tape heater to bottom of liner. Replace if defective.
10. Spillage from ice.	10. Instruct customer on proper use.

Problem: Water on the floor.

Possible cause	*Solution*
1. First pulldown after installing product.	1. This frost-back should stop after first couple of cycles.
2. Overcharge of refrigerant.	2. Purge slightly from high-side charging port. If too much has been purged, evacuate and recharge system. Be sure the freezer coil is not iced as this would cause frost-back.
3. Freezer blower motor inoperative.	3. Check for voltage supply to motor. If none, check wiring. Remember, this motor runs only when the compressor runs.
4. Excessive frost buildup on freezer coil.	4. See Freezing Compartment Too Warm (p. 96).

Problem: Incomplete defrosting or high temperature during defrost.

Possible cause	*Solution*
1. Limit switch.	1. Check bimetal limit switch located near left side of evaporator coil. If limit switch opens too soon, the defrost will be inadequate or incomplete. If it is stuck closed, high freezer temperatures will occur. A loose limit switch will cause the defrost heater to stay on too long. Change limit switch if defective.
2. Inoperative defrost timer.	2. If defrost fails to occur or is incomplete, timer motor may be defective, or switch blades may be distorted, affecting the making and breaking of defrost circuit. Points may stick in defrost position, causing high temperatures. Replace timer if defective.
3. Defective heaters.	3. Defective heaters may cause ice buildup. Check drain heater, shroud heater, and defrost heater. Refer to wiring diagram for location. Replace heaters if necessary.

Problem: Ice accumulation on plate evaporator in refrigerator section (cycle defrost models).

Possible cause	*Solution*
1. Defective heater.	1. Defrost heater attached to connecting tubes may be defective. Replace heater.
2. Poor heat transfer.	2. Defrost heater may not make good contact with connecting tubes. Adjust heater and tubes to make good contact and tape in place.
3. Thermostat.	3. See that thermostat

Possible cause	Solution
	feeler tube is clamped properly to evaporator and that tube does not touch defrost heater. Check calibration. If thermostat cuts in at too low a temperature, change thermostat.
4. Defrost heater inadequate.	4. If ice accumulation continues after above action is taken, add heater to connecting tubes and wire in parallel with existing heater.

Problem: Butter too hard.

Possible cause	*Solution*
1. Butter conditioner selector switch.	1. Check for incorrect setting or wiring connections.
2. Loose or open circuit.	2. Check wiring harness and heater continuity.
3 Incorrect heater.	3. Replace with heater of proper wattage.
4. Heater not properly positioned around conditioner area.	4. Reposition heater so as to direct heat at butter liner area.
5. Butter compartment door not completely closing.	5. Correct door for proper fit.

Problem: Butter too soft.

Possible cause	*Solution*
1. Butter conditioner selector switch.	1. Check for incorrect setting or wiring connections.
2. Incorrect heater.	2. Check heater for proper wattage; replace if necessary.
3. Low usage or warm temperature in fresh-food storage area.	3. Advise colder fresh-food control settings or disconnect butter conditioner heater.

Problem: Foods freeze in the vegetable and meat pans.

Possible cause	*Solution*
1. Meat damper control.	1. Close damper for meat pan duct.
2. Long OFF cycle due to erratic switch.	2. Check operation of door switch, evaporator fan, control, etc.

Problem: Ice buildup at breaker strips.

Possible cause	*Solution*
1. Gasket leak.	1. Check gaskets for proper seal. Replace worn or damaged gaskets. See that freezer is level and on stable surface. Adjust freezer door if necessary.
2. Unsealed liner seams.	2. Check liner for sealed seams. Reseal if necessary.
3. Insulation voids.	3. Check under breaker strips for insulation voids. Reinsulate if necessary.

Problem: Odor in refrigerator.

Possible cause	*Solution*
1. Odorous food.	1. Instruct customer to cover odorous foods when they are stored in the refrigerator. If odor lingers after foods have been removed, clean the refrigerator and rinse with solution of baking soda and water.
2. Dirty drain system.	2. Clean drain system and flush with solution of baking soda and water.
3. Inefficient	3. Check to be sure that

3. *(cont'd.)*

filter if employed.	filter is properly installed. The effective life of a filter is about 1 year. If filter has been in use for more than 1 year, replace it.

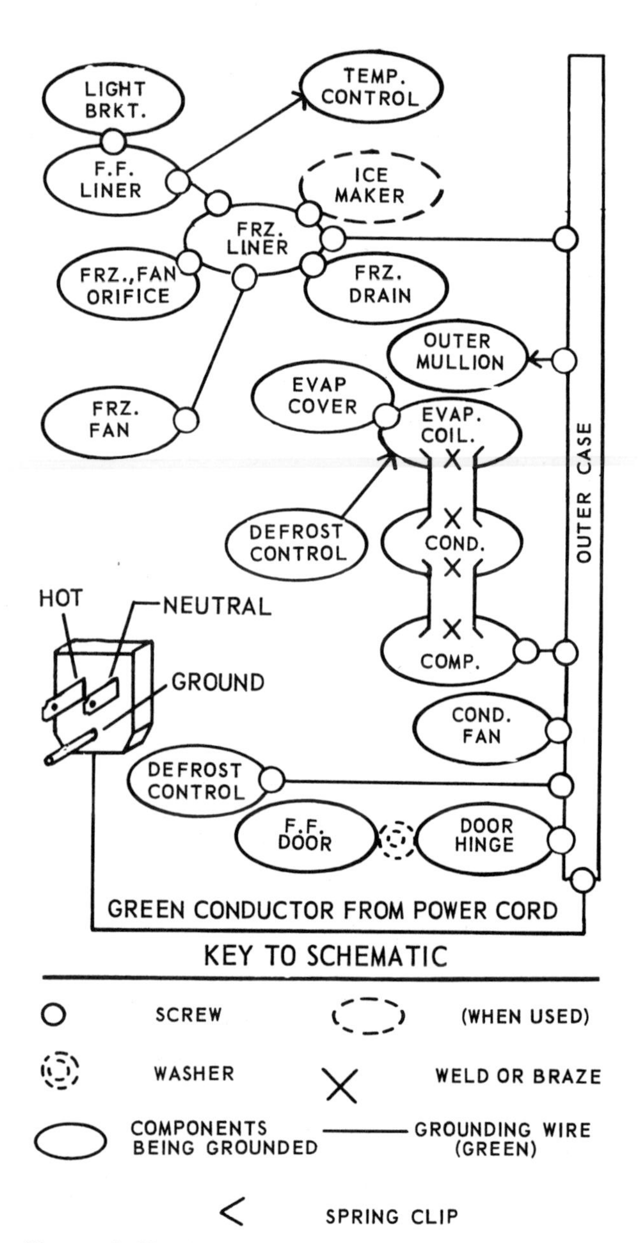

Figure 2-46. Typical grounding system.

Summer Operation

Summer and its hot weather mean a heavier load and harder usage for a refrigerator. Ever since refrigerators have been used in homes, hot summer weather has also meant an increase in performance complaints and service calls. Of the increase in complaints, some are justified in that something is wrong and should be corrected. However, the customer often needs only the assurance that nothing is wrong.

Hot weather complaints cannot be entirely eliminated, but they can be controlled. As soon as the complaint call is received, reference to the use and care booklet may tell the customer what to do to eliminate the complaint and prevent a service call. If the complaint persists, then a service call may be necessary.

A service technician must be tactful and diplomatic when instructing the customer; proper handling of hot weather complaints on the first service call is essential. The expense of later calls can be eliminated when the customer is satisfied on the first call. Summertime operational complaints usually consist of one or more of the following.

1. Food temperature too warm.
2. Refrigerator runs too long.
3. Moisture collects inside.
4. Slow ice cube freezing.
5. Excess frost in freezer compartment.
6. Fresh-food coils will not defrost.

Following are operational checks and procedures which should be made by the service technician.

1. Check for dirty condenser and grilles.
2. Adjust door gasket seal.
3. Measure food temperature.
4. Install temperature and running-time recording instrument.
5. See if control bulb clamp is loose.
6. Check freezer fan and fan switch.
7. Check defrost control and contact with connector block.
8. Check for shorted temperature control circuit.

9. Adjust control calibration or replace control.
10. Instruct the customer on proper usage and realistic expectations.

AUTOMATIC ICE MAKERS

In recent years, many refrigerator models have been featuring special ice cube–making devices. Actually, there are two basic types: autofill and completely automatic ice makers. In the autofill types the trays fill automatically but must be emptied manually. The completely automatic ice maker both fills and ejects the cubes and shuts off automatically. Let us first look at the autofill or self-filling ice tray.

Self-Filling Ice Tray Operation

While there are several different designs of autofill trays on the market, the following description covers the operation of a typical unit. (Always check with the service manual before undertaking any diagnosis of troubles.) The self-filling ice service shelf is similar to those used on models without the self-filling feature. It consists of one large molded plastic tray, an ice storage container, and the wire shelf with the tray pivots. The tray is supported by the shelf pivots on one side and by the tray weigher on the other side.

When the tray is empty, the tray weigher holds one side of the tray approximately $\frac{1}{2}$ in higher than the other side. Closing the freezer drawer pushes the ice shelf up onto its mounting slides. With the ice shelf in place, the top of the plastic handle on the side of the tray will contact the arm of a switch mounting in the front corner of the freezer. The switch then completes the electric circuit to a water solenoid valve which opens and discharges water into the tray through the fill tube mounted in the top of the freezer liner.

When the tray is filled, the tray weigher permits the tray to drop down, thus opening the contacts in the tray switch and cutting off the flow of water. The time usually required to fill the tray is approximately 45 s, and the normal time to freeze one tray of ice is $3\frac{3}{4}$ to 4 h. To eject the ice from the tray, the user opens the drawer, slides the ice tray shelf completely forward, and rotates the tray by lifting the right side of the tray up and over the storage container on the left. Pushing down on the tray at the left rear will rack the tray, ejecting the cubes into the container. Most containers will hold about 100 cubes (7 lb).

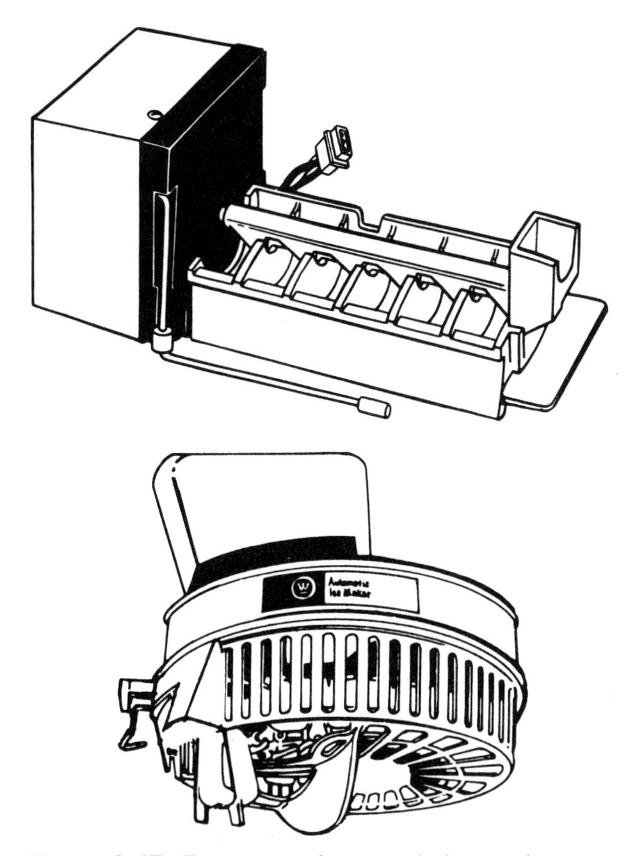

Figure 2-47. Two types of automatic ice service.

Trouble diagnosis of the self-filling ice tray. Here is a trouble diagnosis check list for a typical self-filling ice tray arrangement.

Ice tray does not fill.

1. Refrigerator not plugged in.
2. Water selector switch turned off.
3. Water not plumbed in.

4. Water turned off.
5. Ice tray not installed on pivots on ice shelf.
6. Ice tray shelf pivots worn down.
7. Water fill switch not being actuated by ice tray.
8. Drawer assembly not pushing ice shelf in far enough on closing to operate water fill switch.
9. Water valve or fill tube frozen up. Check valve heater.
10. Water valve not operating.

Ice tray not filling completely or filling too full.

1. Refrigerator not leveled correctly. Tray will have proper level with cabinet when cabinet is tilted back for normal self-closure of door.
2. Ice tray shelf pivots worn down.
3. Tray weigher not at correct vertical position. Empty cubes at pivot end; weigher too low. Empty cubes at movable end; weigher too high. In normal adjustment, movable end of tray should be approximately $\frac{1}{16}$ in low after fill.
4. Ice tray warped or twisted; cube dividers not level.
5. Tray weigher not adjusted for proper fill. To check for proper tray fill, allow to fill, then remove tray from ice shelf and set on level surface. Water level should be $\frac{1}{16}$ to $\frac{3}{32}$ in below cube dividers. Turn adjustment screw "in" to increase fill and "out" to decrease fill.

Water does not shut off, and tray overflows.

1. Tray weigher not functioning, thereby allowing tray to overflow before trip force is reached. Check tray level and tray weigher adjustment.
2. Obstruction under ice tray not allowing ice tray to drop to lower position.
3. Pivot end of ice tray not free, binding resists tripping of tray weigher.
4. Water fill switch not adjusted correctly. Handle on ice tray should move at least $\frac{5}{32}$ in down beyond the OFF position of switch. Adjust switch lever.
5. Combination of fill switch out of adjustment and ice shelf held up by obstruction.
6. Water fill switch does not shut off—check switch lever arm pivot to make sure switch is free to operate. Check lever return spring.
7. Faulty water valve. This may be difficult to identify because cabinet may show signs of overflow with no other faults. Water valve may have stuck open and then closed after overflow because of obstruction clearing inside of valve.
8. Slow water valve leak. This may show up as a small mound of ice on tray under the fill tube.

Water in food (refrigerator) compartment that can be traced directly to autofill unit.

Freeze-up of fill tube and leaks at water valve joints of plumbing line or fill tube. If the freezer drawer is slammed shut, a full ice tray may bounce and momentarily activate the tray fill switch if the switch feeler arm is touching the tray handle or very close to it. When this occurs, a small amount of water may be released into the tube where it may freeze and gradually build up to block the tube. To correct this, make certain there is a clearance between the switch feeler arm and the tray handle when the tray is full. The end of the switch feeler arm can be bent to obtain the clearance necessary. When making an adjustment, be sure to check for correct filling of the tray.

Ice cubes in ice container stick together.

1. Ice container was removed from refrigerator and cubes were allowed to partially melt before it was returned to freezer.
2. Ice tray was ejected before cubes were completely frozen. This may be a problem because it is sometimes difficult to determine when freezing is complete. Watch for moving water bubbles in cubes when tray is tilted; their presence indicates incompletely frozen cubes.

3. Normal cube pressure against other cubes during extended storage will cause sticking.

Ice cubes stick to tray, do not eject.

1. If sticking does occur, it will usually be the corner cubes because they receive the least racking action during ejection. A slight twist of the tray in the direction opposite to the normal will usually clear all cubes. Do not squeeze or push directly on the underside of the cube pockets. A few cubes left in the tray will not affect the next refill.
2. Sometimes a cube will eject but leave a layer of ice or part of the cube stuck to the tray. This is caused by the cracking of the cube when the last portion of the cube freezes. The layer of ice in contact with the plastic surface separates because of expansion force at the center of the cube. The faster-freezing cubes (toward evaporator) will have this tendency more than the slower-freezing cubes.
3. The plastic ice tray can be replaced without replacing the entire ice service shelf assembly.

Frost collects in bottom of ice bucket.

A snowy type of frost will collect in the bottom of the ice bucket after extended use without removal for cleaning. This is moisture trapped in the bucket that cannot migrate to the evaporator. Most of the frost will be at the inside rear of the bucket closest to the evaporator when the cubes are not stirred up or removed.

Slow freezing rate.

1. Normal freezing time for one tray is approximately 4 h.
2. Freezing time for the first tray may be considerably longer if water charge is filled at the start of or during pulldown.

After all adjustments have been completed, the service technician should allow the tray to fill at least twice for a final check. Place ice tray on the shelf and allow the shelf to return to filling position by very slowly allowing the drawer to close. Water should turn on before the drawer gasket contacts the cabinet. Allow the tray to fill before opening the drawer, then inspect water level in the tray without pulling the ice shelf forward or otherwise disturbing the ice tray. All pockets should fill.

Installation of an Automatic Ice Maker

The installation of automatic ice makers can be an important part of a service technician's business. Many people have the makers added after the refrigerator is already installed. For such installation, manufacturer's kits are available. A cold-water supply line must be available for the operation of an automatic ice maker. Use copper tubing, $\frac{1}{4}$-in OD, to connect the refrigerator (or freezer) to the nearest existing cold-water line. Do *not* use plastic tubing or plastic fittings. All installations must be in accordance with local plumbing code requirements. If the refrigerator is to be operated before the water connection has been installed, the ice maker ON-OFF lever should be kept in the OFF position.

To install an automatic ice maker, follow the general procedure given below.

1. Shut off main water supply.
2. Connect the copper tubing to the closest frequently used drinking water line through a shutoff valve. Always locate the valve in an easily accessible spot. If possible, connect into the side of a vertical water pipe. When connecting to a horizontal water pipe, make the connection to the top or side of the pipe, rather than to the bottom to avoid drawing off any sediment from the pipe.
3. After making the connection to the water line, route the copper tubing behind refrigerator, coming through the floor, wall, or an adjacent base cabinet, as close to the wall as possible.
4. Be sure there is sufficient extra tubing (about 8 ft coiled into three turns of 10-in diameter)

to allow refrigerator to move out from wall.

5. If the ice maker is already installed, the pipe can be brought up to make ready for connection. If the unit is not already in place, install the ice-maker mechanism as directed in the kit instruction sheet. But before making the connection to the unit, in either case, turn the water on and flush out the tubing, making certain that all foreign matter is removed from the line. Shut the water off.
6. Connect the copper tubing to the refrigerator with an SAE $\frac{1}{4}$-in flare nut. Fasten the copper tubing to the back of the cabinet with the clamp provided to relieve strain. This is not necessary on models that come with a short length of tubing already clamped to the cabinet. In some localities, sand or other foreign materials may be present in the water supply in such quantities that they may collect in the screen of the valve and tend to reduce the water flow to the ice maker. Where such conditions exist, it is recommended that a filter be installed in the line near the refrigerator. If a screen-type strainer is used, it should be 80 mesh or finer.
7. Turn on the water and check all joints for leaks.
8. Move the refrigerator back to the wall, arranging the coil of copper tubing so that it does not vibrate against the back of the refrigerator or against the wall.
9. Show the user the location of the water shutoff valve, and demonstrate how to shut the water off.
10. Plug the refrigerator power cord into the appropriate receptacle. Set the ON-OFF lever, located on the ice maker housing, to the ON (down) position. Demonstrate the operation and location to the user.

Servicing of an Automatic Ice Maker

Servicing the automatic ice maker is easy for anyone with an understanding of its operation and a working knowledge of simple electrical

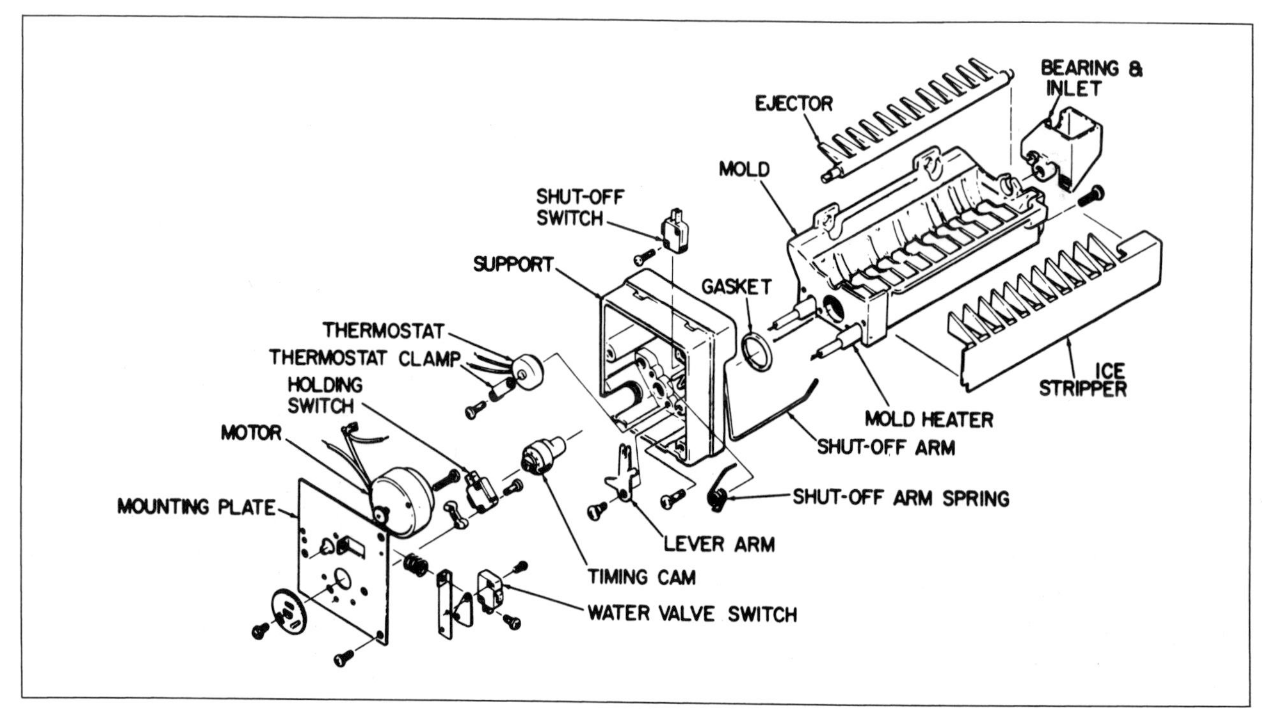

Figure 2-48. Typical ice maker assembly.

wiring diagrams. Repair does not always mean the replacement of parts; at times a slight adjustment of some component may be the only requirement. While there are several different methods of accomplishing automatic ice making, most have the same basic cycle of operation, which can be divided into five parts: freeze, release, ejection, sweep, and water fill. It is extremely important to understand the ice-making cycle before attempting any repair.

A description of a typical cycle of operation begins with water in the mold.

Freeze. Heat is removed from the water in the mold, and freezing takes place. At this time, under normal operating conditions, both the safety thermostat, if used, and the feeler arm switch are closed. The leaf switch contacts are open with the cam positioned as shown in Fig. 2-49. When the cubes are frozen and the temperature reaches about 16°F, the operating thermostat closes.

Release. With the closing of the operating thermostat, a series circuit is completed through the safety thermostat, the feeler arm switch, and the operating thermostat to energize the motor and mold heater. The motor begins rotation but is immediately stalled by the frozen ice. The motor will remain in a stalled condition up to about 2 min until the heater has melted the ice free from the mold. At this time the cam on the motor shaft has advanced sufficiently to close switch No. 1 of the leaf switch, thus completing a "holding circuit" around the feeler arm switch and the operating thermostat.

Once the switches are in a closed position, the cycle will continue, to end with the motor and mold heater energized. Under normal conditions, the operating thermostat will open soon after the mold heater is energized.

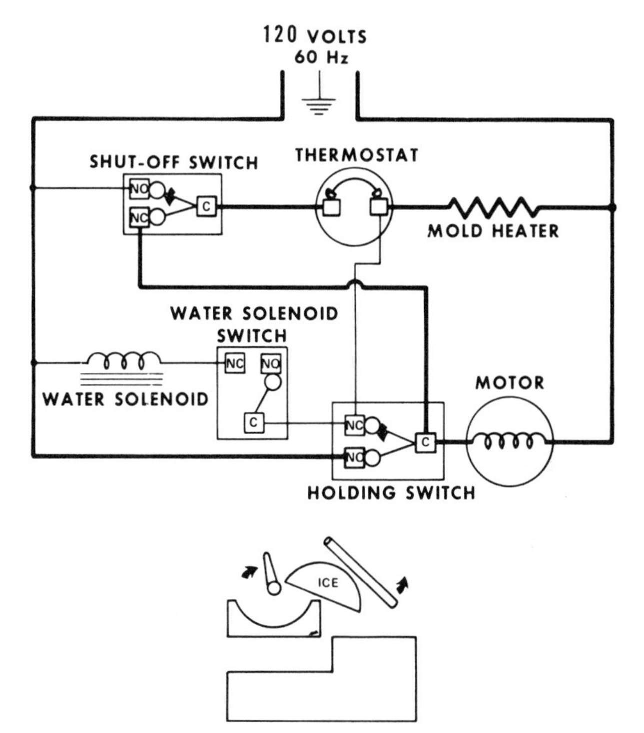

Figure 2-49. Typical ice maker schematic.

Ejection. When the ice has been melted free from the mold, the motor resumes operation. The ejector lever, operated by the cam, raises the pad which pushes the cubes upward. The pad continues to rise until the cubes are fully out of the mold. (Some models use an ejection system which features a conveyor belt removal arrangement. While the system works a great deal differently, the basic theory of operation is the same.)

Sweep. As the pad reaches a maximum height, the feeler arm raises, and the rake sweeps the cubes from the pad into the ice bucket. *Note:* On snap-action types, the rake operation is delayed and, by means of a spring, the cubes are "snapped" from the pad.

While the pad returns to the bottom of the mold, the feeler arm lowers (the feeler arm switch opens if the bucket is full), and the rake returns to its original position against the mold flange. The rake is driven by a fulcrum and a link connected to the rake lever. The rake lever, located between the cam and the motor with one end pivoted on the motor housing, is operated by a pin on the back of the cam. The feeler arm is driven by the actuator which is operated by a link connected to the rake fulcrum.

Water fill. As the pad nears the bottom of the mold, the lobe on the cam closes the switch (water fill switch) which controls water flow, energizing the water valve solenoid. While several types of water valves are used with ice makers, all operate

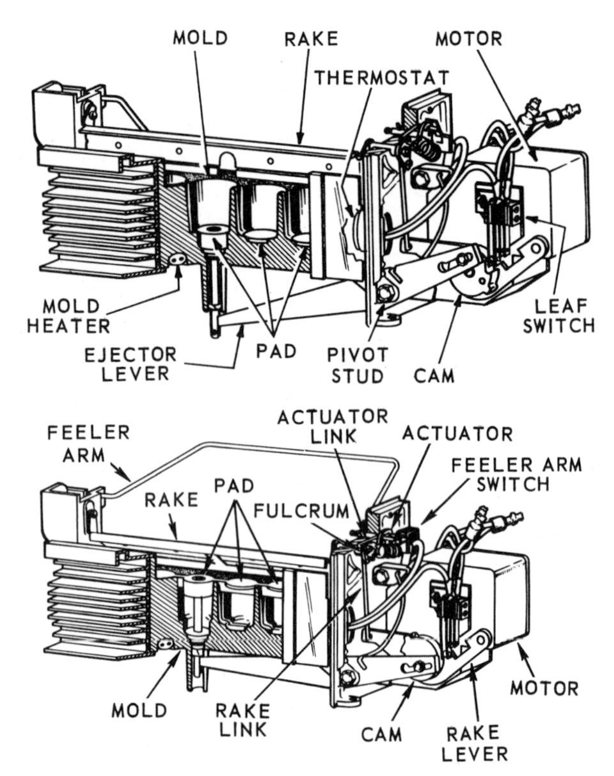

Figure 2-50. Typical ejection (top) and sweep (bottom) operations.

basically be same; that is, the magnetic force of the energized solenoid raises the armature, allowing water to flow through a valve or diaphragm. Water then enters the fill cap for approximately 6 s and is directed into the mold. This furnishes the required amount of water for the next cycle. As a rule, an 80-mesh screen is placed on the inlet of the water valve to prevent foreign matter from entering the valve. Generally, a flow washer "meters" the water flow rate to a constant amount—usually about 12 to 14 cm^3/s for pressures of 15 to 120 lb/in^2.

Cycle termination. The motor continues to rotate the cam, and the pad is seated in the bottom of the mold, until switch No. 1 of the leaf switch is opened by the notch in the cam and returns to its starting position, thus terminating the cycle (the operating thermostat is open at this time).

It may be necessary, on occasion, to test-cycle an ice maker to check its operation. This can be done on the repair bench or while the unit is mounted in the refrigerator. If the ice maker is in an operating refrigerator, take precaution against formation of condensate, by allowing the cold metal components to warm up before removing the front cover. This can be expedited by cycling the assembly with cover in place and the water supply valve closed.

To manually cycle the ice maker, slowly turn the ejector blades clockwise until the holding switch circuit is completed to the motor. When the motor starts, all components except the ice maker thermostat should perform normally to complete the cycle. Then, remove the front cover by prying it loose with the blade of a screwdriver at the bottom of the support housing. If further test cycling is necessary, place the blade of a screwdriver in the slot located in the motor drive gear. Turn it counterclockwise until the holding switch circuit is completed to the motor.

Trouble Diagnosis for Automatic Ice Makers

Here is a summary of conditions which could cause a failure of an automatic ice maker.

Problem: Automatic ice maker fails to start.

Possible cause	*Solution*
1. Feeler or shut-off arm.	1. Make certain that arm is in lowest position. Lower arm if raised.
2. No electric power.	2. Check for power at ice maker. Correct cabinet wiring if defective.
3. Not cold enough.	3. Check mold temperature at a mounting screw. If above 10°F, evaporator is not cold enough.
4. Operating thermostat defective.	4. If mold is below 10°F, manually start ice maker by pushing timing gear. If motor does not start, check thermostat, feeler arm switch, or leaf switch first and, if all right, replace thermostat. If

Possible cause	Solution
	motor does not start, check shutoff switch and motor.
5. Leaf or holding switch faulty.	5. With ejector blades in starting position, check terminals for continuity. Replace switch if open.
6. Feeler arm or shutoff arm switch defective.	6. Check that linkage is proper, adjust if necessary. Check terminals for continuity with arm in lowest position. Replace switch if open.
7. Motor.	7. Check operation with test cord. Replace motor if it fails to start.
8. Safety thermostat defective.	8. Check for continuity. Replace if open.

Problem: Automatic ice maker fails to complete cycle.

Possible cause	*Solution*
1. Leaf or holding switch, if blades are in "ten o'clock" position.	1. With switch depressed, check terminals for continuity. Replace switch if open.
2. Feeler arm or shutoff switch, if blades are in "twelve o'clock" position.	2. Check terminals for continuity. Replace switch if open.
3. Mold heater or operating thermostat, if blades are in "four o'clock" position.	3. Check heater for continuity. Replace heater if open. If heater shows continuity, replace operating thermostat.
4. Motor.	4. Check operation with test cord. Replace motor if it fails to start.

Problem: Automatic ice maker fails to stop at end of cycle.

Possible cause	*Solution*
1. Leaf or holding switch.	1. With ejector blades in starting position, check continuity of terminals. Replace switch if closed.

Problem: Automatic ice maker continues to eject when bin is full.

Possible cause	*Solution*
1. Feeler arm or shutoff switch.	1. Check that linkage is proper. Switch should open when arm is in raised position. Adjust if necessary. Check terminals for continuity with arm raised. Replace switch if closed.

Problem: Automatic ice maker produces undersize ice pieces.

Possible cause	*Solution*
1. Mold.	1. Check for level. Adjust if required.
2. Water supply.	2. Check that supply line and water valve strainer are completely open and that adequate pressure is maintained. Clear restrictions or advise customer accordingly.
3. Water valve switch.	3. Test-cycle and measure water fill. Adjust switch if necessary.
4. Operating thermostat short-cycling indicated by ice sheets or hollow ice in storage bin.	4. Check operating thermostat bond to mold. Ensure good thermal contact with the unit. Check thermostat calibration by replacing with new part.

Problem: Automatic ice maker spills water from mold.

Possible cause	*Solution*
1. Mold.	1. Check for level.

1. (*cont'd.*) — Adjust if required. Check top edge for evidence of siphoning. Prevent capillary action using silicone grease in this area.
2. Water inlet tube. — 2. Check that inlet tube and fill trough fit properly and water does not leak during fill cycle. Adjust fit if required.
3. Water valve. — 3. Check that water does not enter mold after cycle is completed. Replace valve if leaking and water pressure is proper.
4. Water valve switch defective. — 4. Test-cycle and check that water fill does not exceed volume capacity of mold. Adjust switch if this is required. With ejector blades in starting position, check terminals. Replace switch if closed.
5. Unit fails to stop at end of cycle. — 5. Refer to complaint Fails To Stop at End of Cycle above.
6. Operating thermostat short-cycling. — 6. Refer to complaint Undersize Ice Pieces above.

Problem: Water fails to enter mold.

Possible cause	*Solution*
1. Water supply.	1. Check that water line, valve, and valve strainer are open. Remove restriction, open valve, or instruct customer accordingly.
2. Water valve.	2. Observe for ice in inlet tube and fill trough. If obstructed with ice, check water valve for slow leak. If valve leaks, check for proper water line pressure. Replace water valve if pressure is within specifications.
3. Defective valve solenoid coil.	3. Check terminals for continuity. Replace coil if open or shorted.
4. Defective water valve switch.	4. With plunger out, check terminals for continuity. Replace switch if open.

Commercial Ice Makers

With the increased use of commercial ice cube makers in supermarkets, roadside stands, motels, etc., there is a demand for their servicing. In this book, it would be impossible to discuss all the types and designs of commercial ice makers, or even to generalize enough to cover the subject adequately. However, with the proper service manual and the general information given in this chapter, the average service technician should be able to cope with any servicing problems on small commercial ice cube makers.

Some special equipment items such as chilled water dispensers, ice cube dispensers, and other so-called custom items have not been described in this chapter. Most of these items are in one model only and made by only one manufacturer. They also are only featured for a year or so. Therefore, if you must service such a custom item, check the manufacturer's service manual and follow those instructions to the letter. In addition, many of these items require direct replacement; thus, if they go bad, the entire faulty unit must be removed and replaced with a completely new part.

Freezers

CHAPTER

3

As a service technician, you may be called upon to service a separate or so-called free-standing freezer unit. Actually, free-standing freezers are of two general types, chest and upright. In the upright freezer, the common practice is to fasten the coils below some or all shelves, as well as at the top, back, and side walls of the unit. This arrangement tends to increase the speed of freezing since more food comes into direct contact with the refrigerant. The chest-type freezer, on the other hand, may have a single compartment or a separate, fast-freezing section. While the chest type is considered more economical to operate, upright freezers seem to be more popular with users.

OPERATION OF REFRIGERATION SYSTEM

The refrigeration system of any freezer is the simplest form of refrigeration and usually consists of only five basic parts, the compressor, a condenser, a drier, a heat exchanger, and an evaporator. The only variation is on those units using the large compressor; either a fan will be used in the machine compartment to cool the compressor or an oil cooler will be used. If an oil cooler is used, it generally consists of the top seven passes of the condenser, and thus the condenser becomes a combination oil cooler and condenser.

Refrigeration then takes place by the following procedures in the same way as was described in Chap. 1. That is, hot high-pressure gas is pumped from the compressor to the top seven passes of the condenser which are used as an oil cooler on those models having a large compressor, or it is pumped directly into the condenser on those models not having the large compressor. On models using an oil cooler, it tends to cool the hot high-pressure gas as it passes through the continuous pass of tubing in the compressor shell, and the hot high-pressure gas continues back to the lower portion of the combination oil cooler

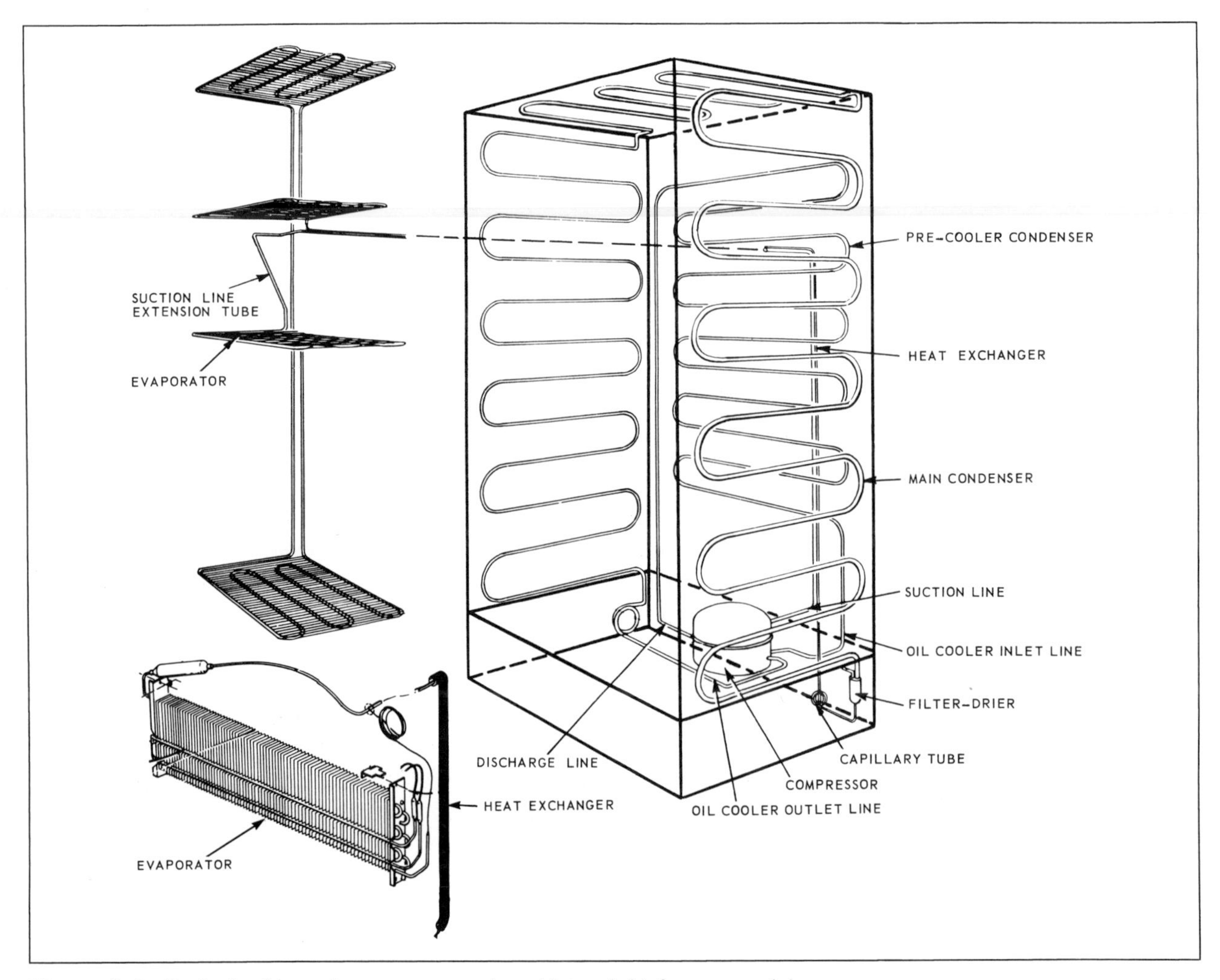

Figure 3-1. Typical refrigeration system employed in upright freezer models.

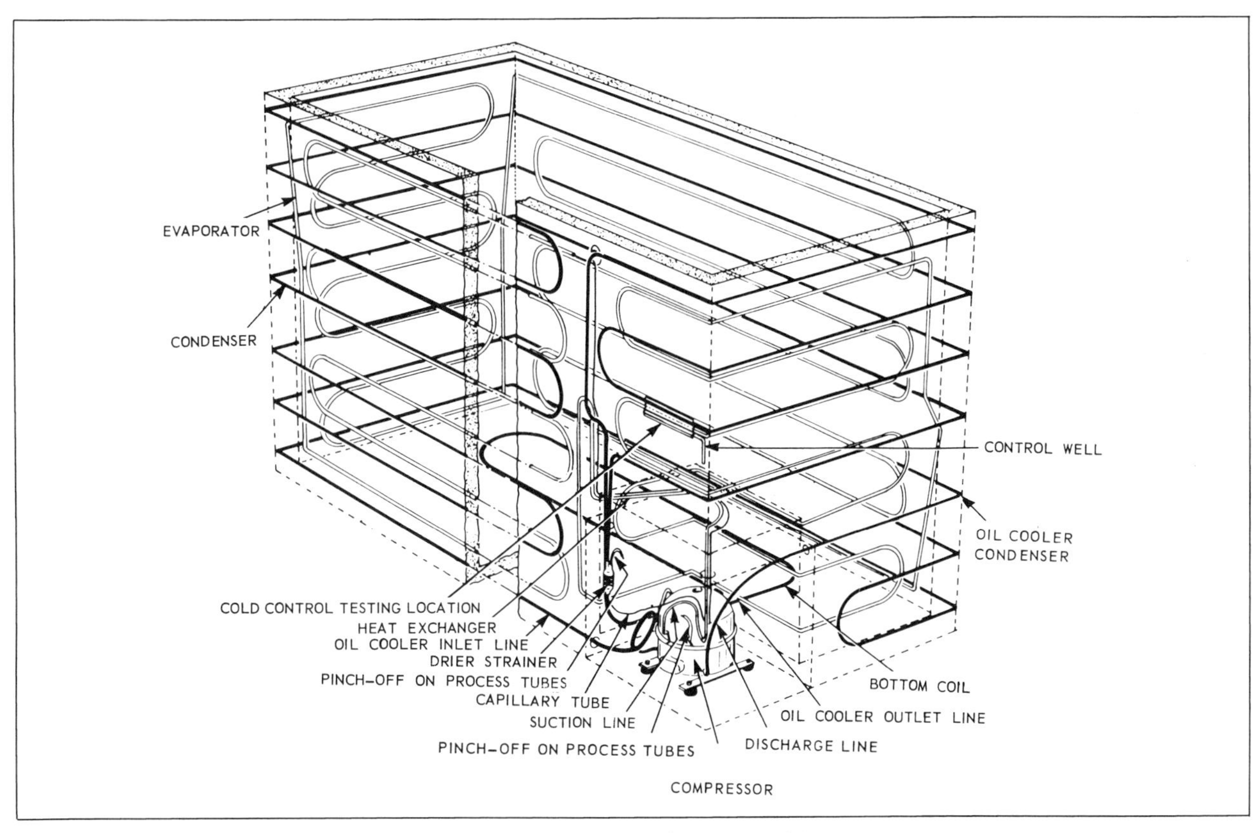

Figure 3-2. Typical refrigeration system employed in chest freezer models.

and condenser. Partially condensed refrigerant from the oil cooler is superheated in the compressor and therefore serves to cool the oil in the compressor. The refrigerant then passes out the oil cooler tubes in the compressor back to the regular condenser, where the hot high-pressure gas is turned to a liquid. This liquid then passes through the drier where any moisture or impurities are entrapped. From the drier the liquid refrigerant passes through the capillary tube into the evaporator, where it gradually changes from a liquid to a low pressure gas. In changing from a liquid to a low-pressure gas, heat is absorbed, thereby cooling the refrigerator. A low-pressure gas then passes to the suction line back to the heat exchanger and down to the compressor. The heat exchanger serves the purpose of gaining some efficiency from the unit by the fact that the cool gas coming down the suction line will have a tendency to cool the hot liquid going up the capillary tube, since the two lines are soldered together, thereby gaining more efficiency.

To sum up, the refrigerant flow through the freezer system is as follows.

From the...	*As a...*	*To the...*
Compressor	Hot gas	Oil cooler condenser
Oil cooler condenser	Precooled gas	Compressor (oil cooler tubes)
Compressor (oil cooler tube)	Hot gas	Condenser
Condenser	Liquid	Drier
Drier	Liquid	Capillary tube
Capillary tube	Liquid	Evaporator
Evaporator	Liquid and gas	Suction line
Suction line	Cool gas	Compressor

Airflow. In all conventional non-frost-free freezers, airflow is by convection; that is, the cold air falls from the top of the freezer where it is warmed on its way down and rises back to the top because it is warmer than the air coming down from the top. Some models, as an alternative to convection, use forced air, by means of a fan. This design has been used satisfactorily in upright freezers, for instance, for a number of years and creates no mechanical problems with the possible exception of causing some confusion on behalf of the user if all the shelves do not frost up evenly. Sometimes the airflow down through the shelves on the refrigerated shelf-type freezer will be in such a manner that the top shelf, or perhaps the top two shelves, will not frost over entirely. The service technician must take particular caution not to imply to the user that this is a problem but should, however, attach suitable temperature-sensing devices to any shelves that are not frosted over as well as to the shelves that are frosted over and demonstrate to the customer's satisfaction that all shelves are refrigerating at approximately the same temperature. Actually, you will sometimes find the unfrosted shelf to be running colder than the frosted shelf.

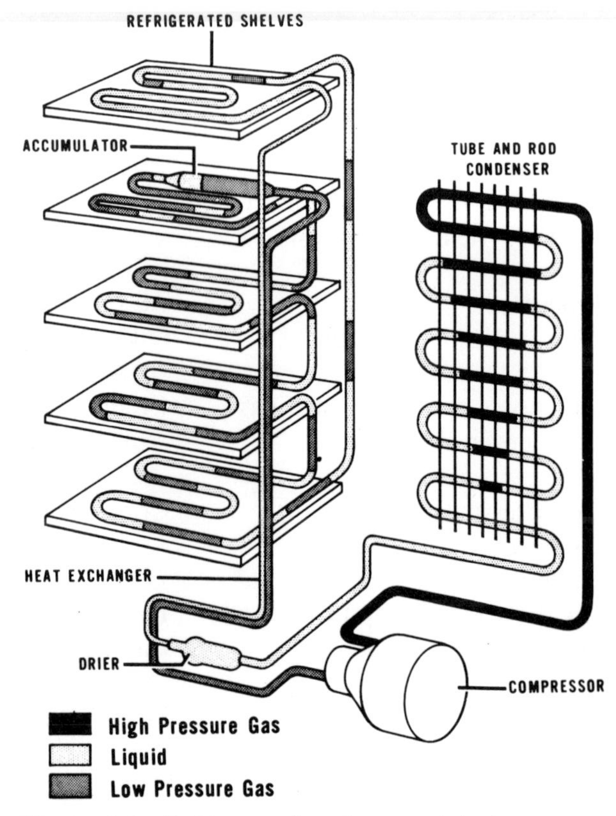

Figure 3-3. Refrigerant flow in an upright freezer.

The airflow in an upright no-frost or frost-free freezer is as follows. Air is pulled through the air inlet, which is a part of the bottom breaker, and over the evaporator which is beneath the bottom of the food liner by the interior fan. The air is then blown out of one of the fine air outlet openings where it is circulated through the freezer cabinet over the inner door shelves and back down to the air inlet grille, where the cycle is repeated. The fan does not run when the unit is off, the door is open, or the freezer is in defrost.

Defrost cycle. The frost-free or no-frost freezer coil is defrosted by an electrical heater which heats the coil. This heater, often called a flash defroster, is activated by a timer and terminated by a thermodisk-type thermostat. The timer is not, as a rule, in series with the unit control, and therefore the freezer defrosting is not tied in with the compressor running time. The freezer then will usually defrost every 12 h in that the timer will be advanced to a point where the switch contacts are made to the defrost heater. Once the ice is removed from the coil and a predetermined temperature is reached at the defrost termination thermostat, the defrost heater will be de-energized, but the compressor and freezer fan will not start to run until the contacts in the timer have been reset. The timer contacts are reset only by the timer motor advancing the timer out of defrost. For example, the total time length for this to be accomplished may be 30 min from the moment the timer originally advanced the contacts to the defrost position, not 30 min from the moment the defrost termination thermostat de-energized the defrost heater. With many freezers, the length of defrost time, usually about 30 min, is fixed, and cannot be increased or decreased.

With nonautomatic types, manufacturers do

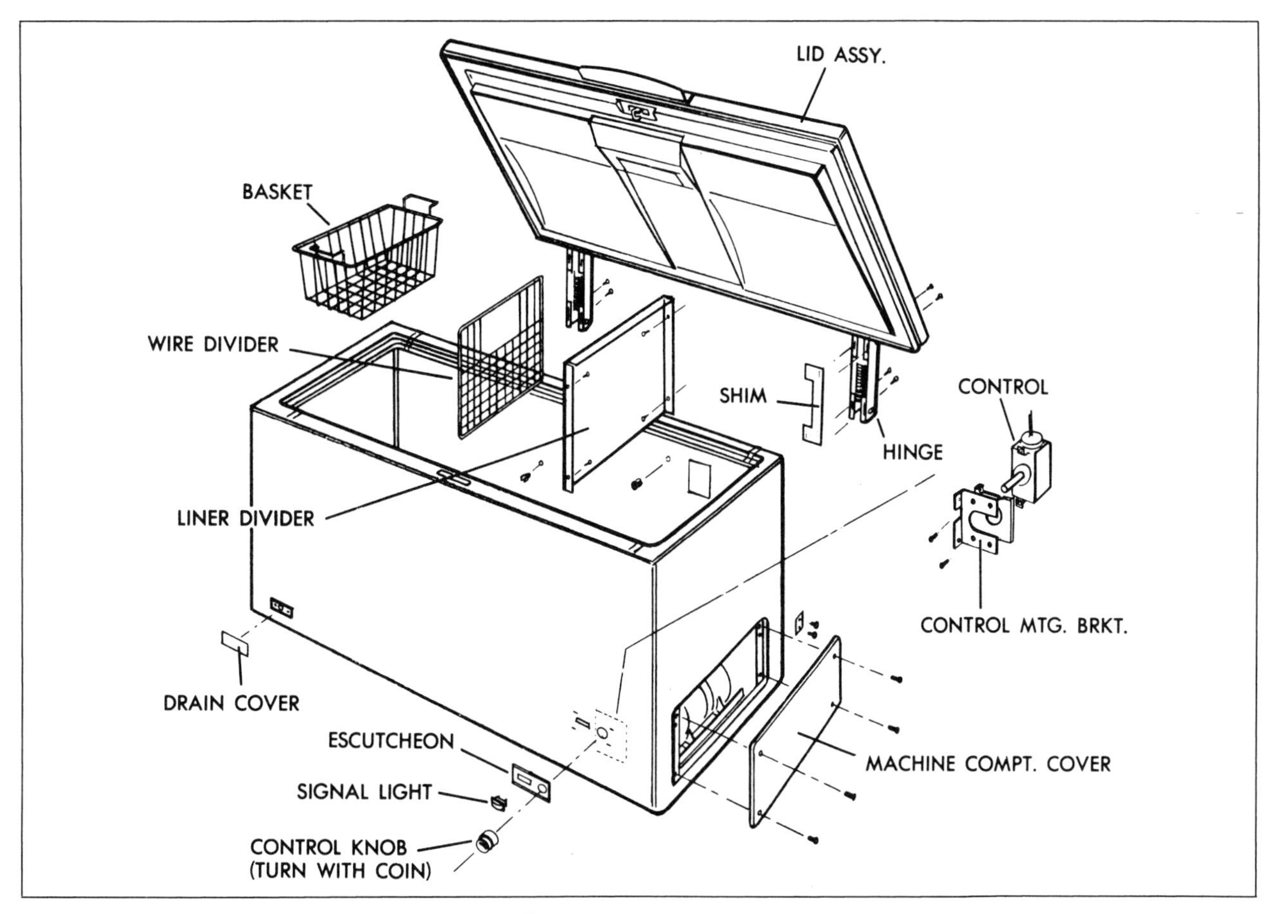

Figure 3-4. Exploded view of a typical chest freezer.

sometimes attach a hot gas valve and sufficient tubing to use the hot gas to defrost a freezer. When this is done, it is a simple matter of attaching the hot gas valve so that the gas passes out of the compressor to the valve (when the valve is closed) on to the condenser and through the rest of the refrigeration system to produce refrigeration. During a defrosting cycle, the user throws a switch energizing the valve which pulls a plunger up, thereby bypassing the condenser. The gas out of the compressor goes through a hot gas line rather than to the condenser, and the hot high-pressure gas is pumped directly into the evaporator (freezer shelves) and melts the ice from the evaporator.

Drain system. Some freezers have a system to drain water from them. Sometimes drain systems are also used on non-frost-free freezers. These are simple arrangements primarily of a hose running from the food liner through the insulation out the bottom of the cabinet and clamped to the bottom of the shell with a small bracket. These hoses normally must be placed in a suitable container to catch the water. The likelihood of any of these hoses causing a problem is not too great.

ELECTRIC COMPONENTS CHECKUP

The electrical controls are basically the same as for a refrigerator. For instance, a thermostat or temperature control maintains the selected temperature; its operation is based on the expansion of fluids according to the temperature. The thermostat capillary tube contains a liquid with

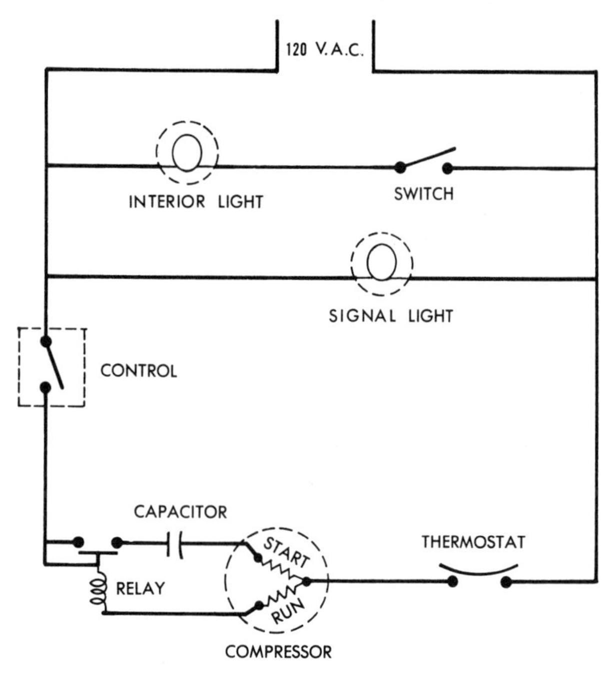

Figure 3-5. Schematic diagram of a typical non-automatic defrost model.

physical characteristics of a prefixed temperature; one side is connected to the upper refrigerating shelf (in an upright model) to a small bellows located in the thermostat. Variations in the temperature of the refrigerating shelves are transmitted to the thermostat by means of springs and levers and will open and close the thermostat contacts. The range is usually from MIN to MAX, with the MAX position the coldest. (Some models are marked with numbers, the highest number being the coldest.) To defrost a nonautomatic freezer, set the thermostat at STOP; the compressor will stop and restart only when the thermostat is set at one of the operating positions.

The thermostat can lose its calibration, and this is indicated by a gradual lengthening of the period between stopping and starting. This fault can be corrected by turning the thermostat knob toward a colder setting. To check whether the thermostat is faulty, proceed as follows.

1. Remove the plug from the socket.
2. Turn the thermostat knob toward an operating setting. Insert the two terminals of the tester on the thermostat terminals. If the instrument does not show continuity, the thermostat contacts are open and the thermostat must be replaced. The thermostat contacts may also be stuck together because of oxidation caused by the electric arc on the contacts. In this case the compressor will run continuously.
3. Whatever the cause of the inconvenience, it will be necessary to replace the thermostat. Do not try to repair the thermostat or to change the calibration point, as this will automatically annul the guarantee. Calibration will always be subjective if suitable testing instruments are not utilized.

Compressor. A starting relay, mounted on the compressor body, controls the electrical operation during starting, interrupting the starting

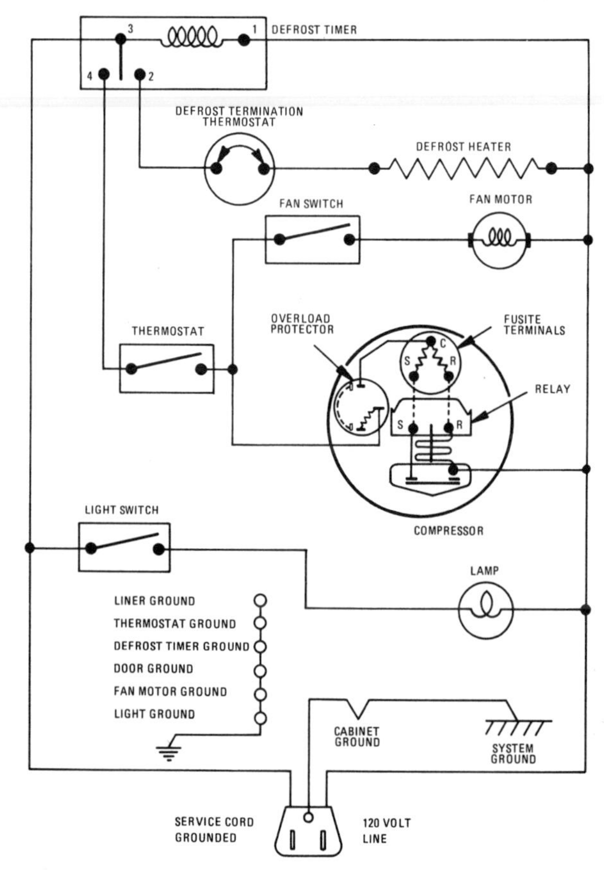

Figure 3-6. Schematic diagram of a typical automatic defrost model.

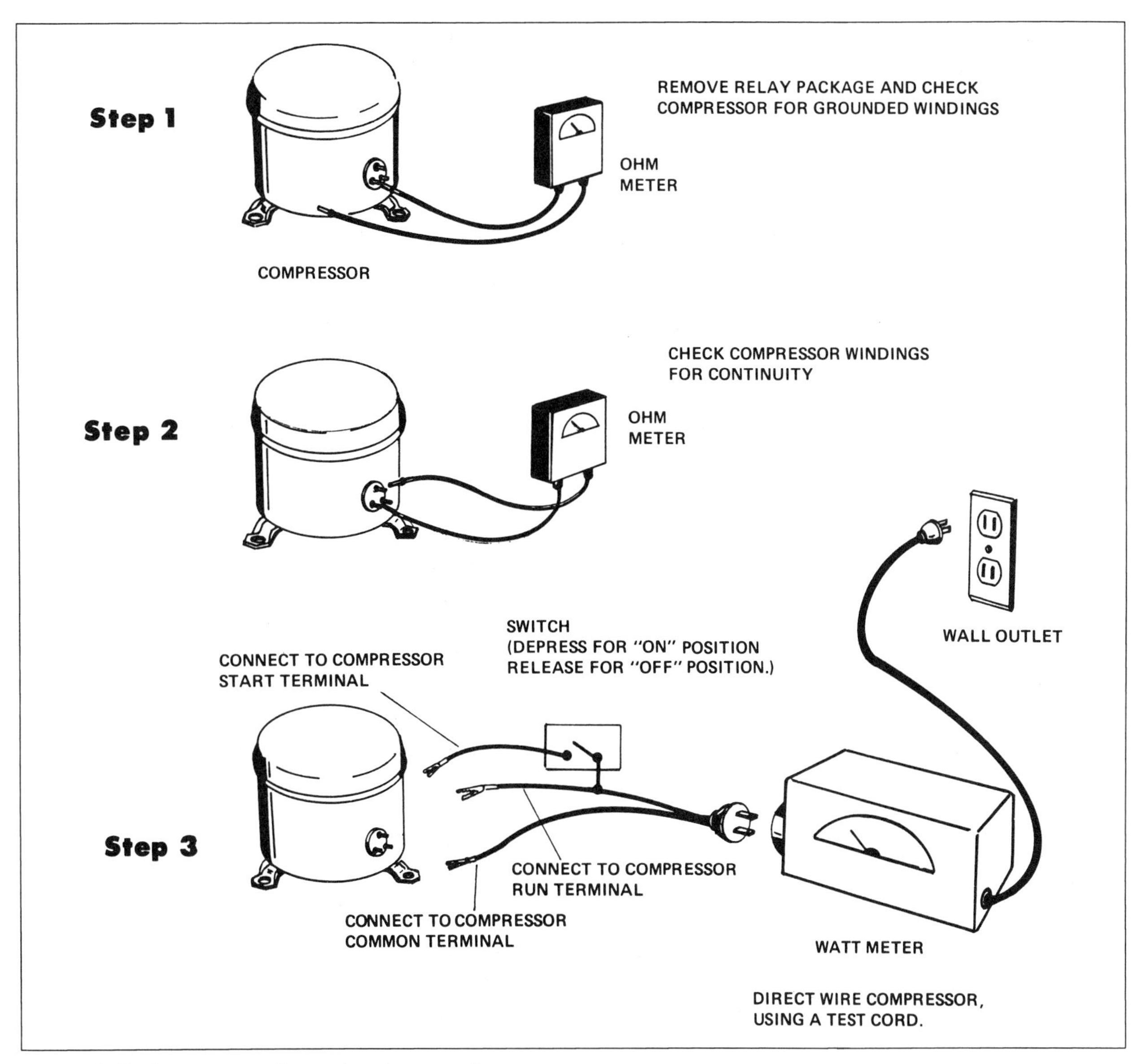

Figure 3-7. Compressor diagnosis prior to replacement.

circuit when the motor has reached its normal rotating speed. A thermic protector on the hermetic compressor interrupts the operation when there are critical thermal conditions (or electrical conditions such as variation of voltage or overheating). For a proper electrical check of the compressor, carry out the following operations in order to establish which cooling system components are faulty and must be replaced.

With a conventional voltmeter check the voltage of the freezer; it should have 10 percent tolerance with respect to the nameplate values. Remove the terminal box cover mounted on the compressor, exposing the compressor terminals, the starting relay, and the overload protector to view. Check whether the compressor terminal ends are fed with current. Touch the two compressor terminal ends with the circuit test lamp in the following order.

1. Touch COMMON and START terminals.
2. Touch COMMON and RUN terminals.

If the lamp does not light on either of the positions, replace the compressor. Touch the compressor crankcase with the circuit test lamp, in series with the lamp. If the lamp does not light, replace the test lamp terminals and repeat the test. If the lamp lights up on both positions, one of the warning lamp terminals can be used to touch the compressor terminals as follows.

Touch the compressor body with one terminal end and the compressor terminals with the other terminal end in sequence. If the light goes on, the compressor is grounded and must be replaced. Install a 15-A fuse in the circuit and start the compressor. If the fuse does not blow out, connect an ammeter in series to the circuit and check the amperage of the compressor. If the amperage is over 15 percent of the value printed on the compressor nameplate, there is probably a short circuit inside the electric windings of the compressor, or a mechanical defect which prevents the starting of the compressor. In both cases the compressor must be replaced.

Starting relay. The starting relay has the task of aiding the compressor in starting. When the compressor starts, a high starting current flows into the relay, energizing the solenoid which causes the closing of the movable contacts and the subsequent closing of the circuit. As the compressor reaches its normal rotating speed, the current requirement decreases to the point at which the solenoid is no longer able to attract the movable contacts. When relay contacts open, the starting winding is no longer energized, and the compressor continues to operate with the running winding only. To test the starting relay, proceed in the following order.

1. Disconnect the feeding cord plug.
2. Set the thermostat knob at STOP.
3. Insert the circuit tester plug in the socket.
4. Touch the relay terminal, in series with the line, with the circuit lamp tester and the relay terminal corresponding to the wire originating in the overload cut-out with the circuit lamp tester. If the lamp does not light, the relay coil is broken, and it is therefore necessary to replace the relay. If the lamp goes on, try to join the two relay terminals with a wire junction, with the freezer plugged in and the thermostat knob toward an operating position. If the compressor starts, it is evident that the movable part of the relay is blocked with the contacts open, which requires the replacement of the relay.

If, on the other hand, the compressor operates with excessively short cooling cycles and you are sure that the thermostat is normal, the relay can be checked as follows.

1. Insert the freezer supply cord plug in the wall socket.
2. Set the thermostat at STOP.
3. Connect the circuit tester ends to the relay terminal to which the current arrives and the end where the conducting wire from the thermostat is located.
4. Turn the thermostat knob to an operating position and watch the test lamp. If it remains unlit for a few seconds and then lights up and does not go off any longer, it means that the relay is blocked with the contacts closed and must be replaced.

Such items as the defrost control and overload switch are checked in the same manner as detailed for the refrigerator in Chaps. 1 and 2.

REFRIGERANT SYSTEM REPAIR

Methods of evacuating a freezer's refrigerant system, checking for a leak, cutting suction tube, recharging a system, replacing a compressor, etc., are the same as for a refrigerator's refrigerant system (see Chap. 2). The most common refrigerant problem with a freezer that the service technician has to face is the unit running but not refrigerating.

When there is an obstruction in the liquid line and/or capillary of the refrigerating unit due to moisture within the unit freezing into ice or due to other foreign matter in the system, the following conditions will result.

1. At low room temperatures the unit will run continuously as long as the obstruction is present. At high room temperatures the unit will cycle on its thermostat.
2. The evaporator will warm up (over its entire surface if the flow of liquid is entirely shut off and over lesser portions if some liquid continues to leak through), and the frost will melt.
3. Under this condition, operation may be noisy, and the compressor motor may cycle on the thermostat.
4. If the liquid line has been blocked by ice, this ice will melt after the refrigerator has warmed up enough. The unit will then begin to refrigerate normally and will continue to do so until the ice builds up again. If the unit freezes up so frequently that it causes customer dissatisfaction, it should be repaired.

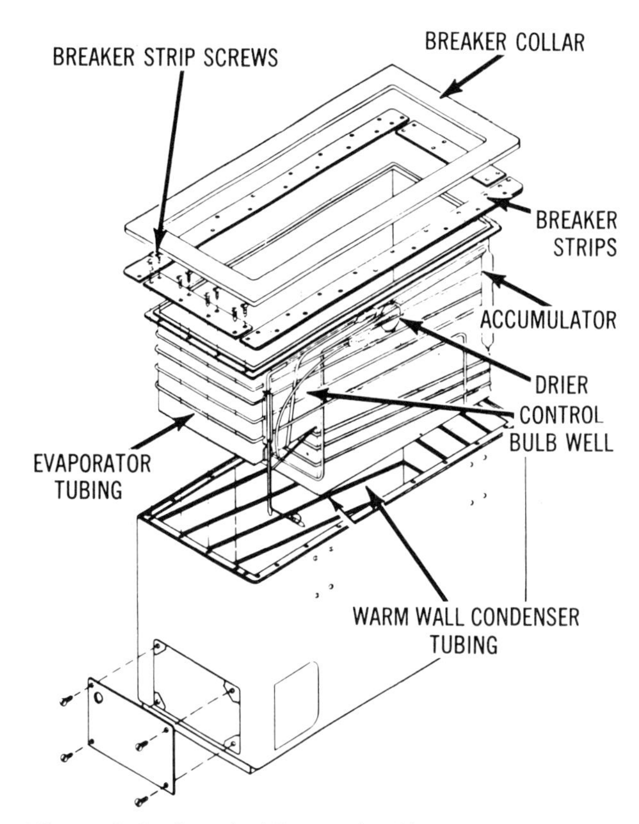

Figure 3-8. A typical liner and cabinet.

If the refrigerating unit is undercharged or some of the refrigerant gas has leaked out, the unit will provide some refrigeration. To test for undercharge, check the shelf with an accumulator to see if it is frosted up.

Where the compressor valves are broken, there will be no refrigeration, and the evaporator will warm up and defrost. The unit will run continuously and will draw very low wattage. The unit will run exceptionally quietly. The compressor will "coast" to a stop when the unit is turned off, if the valves are broken. (Unit will start immediately after shutting off without the benefit of equalization.)

The need for servicing the evaporator or the condenser of a chest-type freezer is very rare since both consist of a simple length of tubing. The evaporator as stated previously is tubing welded to the back of the liner. The (warm wall) condenser is a length of tubing attached to the inside surface of the outer cabinet. During operation, you can feel the warmth of the condenser on the cabinet sides. This reduces the possibility of condensation on the outer surface. Any sealed system entry will most likely be to the condensing unit or at the capillary tube and suction line joints.

The evaporator and condenser coils of freezers using fiberglass (spintex) or rock wool are accessible if needed. Take off the breaker collar and strips (Fig. 3-8). Pull the thermostat sensing bulb down into the unit compartment. Lift the liner out of the cabinet high enough to rest on two boards placed across the top of the cabinet. Remove the insulation to check the condenser coils on the side walls or the precooler on the cabinet bottom.

Evaporator and condenser coils of thin wall freezers with foamed-in-place insulation are not accessible. The walls of these models are about 2 in thick. If the coils are defective within warranty, the customer receives a new freezer and the old freezer is returned to the factory. This is an very rare occurence.

The most likely need for service to the sealed system of a foamed-in-place liner will be in the compressor compartment or at the capillary tube and suction line connections. These are the locations of solder joints and serviceable operating components. In the event of a sealed system problem, the technician must first thoroughly check out the obvious problem areas.

The very last components to suspect are the evaporator and condenser. If a freezer is returned claiming defects in either of these parts and it is found that the problem was actually in a serviceable area, the freezer will be returned to the sender and charged for repairs and for shipment both ways. When servicing the heat exchanger (capillary tube and suction line connections) or checking for leaks, peel back the

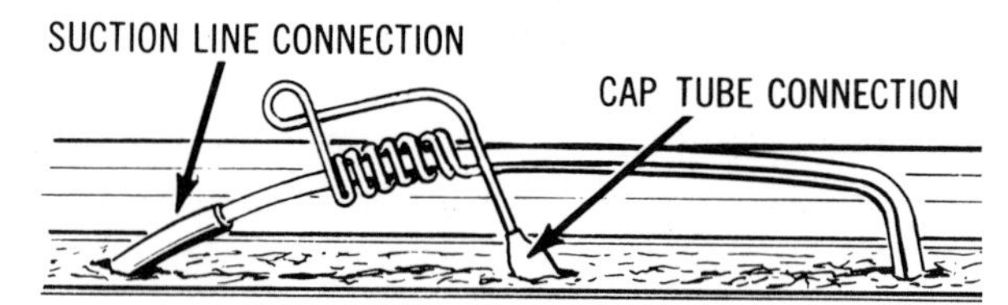

Figure 3-9. Method of exposing a heat exchanger.

Table 3-1. Failure and disposition chart.

Failure	Disposition							
	Replace compressor	Replace heat exchanger	Replace liner assembly	Replace condensor	Repair by brazing	Replace process tube	Install drier	Evacuate recharge
A. Compressor								
1. Noisy	×					×	×	×
2. Stuck	×					×	×	×
3. Inoperative	×					×	×	×
4. Burned out*	×	×				×	×	×
5. Leak	×					×	×	×
B. Heat exchange								
1. Restriction		×				×	×	×
2. Broken capillary		×				×	×	×
C. Liner								
1. Aluminum tube leak			×			×	×	×
D. Condenser								
1. Internal leak				×		×	×	×
2. Radiant condenser					×	×	×	×
E. Refrigerant Leak								
1. Copper to copper tube joint					×	×	×	×
2. Copper to steel tube joint					×	×	×	×
3. Steel to steel tube joint					×	×	×	×
4. Brass to copper tube joint					×	×	×	×
5. Brass to steel tube joint					×	×	×	×

* System must be thoroughly cleaned and flushed.

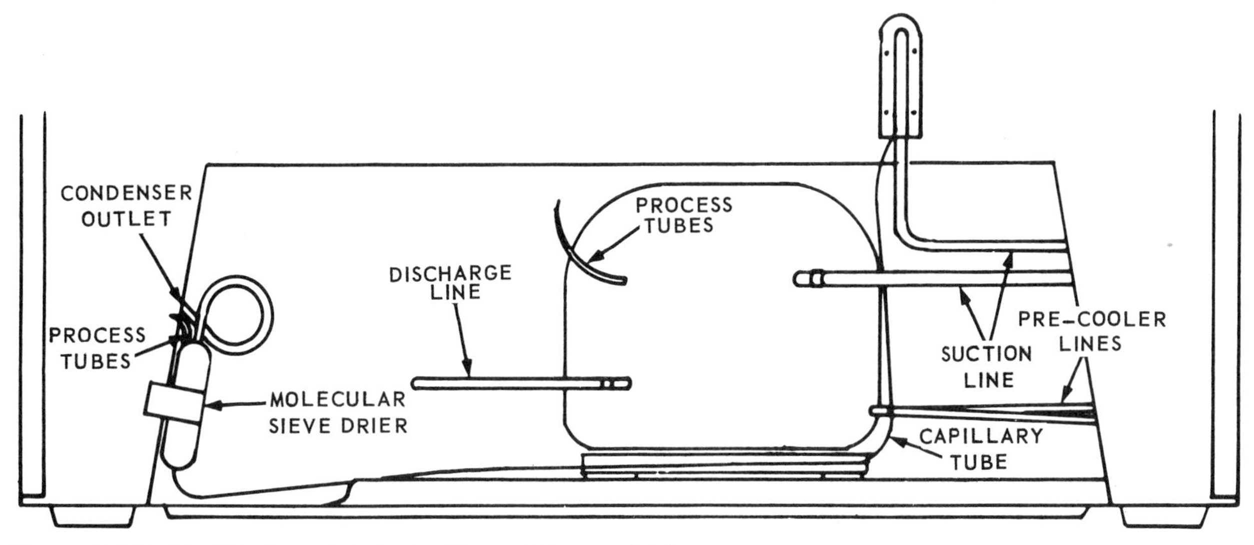

Figure 3-10. Identification of typical refrigerant lines and tubes.

vinyl breaker trim collar on the hinge side to expose the insulated space. Scrape off the top of the insulation ($\frac{1}{2}$ in or so) to locate the heat exchanger. Pull it upward to free it from the foam (Fig. 3-9). Check these joints with a bubble solution only. Do not use a halide or electronic leak detector near the foam. Refrigerant 11 is used as an expanding agent in the foaming process. It will react to a leak detector giving a false reading. The halide or electronic detector must still be used in the unit compartment.

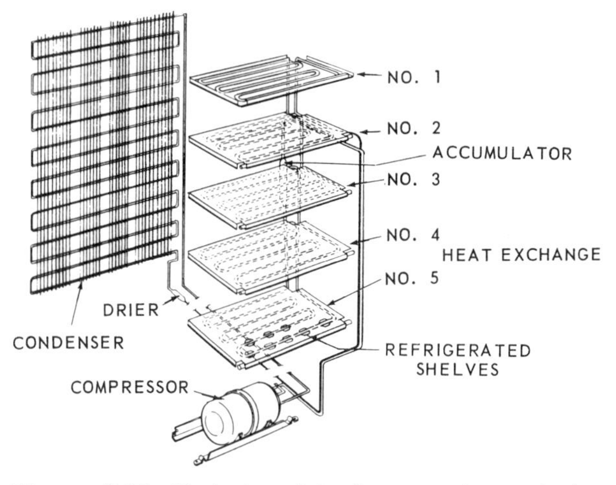

Figure 3-11. Typical upright freezer unit employing refrigerated aluminum shelves.

TROUBLE DIAGNOSIS

The most important part of trouble diagnosis of any appliance, as mentioned previously in the other books of this series, is to first determine exactly what the appliance is designed to do and how it is designed to perform. Once you know how the appliance is supposed to operate, trouble diagnosis becomes a routine elimination of those components, parts, or problems that prevent the appliance from operating in the manner it was designed to do. The following are the major problems one faces when undertaking the servicing of free-standing freezers—both upright and chest.

Problem: Unit will not operate.

Possible cause	*Solution*
1. Service cord pulled out of wall receptacle.	1. Plug in service cord.
2. Blown fuse in feed circuit.	2. Check wall receptacle.
3. Bad service cord plug,	3. If wall receptacle is "live," check circuit

3. (*cont'd.*) loose connection, or broken wire.	and make necessary repairs.
4. Inoperative temperature control or thermostat.	4. Power element may have lost charge, or points may be dirty. Check points. Short out thermostat. Repair or replace thermostat.
5. Inoperative relay.	5. Replace.
6. Defective capacitor.	6. Replace.
7. Stuck or burned out compressor.	7. Replace compressor.
8. Low voltage.	8. Check voltage. It must be 120 V ±10 percent at moment of starting with all other loads on the line.
9. Inoperative overload protector.	9. Replace.
10. Inoperative defrost timer (if used).	10. Replace.

Problem: Unit runs all the time.

Possible cause	*Solution*
1. Thermostat or temperature control adjustment.	1. Readjust or change thermostat.
2. Restricted air flow.	2. Provide proper clearances around cabinet. (Allow at least 6-in clearance above and 4 in at sides and back of upright freezers. Clearances of 2 in at the ends and 4 in at the rear are required for chest freezers.)
3. Short refrigerant charge.	3. Cabinet or chest temperatures abnormally low in lower section. Not enough refrigerant to flood evaporator coil at outlet.
4. Inefficient compressor.	4. Replace compressor, if necessary.
5. Improper wiring.	5. Check wiring and correct as necessary.
6. Overcharge of refrigerant.	6. Check suction pressures and frost line. Take necessary corrective action.
7. Partial restriction.	7. Locate restriction and remove. Then evacuate and recharge system.

Problem: Unit short cycles.

Possible cause	*Solution*
1. Thermostat erratic or out of adjustment.	1. Readjust or change thermostat.
2. Cycling on the overload relay.	2. This may be caused by low or high line voltage which varies more than 10 percent from 120 V. It may also be caused by high discharge pressure caused by air or noncondensable gases in the system. Correct either condition, as necessary.
3. Loose electric connection.	3. Tighten connection.
4. Faulty or incorrect overload.	4. Replace overload.
5. Tight compressor.	5. Check with no load. If compressor is faulty, replace.
6. Refrigerant line clogged.	6. Locate and repair.

Problem: Unit runs too much.

Possible cause	*Solution*
1. Freezer overloaded with unfrozen food.	1. Instruct customer to freeze no more than 10 percent of freezer capacity at one time.
2. Thermostat or temperature control set too cold.	2. Move dial to warmer setting.
3. Low refrigerant charge.	3. Unit will run longer to remove the necessary amount of heat and will operate at a lower than normal suction pressure. Put in normal charge and check for leaks.
4. Overcharge of refrigerant.	4. Excessively cold or frosted suction line results in lost refrigeration effort. Unit must run longer to compensate for the loss. Purge off excessive charge.
5. Temperature control or thermostat out of calibration.	5. Recalibrate or replace.
6. Low compressor capacity.	6. Replace compressor.
7. High room temperature.	7. Any increase in temperature around the cabinet will increase the refrigeration load and will result in longer running time to maintain cabinet temperature. Advise customer to move freezer to cooler area.
8. Restricted airflow over condenser.	8. This can be an improperly designed built-in cabinet, causing obstruction to the airflow around cabinet shell. Can also be caused by air or noncondensable gases in system, resulting in a higher head pressure which produces more reexpansion during the suction stroke of the compressor, and, consequently, less suction vapor. Increased running time must compensate for loss of efficiency. Correct the condition.

Problem: Unit operates improperly on thermostat.

Possible cause	*Solution*
1. Low voltage.	1. Check voltage. It must be 115 V ± 10 percent at moment of starting with all other loads on the line.
2. Starting relay stuck closed.	2. Check operation of starting relay on compressor.
3. Poor air circulation through condenser.	3. Check for obstruction.
4. Pressures in system not equalized.	4. Wait until unit stops before trying to start it again.
5. Liquid line restricted.	5. Locate and repair.
6. Thermostat out of calibration.	6. Replace thermostat.
7. Defective compressor motor.	7. Replace compressor.
8. Poor cabinet door seal.	8. Adjust door for proper seal.
9. Cabinet light on continuously.	9. Check operation of light switch. Check to see if door hits light button.

Problem: Compressor runs but does not refrigerate (see p. 118).

Possible cause	*Solution*
1. Refrigerant leak or shortage of refrigerant.	1. Locate leak and repair. Recharge system.
2. Restriction or moisture in refrigerant line.	2. Locate and repair.
3. Clogged tubing.	3. Check tubing for completely closed kinks. Replace if necessary.
4. Broken compressor valves or internal tubing.	4. Replace compressor.

Problem: Freezer noisy or vibrates.

Possible cause	*Solution*
1. Tubing rattles.	1. To prevent rubbing as well as noise, adjust tubes so they do not touch.
2. Unit not properly leveled.	2. Check alignment and correct position of freezer on the floor.
3. Food shelves, dividers, or baskets rattle.	3. Adjust to fit.
4. Noisy compressor.	4. Replace compressor.

Problem: Storage temperature too cold.

Possible cause	*Solution*
1. Temperature control set too cold.	1. Set warmer.
2. Thermostat is out of adjustment.	2. Readjust or change the thermostat.
3. Thermostat feeler bulb contact bad.	3. If the bulb contact is bad, the bulb temperature will lag behind the cooling coil temperature. The unit will then run longer and make the freezer too cold. See that bulb makes good contact with bulb well and that it is in its proper position.

Problem: Storage temperature higher than normal.

Possible cause	*Solution*
1. Thermostat selector knob set too warm.	1. Set colder.
2. Thermostat contact points dirty or burned.	2. Clean points or replace thermostat.
3. Thermostat out of adjustment.	3. Readjust or change thermostat.
4. Loose electric connection.	4. This may break the circuit periodically and cause the freezer to become warm because of irregular or erratic operation. Check circuit and repair or replace parts.
5. Incorrect wiring.	5. Wire correctly (see wiring diagram).
6. Excessive service load or abnormally high room temperature.	6. Instruct customer to freeze no more than 10 percent of freezer capacity at one time. Also advise on location of unit.
7. Restricted air circulation.	7. Allow for proper clearance (see Solution No. 2 under Unit Runs all the Time, p. 120).
8. Loss of refrigerant charge.	8. Locate leak and repair. Recharge.
9. Poor door seal.	9. Adjust door for proper seal.

10. Excessive frost accumulation on refrigerated shelves (upright manual-defrost models).	10. Remove frost.
11. Excessive frost accumulation on evaporator.	11. On no-frost models, check defrost heater, termination thermostat, and defrost timer. Also check for inoperative fan motor, improper fan position, or inoperative door switch. Repair or replace as needed.
12. Compressor cycling on overload protector.	12. Check protector and line voltage at compressor.

Problem: Large accumulation of ice beneath evaporator.

Possible cause	*Solution*
1. Plugged drain system.	1. Remove obstruction.
2. Drain-pan heater open.	2. Replace heater.
3. Drain-pan heater not in good contact with bottom of drain pan.	3. Adjust heater so that good contact is made with bottom of pan.
4. Loose or broken electric connections.	4. Repair or replace connections.
5. Freezer located in ambient temperature of below 35°F.	5. Instruct customer to move freezer to warmer location.

Problem: Defrost water runs onto floor, or excessive moisture in insulation in the bottom of the freezer.

Possible cause	*Solution*
1. Defrost tray not level.	1. Level tray.
2. Poor door seal causing excessive ice buildup.	2. Adjust door for positive seal.
3. Freezer installed in area of extremely high humidity.	3. Instruct customer to change location.
4. Vapor leak in door or cabinet.	4. Locate vapor leak and reseal. Replace wet insulation.
5. Bottom pan not sealed.	5. Make certain pan and breaker strip are sealed watertight by laying a bead of mastic on flange of pan.

Problem: Unit will not defrost.

Possible cause	*Solution*
1. Defective thermostat.	1. Check and replace if faulty.
2. Open heating element.	2. Check and replace if faulty.
3. Loose connection.	3. Check and repair as necessary.

Problem: Exterior of cabinet or chest is warm or hot.

Possible cause	*Solution*
1. Normal temperature will vary depending on load and ambient temperature.	1. Instruct customer of this fact.
2. Poor door seal (will increase load).	2. Adjust door for positive seal.

PERFORMANCE DATA AND THEIR USE

Performance data for a given refrigerator or freezer are usually given in the service manual. Such data help the service technician to determine whether a unit is operating up to full

MODEL NO.	UNIT CONTROL CUT ON and CUT OFF at settings					DEFROST CONTROL	DEFROST WATTAGE	REF. CHARGE (IN OZS.)	FOOD COMPT. TEMPERATURE (at #3 setting)			CRISPER TEMPERATURE (at #3 setting)			FREEZER AIR (at Cut Off)			RUNNING WATTS (at Cut Off)			KWH/24 CONSUMPTION			% RUNNING TIME			SUCTION PRESSURE		
	#1	#2	#3	#4	#5				70°	90°	110°	70°	90°	110°	70°	90°	110°	70°	90°	110°	70°	90°	110°	70°	90°	110°	70°	90°	110°
FHK150			0/-14					11.0							2.0	1.0	0	235	235	235	2.1	3.2	5.2	39	62	100	2	2	2
FHK170			0/-14					11.0							2.0	1.0	0	235	235	235	2.2	3.4	5.2	40	64	100	2	2	2
FHK210			0/-16					12.0							2.0	1.0	0	300	300	300	2.8	4.4	6.7	44	66	100	1	1	1
FHK211			0/-16					12.0							2.0	1.0	0	300	300	300	2.8	4.4	6.7	44	66	100	1	1	1

Figure 3-12. Typical performance data chart.

performance. A typical example of a performance data table is given in Fig. 3.12. But remember that the performance data given are usually no-load figures and should be used to gain a general approximation of performance of freezer. No-load can be approximated by using a time and temperature recorder over a 24-h period. Incidentally, the data as given below apply in general to refrigerator performances.

Control calibration. All control calibration figures are written with cut-on temperatures over the cut-off temperatures, i.e., cut-on/cut-off. Also all control calibration figures are nominal. You must remember that they have a tolerance of +20°.

Refrigeration charges. All charges are applicable if a separate vacuum pump is used (see p. 24).

Temperatures. Measure temperatures by wrapping a thermistor in a 1-in ball of permagum and placing the ball in the approximate center of the area to be tested.

Wattages. All wattages are given with assumption that the voltage is 120 V ± 10 percent. All readings are accurate within ±10 percent. Wattage, a true measure of power, is the measure of the rate at which electric energy is consumed. Therefore, wattage readings are useful in determining compressor efficiency, proper refrigerant charge, whether a restriction exists, or whether there is a malfunction of an electric component.

Amperes, measured with an amprobe, multiplied by the voltage, are not a true measurement of power in an ac circuit. Such a measurement gives only voltamperes or apparent power. This value must be multiplied by the power factor (phase angle) to obtain the true or actual ac power. The actual power is indicated by a wattmeter. See p. 115 for the method of measuring wattage with a wattmeter.

Cycles per day. You can determine ON and OFF times as follows.

$$\text{ON time} = \frac{1{,}440}{\text{cycles per day}} \times \text{percent of run time}$$

$$\text{OFF time} = \frac{1{,}440}{\text{cycles per day}} \times (100 - \text{percent of run time})$$

This will give you an approximation only because run time will vary with the number of door openings.

Percent of run time. This figure will vary with the number and length of door openings. Therefore, figures are not final, but should be used only as an approximation.

Pressures. Applicable just before cut-off.

Room air conditioners

CHAPTER

4

The room air conditioner with its controlled cooling and drying effects on the room air is often referred to as a *comfort conditioner*. Since comfort is the ultimate goal of air conditioners, let us first define comfort. Idealistically, comfort is a state of being totally unaware of the surrounding atmosphere. You are not too cold, nor too hot; not too dry, nor too humid; not too drafty, nor too stuffy.

To meet all these conditions, the heat load of the area to be cooled must be carefully calculated. This calculation, if done on a form approved by the AHAM (Association of Home Appliance Manufacturers), will ensure the selection of the correct unit. An air conditioner which bears the "AHAM approved" sticker must be sized to AHAM specifications. Instead of the capacity being rated in tons or horsepower (two terms which were abused in the past), AHAM standards call for the rating to be shown in Btu for cooling and pints of water per hour for moisture removal.

Installing either too small or too large a unit is a false economy. If an air conditioner is too small, it will run continuously, do little or no cooling, and do nothing more than act as a dehumidifier. Room temperature will remain high. Too large a unit may bring temperature in the immediate area down very quickly and then shut off too soon (if operating on a thermostat). Not only will the room temperature fluctuate greatly, but little dehumidification will take place because the unit did not run long enough to remove sufficient moisture. High humidity can be more uncomfortable than high temperatures.

HOW TO CALCULATE A ROOM LOAD

It is important as a service technician to know how to calculate the heat loads for room air conditioners. Many times troubles are caused by the use of units of improper size. Therefore, let us look at the factors influencing the correct heat calculations. Since heat always travels from the warmer to the colder object or air, air-conditioning engineers must be concerned with the "heat gain" into the building or room. They are working with an average of a 20°F temperature difference between the inside and the outside of the building. Part of this heat gain is by conduction through the walls, windows, and roof. Radiation from the sun is an additional source of heat gain to be reckoned with; it can easily add another 10 percent or more to the heat load. The sun's heat not only affects the the cooled area directly but also has a flywheel effect. The heat absorbed by the exterior walls of a building from the sun will still be making itself felt long after the sun has gone down.

Heat gain not only is due to transfer of heat through the walls, windows, etc., but also is the result of infiltration of warm air through cracks around doors, windows, and building framework as well as the opening and closing of doors. The amount and speed of this heat gain can be lessened by the use of double glass or storm windows, insulated walls and ceilings (or roofs), and awnings or other means of shade. Light-colored outside surfaces and roofs reflect heat. A vapor barrier type of wall insulation will also reduce infiltration of moisture as well as warm air through the walls.

Almost all the heat leakage into the air-conditioned area from the outside can be measured in degrees Fahrenheit and is called *sensible* heat (see Heat in Chap. 1).

The heat loads generated inside the living area that air conditioners must cope with are latent or hidden sources. These internal sources of heat are lights, appliances, cooking, people, baths and showers, etc. The amount of heat given off by lights and appliances can be found by multiplying their wattage by 3.4. For instance, three 100-W light bulbs in a room add 1020 Btu, all sensible heat, to the load the air conditioner must handle (300 watts × 3.4 = 1020 Btu). The AHAM cooling load estimate form drops the 0.4 and asks you to multiply by 3. Cooking also releases a large amount of heat and moisture into a room (both sensible and latent heat). A kitchen range with all burners going releases enough heat to use up the effect of a 3-ton air conditioner. (This situation can be helped by use of an exhaust fan.)

The "people load" is often underestimated when selecting an air conditioner. One person adds an average of 600 Btu/h to the load; while doing moderately heavy work, 1000 Btu;

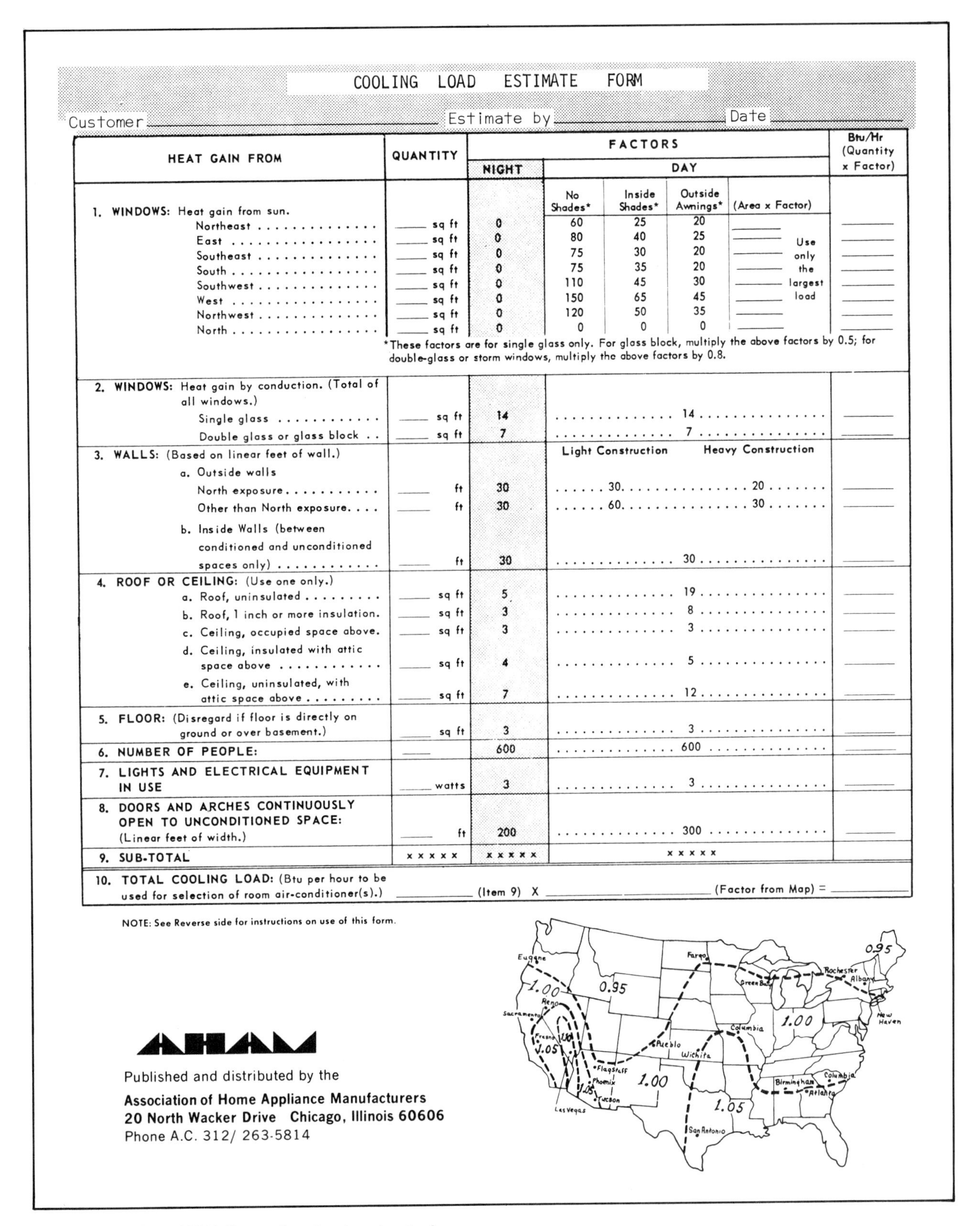

COOLING LOAD ESTIMATE FORM

Customer ______ Estimate by ______ Date ______

HEAT GAIN FROM	QUANTITY	FACTORS: NIGHT	FACTORS: DAY				Btu/Hr (Quantity x Factor)
			No Shades*	Inside Shades*	Outside Awnings*	(Area x Factor)	
1. WINDOWS: Heat gain from sun.							
Northeast	____ sq ft	0	60	25	20	____ Use only the largest load	____
East	____ sq ft	0	80	40	25	____	____
Southeast	____ sq ft	0	75	30	20	____	____
South	____ sq ft	0	75	35	20	____	____
Southwest	____ sq ft	0	110	45	30	____	____
West	____ sq ft	0	150	65	45	____	____
Northwest	____ sq ft	0	120	50	35	____	____
North	____ sq ft	0	0	0	0	____	____
*These factors are for single glass only. For glass block, multiply the above factors by 0.5; for double-glass or storm windows, multiply the above factors by 0.8.							
2. WINDOWS: Heat gain by conduction. (Total of all windows.)							
Single glass	____ sq ft	14	14				____
Double glass or glass block	____ sq ft	7	7				____
3. WALLS: (Based on linear feet of wall.)			Light Construction		Heavy Construction		
a. Outside walls							
North exposure	____ ft	30	30		20		____
Other than North exposure	____ ft	30	60		30		____
b. Inside Walls (between conditioned and unconditioned spaces only)	____ ft	30	30				____
4. ROOF OR CEILING: (Use one only.)							
a. Roof, uninsulated	____ sq ft	5	19				____
b. Roof, 1 inch or more insulation.	____ sq ft	3	8				____
c. Ceiling, occupied space above.	____ sq ft	3	3				____
d. Ceiling, insulated with attic space above	____ sq ft	4	5				____
e. Ceiling, uninsulated, with attic space above	____ sq ft	7	12				____
5. FLOOR: (Disregard if floor is directly on ground or over basement.)	____ sq ft	3	3				____
6. NUMBER OF PEOPLE:	____	600	600				____
7. LIGHTS AND ELECTRICAL EQUIPMENT IN USE	____ watts	3	3				____
8. DOORS AND ARCHES CONTINUOUSLY OPEN TO UNCONDITIONED SPACE: (Linear feet of width.)	____ ft	200	300				____
9. SUB-TOTAL	xxxxx	xxxxx	xxxxx				

10. TOTAL COOLING LOAD: (Btu per hour to be used for selection of room air-conditioner(s).) ______ (Item 9) X ______ (Factor from Map) = ______

NOTE: See Reverse side for instructions on use of this form.

AHAM

Published and distributed by the
Association of Home Appliance Manufacturers
20 North Wacker Drive Chicago, Illinois 60606
Phone A.C. 312/ 263-5814

Figure 4-1. The AHAM's cooling load estimate form.

INSTRUCTIONS FOR USING COOLING LOAD ESTIMATE FORM FOR ROOM AIR CONDITIONERS

(FROM AHAM STANDARD CN1)

A. This cooling load estimate form is suitable for estimating the cooling load for comfort air-conditioning installations which do not require specific conditions of inside temperature and humidity.

B. The form is based on an outside design temperature of 95 F dry bulb and 75 F wet bulb. It can be used for areas in the continental United States having other outside design temperatures by applying a correction factor for the particular locality as determined from the map.

C. The form includes "day" factors for calculating cooling loads in rooms where daytime comfort is desired (such as living rooms, offices, etc.), as well as "night" factors for calculating cooling loads in rooms where only nighttime comfort is desired (such as bedrooms). "Night" factors should be used only for those applications where comfort air-conditioning is desired during the period from sunset to sunrise.

D. The numbers of the following paragraphs refer to the correspondingly numbered item on the form:

1. Multiply the square feet of window area for each exposure by the applicable factor. The window area is the area of the wall opening in which the window is installed. For windows shaded by inside shades or venetian blinds, use the factor for "Inside Shades." For windows shaded by outside awnings or by both outside awnings and inside shades (or venetian blinds), use the factor for "Outside Awnings." "Single Glass" includes all types of single-thickness windows, and "Double Glass" includes sealed air-space types, storm windows, and glass block. Only one number should be entered in the right-hand column for item 1, and this number should represent ***only the exposure with the largest load.***

2. Multiply the total square feet of *all* windows in the room by the applicable factor.

3a. Multiply the total length (linear feet) of all walls exposed to the outside by the applicable factor. Doors should be considered as being part of the wall. Outside walls facing due north should be calculated separately from outside walls facing other directions. Walls which are permanently shaded by adjacent structures should be considered as being "North Exposure." Do not consider trees and shrubbery as providing permanent shading. An uninsulated frame wall or a masonry wall 8 inches or less in thickness is considered "Light Construction." An insulated frame wall or a masonry wall over 8 inches in thickness is considered "Heavy Construction."

3b. Multiply the total length (linear feet) of all inside walls between the space to be conditioned and any unconditioned spaces by the given factor. Do not include inside walls which separate other air-conditioned rooms.

4. Multiply the total square feet of roof or ceiling area by the factor given for the type of construction most nearly describing the particular application. (Use one line only.)

5. Multiply the total square feet of floor area by the factor given. Disregard this item if the floor is directly on the ground or over a basement.

6. Multiply the number of people who normally occupy the space to be air-conditioned by the factor given. Use a minimum of 2 people.

7. Determine the total number of watts for lights and electrical equipment, except the air conditioner itself, that will be *in use* when the room air-conditioning is operating. Multiply the total wattage by the factor given.

8. Multiply the total width (linear feet) of any doors or arches which are continually open to an unconditioned space by the applicable factor.

   NOTE—Where the width of the doors or arches is more than 5 feet, the actual load may exceed the the calculated value. In such cases, both adjoining rooms should be considered as a single large room, and the room air-conditioner unit or units should be selected according to a calculation made on this new basis.

9. Total the loads estimated for the foregoing 8 items.

10. Multiply the sub-total obtained in Item 9 by the proper correction factor, selected from the map, for the particular locality. The result is the total estimated design cooling load in Btu per hour.

E. For best results a room air-conditioner unit or units having a cooling capacity rating (determined in accordance with the AHAM Standards Publication for Room Air Conditioners, CN 1-1967) as close as possible to the estimated load should be selected. In general, a greatly oversized unit which would operate intermittently will be much less satisfactory than one which is slightly undersized and which would operate more nearly continuously.

F. Intermittent loads such as kitchen and laundry equipment are not included in this form.

RAC-1

Printed in USA

when bowling, 1200 Btu; while sleeping, 300 Btu. It can be seen that 10 people in a room can add 6000 Btu or a load equal to a 1/2 ton of air conditioning. This load is both sensible and latent heat. In fact, more than half is latent heat caused by evaporation of body moisture. Mopping and cleaning, air-drying clothes, ironing clothes, and using baths and showers add both sensible and latent heat to the load if performed in the air-conditioned area. This heat must be removed.

The most accurate way to calculate a heat load and to select a room air conditioner of the proper size is with the use of the cooling load estimate form (see Fig. 4-1). This form lists all the components of a building, windows, walls, etc. Heat transfer properties of these components have been predetermined, and a multiplying factor furnished. By multiplying the quantity (square foot, etc.) times the factor, the heat gain in Btu/h from each component is easily computed. Just follow the directions on the back of the form. As an example, let us figure a heat load on a room 15 by 20 ft. It has two 3 by 4 ft windows, unshaded, and is located on the west side of the building. The windows are of single panes of glass. The ceiling is uninsulated with an attic space above. Two people ordinarily occupy this room, and lighting is by two 100-W bulbs. The figures necessary in the calculation are the following.

Item No. 1 is heat gain from the sun by radiation. Our west windows are 24 ft^2, have no shades, and are of single panes of glass, so 24×150 (factor) = 3600 Btu. Item No. 2 is heat gain through the window by conduction. So 24 $ft^2 \times 14$ (factor) = 336 Btu. Item No. 3 is one outside wall of 20 linear ft: 20 ft $\times$ 60 (light construction factor) = 1200 Btu. Three inside walls next to unconditioned space, consisting of two 15-ft walls and a 20-ft wall equal 50 linear ft ($50 \times 30 = 1500$ Btu). Item No. 4 is the ceiling, uninsulated with attic space: 300 $ft^2 \times 12$ (factor) = 3600 Btu. Item No. 6 is two people: 2×600 (factor) = 1200 Btu. Item No. 7 is the lights: 200 W $\times$ 3 (factor) = 600 Btu. Adding these quantities together results in 12,036 Btu subtotal. For item No. 10 check the map below, then multiply the subtotal by the map factor determined by the geographical location of our building. This will give us the total load. In Fargo, N.D., we would multiply by 0.95; in Chicago, by 1.00; in San Antonio, by 1.05.

For best results, an air conditioner having an AHAM-rated cooling capacity as close as possible to the total estimated load should be selected. In general, a greatly oversized unit which would operate intermittently would be much less satisfactory than one which is slightly undersized and which would operate more nearly continuously.

AIR-CIRCULATION SYSTEM

A properly designed room air conditioner should perform the following tasks.

1. Regulate and control temperature.
2. Regulate and control humidity.
3. Furnish proper ventilation.
4. Recirculate the contained air within the conditioned space.
5. Filter and clean the air.

Modern room air conditioning is accomplished by the application of the principles of mechanical refrigeration to an air recirculation system arranged to absorb heat from the air within a contained space (a room) and transfer the heat to the outside of the structure. Thus there are only two basic systems in a room air conditioner: the refrigeration and the air circulation systems. We shall begin with the air system. There are two kinds.

System 1. The air system on the outside includes the condenser fan and dissipates the heat taken from the cooled area to the outside atmosphere. The condenser fan also picks up and dissipates into the outside atmosphere the condensate (water) which has been removed from the air in the cooled area. The slinger

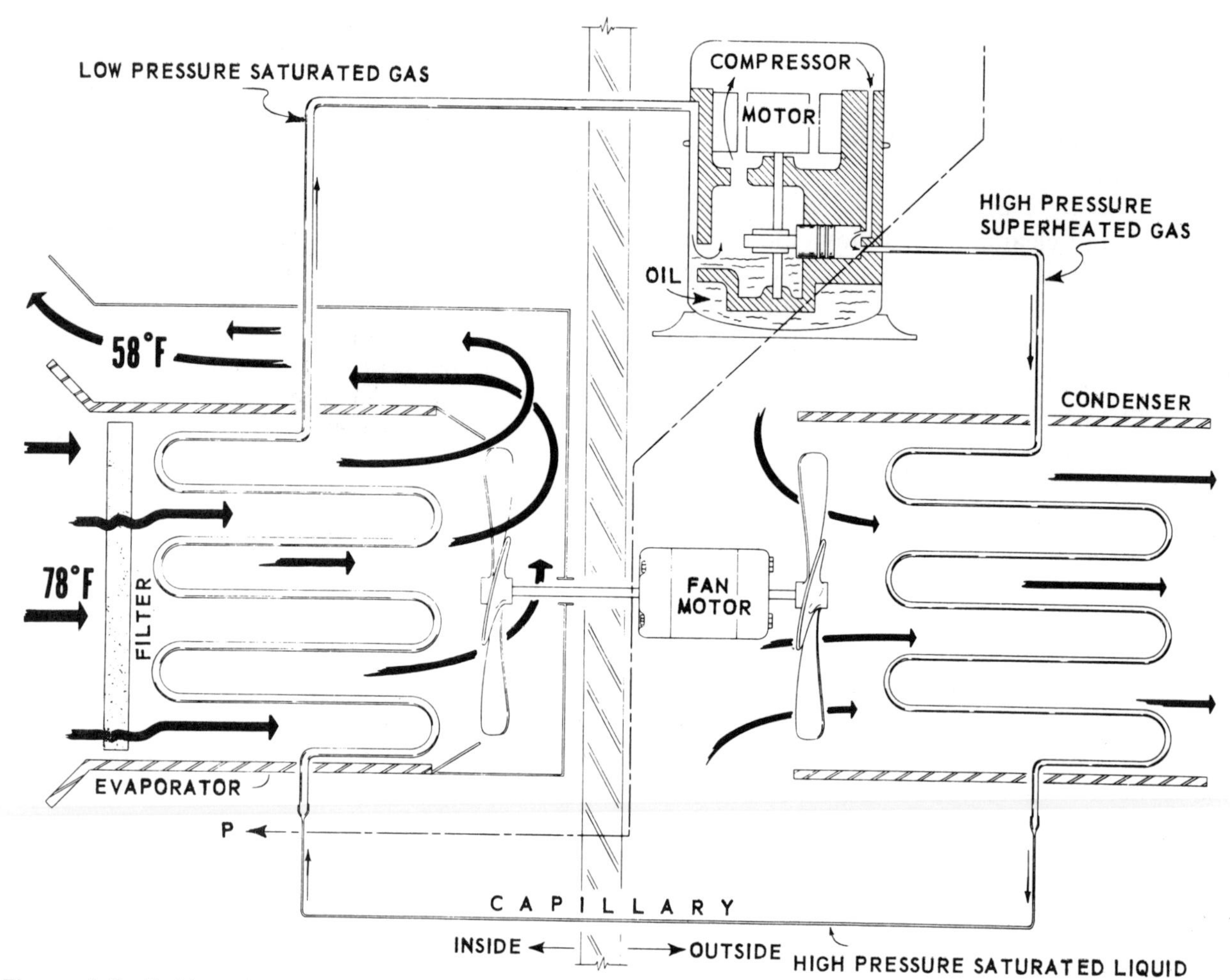

Figure 4-2. Refrigeration cycle of a room air conditioner.

ring on the outer periphery of the fan performs this function.

System 2. The air system within the cooled area circulates room air over the cooling unit. In so doing, the air loses sensible heat to the cooling unit. The cooling unit also condenses water vapor (latent heat) from the room air. A solid insulated partition (bulkhead) inside the air conditioner separates these two air systems.

These systems are so arranged that the cooling coils (which absorb heat) are located on the room side of the window and the condensing unit (which rejects heat) is on the outside of the window. It is important that no intermingling of the hot condenser air and cooled room air takes place. A partition (bulkhead) inside the air conditioner separates the room air from the condenser air. The rubber seals and panels at the window prevent exchange of inside and outside air. Tight installation at the window is of utmost importance.

Fan blades. The evaporator fan and condenser fan blades attach to each end of the motor fan. As a rule, the squirrel-cage type evaporator fan is employed because it has a low sound level and a high air-pulling capacity. It must pull the room air through the evaporator coil and expel the cooled dry air back into the room. A blade-type fan may be found on some small compact conditioners where blower space is limited.

On the outside end of the motor shaft is the condenser fan. This blade-type fan circulates

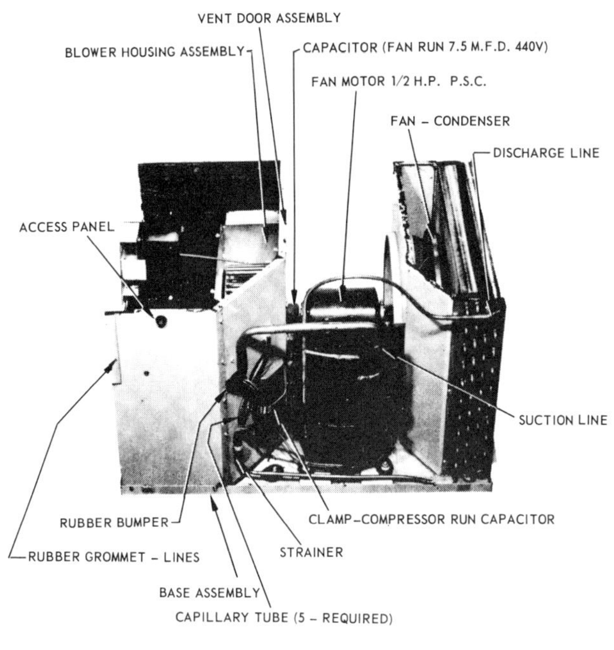

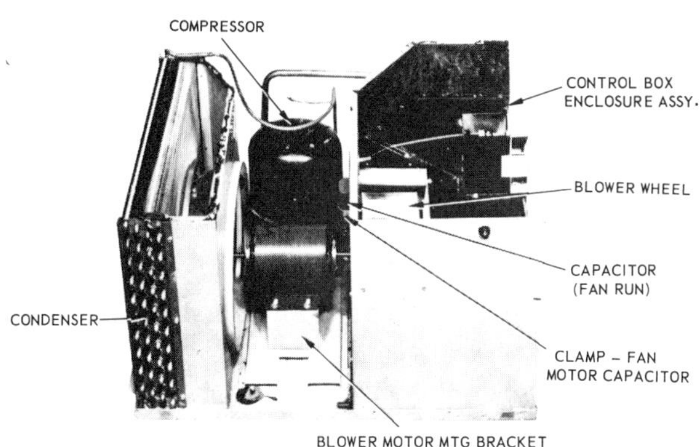

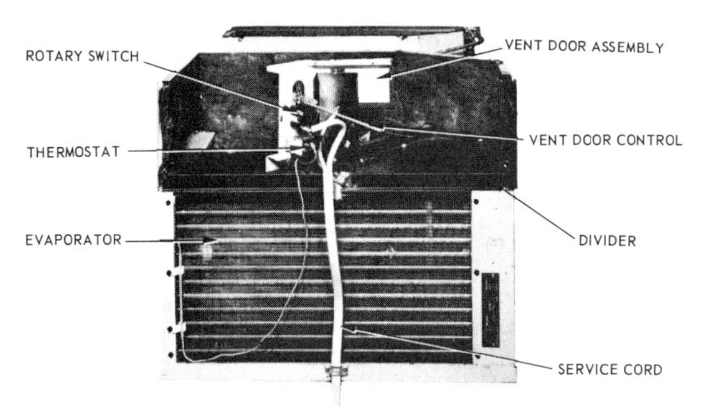

Figure 4-3. Major working parts of a typical room air conditioner.

outdoor air over the compressor and out through the condenser.

If you suspect trouble with the blades or fan motor, take extreme care to keep clear of the spinning rotor and blades. Be sure to set the switch in the OFF position when examining clearance of both the evaporator fan and condenser fan blades. Rotate the blades slowly by hand and listen for a scraping or grating noise. If blades are hard to turn, or if you feel or hear any scraping or grating noise, the air conditioner chassis must be removed from the cabinet to permit a complete examination to determine the cause of the trouble and its proper correction.

Occasionally a fan blade setscrew becomes loosened and emits a clicking sound between the fan blade hub and the motor shaft. Loosened setscrews can be detected by rocking the blades of the evaporator fan. As the fan blades and motor shaft assembly change direction of rotation, the loosened setscrew will bump on the shaft. Before a loosened setscrew can be tightened, the air conditioner chassis must be removed from the cabinet.

Fan motor. The fan motor may be either a single-speed or multispeed type, according to the model. It may be either a shaded-pole or a permanent split capacitor (PSC) type. All quality room air conditioners are equipped with built-in overload switches. These switches may not be shown on wiring diagrams because of space limitation, but have always been in use.

Single-speed fans are found on a few low-capacity conditioners. Two-speed and three-speed fan motors are used on all the others. The multispeed shaded-pole motors used in the early 1960s by most manufacturers were equipped with a reactor in series with the motor winding to achieve a slower speed. The reactor reduces the voltage applied to the motor. Actually, this reactor is a coil of wire similar to a ballast or choke coil. It may have more than two wire taps if used for more than two speeds. Each tap furnishes a different resistance to current flow.

In recent years, speed changes in shaded-pole motors were made through the motor windings. The highest motor speed is through the main winding. A second winding is placed in series with the main winding for a slower motor speed. A third winding is used in three-speed fans. With the three windings in series, the fan motor turns at its slowest speed.

The greatest number of fan motors, particularly in the larger-capacity air conditioners, are the PSC type. These motors have two windings: a run or main winding and a phase (sometimes called *start*) winding. Both windings are wound with about the same size and length of wire. A run capacitor is in series with the phase winding. The capacitance causes the electrical flow through this winding to shift out of phase with that of the run winding. A rotating field is thus set up causing the rotor to turn. Remember that the PSC motor is more efficient than the shaded-pole type. It has more power and draws less amperage.

Multispeed PSC fan motors contain additional main or run windings. The phase winding is in series with a capacitor and in parallel with the main winding. For high-speed operation only, the phase and main windings are energized. For a slower speed, the high-speed winding is placed in series with a second run winding. A three-speed motor will have a third winding in series.

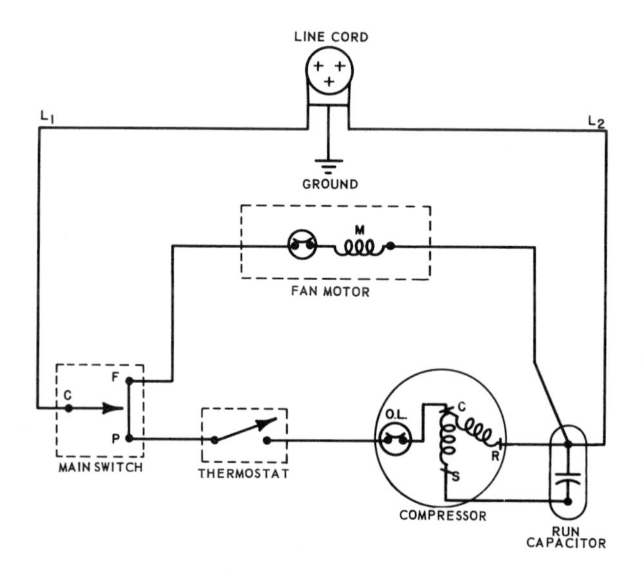

Figure 4-4. Schematic of a typical room air conditioner.

If you suspect a problem with the fan motor, first study the schematic wiring diagram, then make sure that all electric connections between power source and the fan motor are good. Most fan motors are equipped with a capacitor which may be tested in the same manner as described later in this chapter.

To test the fan motor windings, remove the motor from the terminal connections and check the continuity of the windings with a test light or an ohmmeter. An "open" circuit indicates a defective winding, a broken connection, or a bad motor protector inside the motor. Continuity between the windings and the frame indicates an internal short or ground, and the motor must be repaired or replaced.

Another testing procedure for a fan motor is to disconnect the motor wires and energize the motor independently with a test cord. With a test cord properly connected, the motor should operate if the motor is in good condition.

Many fan motors are equipped with oiler tubes. SAE-20 weight of a nondetergent type of oil is recommended. (Since most fans have Oilite bearings, a detergent oil can ruin the oil-absorbing quality of bearings.) If the motor does not have oiler tubes, it has a lasting lubricant and therefore requires little attention.

Do not overoil as the reservoir may overflow and damage the motor. Add five or six drops of oil to each bearing. Be sure to replace the oiler plugs to keep water and dirt from getting into the fan motor bearings.

REFRIGERATION SYSTEM

For a better understanding of the application of the mechanical refrigeration machine for the air conditioner, the service technician should be familiar with certain thermodynamic facts dealing with heat and its transfer from one form of energy to another and from one medium to another. For instance, the refrigeration cycle is a series of changes of state of the refrigerant. In this process, as described in Chap. 1, the

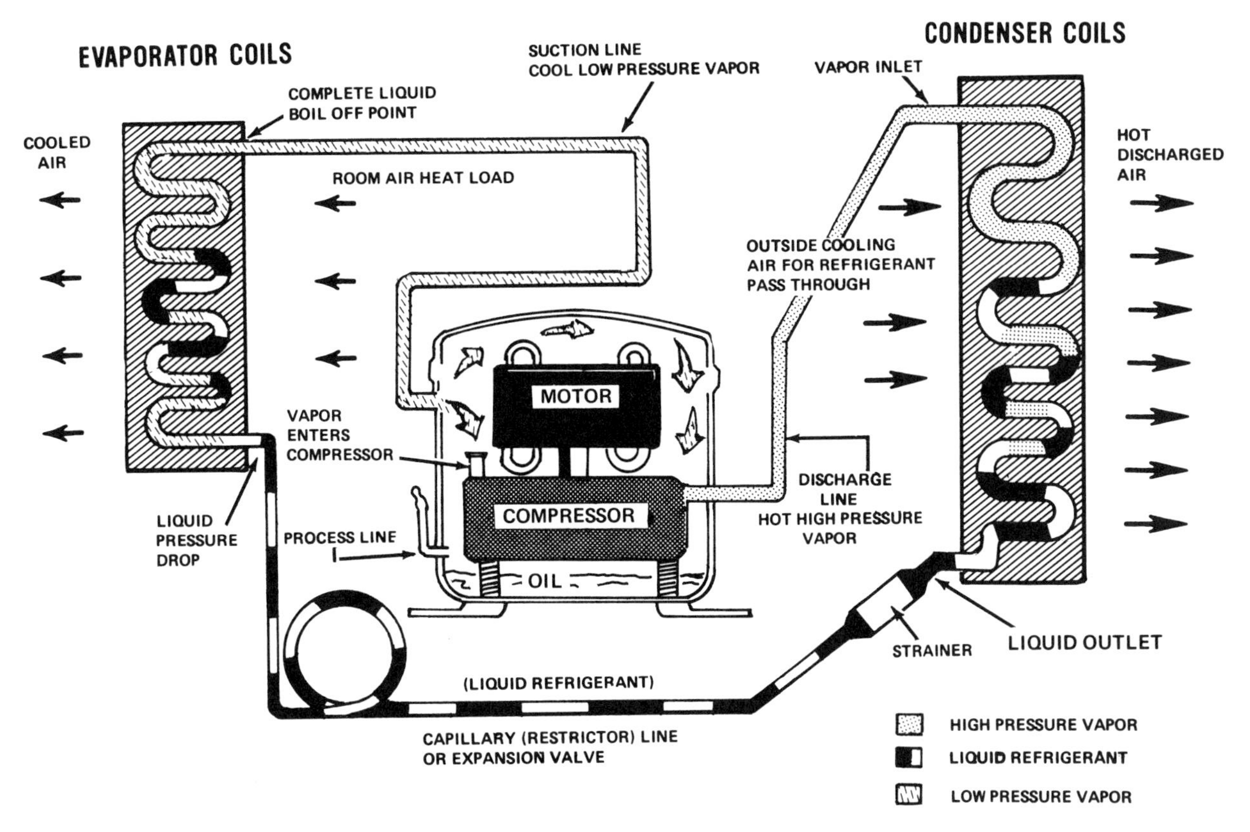

Figure 4-5. Cycle of refrigeration of a room air conditioner.

refrigerant is changed from a liquid to a vapor and restored to a liquid. The energy (heat) changing the liquid refrigerant to a vapor is the heat extracted from the air in the conditioned space. The complete cycle consists of four processes.

1. Heat gain in the evaporator
2. Pressure rise in the compressor
3. Heat loss in the condenser
4. Pressure loss in the capillary

Basically the refrigeration cycle consists of two heat transfer processes and two pressure change processes.

Heat gain in the evaporator. The heat-laden air inside the house is constantly recirculated through the coils of the evaporator. Heat from the air is transferred to the coils of the evaporator, resulting in the *heat gain* in the evaporator.

Pressure rise in the compressor. When the compressor is started, the pressure in the evaporator is reduced, causing the liquid refrigerant in the evaporator to vaporize or boil. Since the pumping capacity of the compressor is constant, the pressure in the low side is balanced between the inflowing liquid from the capillary and the compressor intake. This process results in the use of the heat in the air to bring about a change of state—liquid to vapor—of the refrigerant; during this change each pound of liquid refrigerant which is "boiled off" will absorb a considerable amount of heat, and the heat is carried out of the evaporator with the vapor.

The vapor being pumped out of the evaporator is termed a *low-pressure saturated gas*; that is, it is said to be saturated with the heat that came out of the air in the house. When heat is held in a vapor, the heat energy is called *latent heat of evaporization.* When the low-pressure saturated gas enters the compressor, it is compressed and forced into the condenser coil under high pressure.

The process of compression requires power from the compressor-motor; this power is expended to compress the gas. As the low-pressure gas enters the compressor at 40 to 50°F, the temperature is increased by the compression action to 150 to 180°F, and the pressure evaluated to 225 to 250 lb/in^2 gauge. After the gas leaves the compressor, it is referred to as *high-pressure superheated gas.*

Heat loss in the condenser. As the hot gas passes into the condenser, heat flows from the gas to the tubing and to the fins where the heat is transferred to the cool air flowing through the coil. Under the high-pressure condition the heat-saturated gas (refrigerant) undergoes another change of state, vapor to liquid. As the process of condensation takes place, the refrigerant gives up all the heat it absorbed in the evaporator and also in the compressor. When this heat is released to cause the vapor to change back to a liquid, the heat involved in the process is also referred to as *latent heat of evaporization.*

Heat flows from a warm to a cool body; therefore, in order to handle the heat being removed from the condenser coil, the coil is situated in the rear end of the unit and is isolated from the air inside the house by the weatherproof bulkhead. The fan circulates outside air through the coil, and the heat is transferred to the cooler air and is carried away from the building.

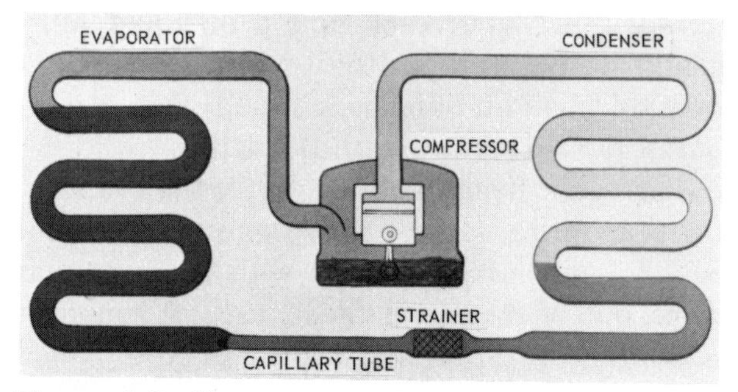

Figure 4-6. Piston-type compressor system.

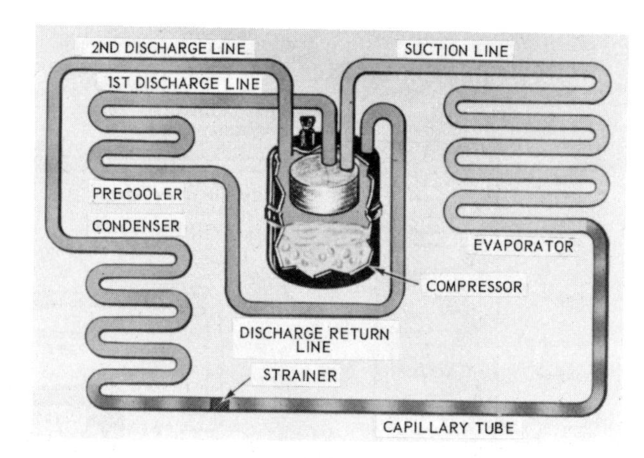

Figure 4-7. Rotary compressor initial startup.

The liquid refrigerant flows to the lower area of the condenser coil where it is accumulated and impounded ahead of the capillary tube.

Pressure loss in capillary. The capillary is a small tube which serves to change the high-pressure liquid to a low-pressure liquid. This is achieved by the flow resistance of the liquid refrigerant through the small bore of the capillary tube.

Compressor. The sealed compressor system is similar to other refrigeration systems. R-22 is usually used as the refrigerant. The capillary tube has a larger diameter than that of a freezer or refrigerator to operate at a high evaporator temperature and pressure.

As with refrigerators, two basic types of compressors are used in room air conditioners: rotary and piston. The piston type (Fig. 4-6) is most popular and is often called a *low-side pump.* The area inside the shell around the motor and pump is under low-side pressure. The pump pulls from the interior of the shell. High-pressure gas is discharged directly to the condenser. For more on piston-type compressors, see Reciprocating Compressors in Chap. 1.

Some makes of room conditioners are equipped with a rotary compressor (Fig. 4-7) much like an oversized refrigerator compressor. The interior of the dome is in the high-pressure side of the system. It has four tubing connections: a suction, a first discharge, a discharge return, and a second discharge. The first discharge tube leads to a precooler coil in the condenser assembly. The precooler functions also as an oil separator. As the hot gas and oil vapor pass through the precooler, the oil, with its higher condensation temperature, will cool to its dew point and form droplets. Upon reentering the compressor at the discharge return, the oil drops will settle into the crankcase. The refrigerant vapor gives off some of its heat and continues out the second discharge tube to the condenser.

Since the rotary compressor has a high-side crankcase, it will take a few minutes longer than the piston-type compressor to start cooling on initial startup. This condition will probably be more noticeable to the technician than to the customer. During an extended shutdown of this unit, much of the refrigerant liquid will migrate into the compressor crankcase and mix with the oil. It may take 7 to 10 min on initial startup for the refrigerant to leave the crankcase and balance itself in the system.

A comparison of reciprocating piston pumps and rotary vane pumps will show the advantages of each (Fig. 4-8). When operating under normal loads and ambient temperatures, the two types will draw about equal amperage. As temperatures rise with the corresponding increase in refrigerant pressures, the piston pump will lose comparative efficiency because of the

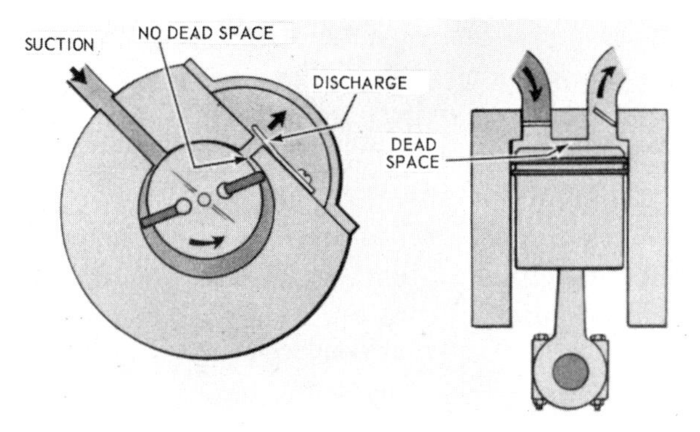

Figure 4-8. (Right) A comparison of reciprocating piston pump and (left) the rotary vane pump.

space between the piston and the cylinder head. The higher the pressures, the more vapor remains to reexpand in the cylinder cavity between piston strokes. The capacity of the pump is reduced.

The rotary compressor pump, with its wiping-vane action, will exhaust the entire cylinder vapor load to the high side. When load pressures increase in high ambient temperatures, the rotary pump efficiency actually increases. The piston pump, on the other hand, will initially pull down a couple of minutes faster than the rotary pump. The rotary, however, draws less amperage at start. This is important in areas of marginal voltage or brownouts.

The capillary tube in the compressor system is a restrictor for metering the flow of refrigerant from the condenser to the low side in the evaporator. A pencil-type strainer is often placed ahead of the capillary tube. It is designed to catch any solid foreign material in the system before it can enter or plug the capillary tube.

Compressors equipped with an internal overload have been in use on room air conditioners for several years. They are usually found on models of 21,000 Btu and larger. Thus, in the event the condenser fan stops or the condenser airflow is restricted, the high-side pressure will rise. The extra load on the compressor will cause the windings to heat up and actuate the internal overload. Once this happens, a waiting period of $\frac{1}{2}$ to 2 h is necessary before it will reset. The waiting time depends on ambient temperatures and the residual heat in the copper winding.

Manufacturers are now building an internal pressure relief (IPR) valve into some of their compressors. It is located between the high-pressure discharge chamber and the low-pressure suction chamber of the compressor pump. The IPR valve is usually a spring-loaded, ball-type. If the pressure difference between the high and the low side in the pump becomes greater than approximately 550 lb because of fan stoppage or blocked airflow, the valve will open. This will permit high-pressure discharge gas to flow directly into the suction side of the compressor.

The valve opening will cause the compressor pumping sound to change, similar to the change in sound of the automatic transmission in your car when shifting. The internal pressure will rise to the valve opening pressure in about $1\frac{1}{2}$ min. Once the valve opens, it will not close until the compressor goes off on overload. Remember that the internal overload switch will still open on either excessive motor winding temperatures or high amperage draw. The action of the IPR valve and overload switch will continue until the cause of the high head pressure is corrected. However, the number of overload cycles will be fewer since the IPR valve will bypass the high-pressure gas, reducing the tendency for winding temperatures to become extremely high and cause an overload trip.

Compressor motor. The compressor motors are generally permanent split capacitor (PSC) types. They operate in the same manner as the PSC fan motor, but are of higher horsepower and use capacitors of higher microfarad rating. Also the run capacitor, since it is in the circuit continuously, is usually filled with an oil which acts as a coolant. The oil was removed and the can cut away for this picture to illustrate the tightly wrapped foil and paper layers making up the plates of the capacitor. The marked terminal may be indicated with either a dimple or a spot of paint. It means that this terminal is connected to the capacitor plate closest to the can or outer case of the capacitor.

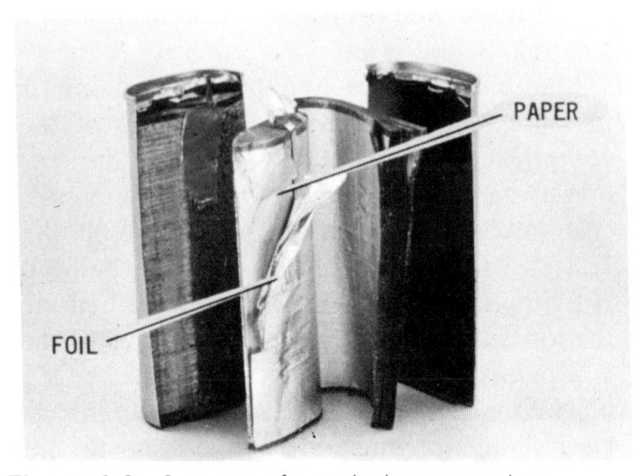

Figure 4-9. Cutaway of a typical run capacitor.

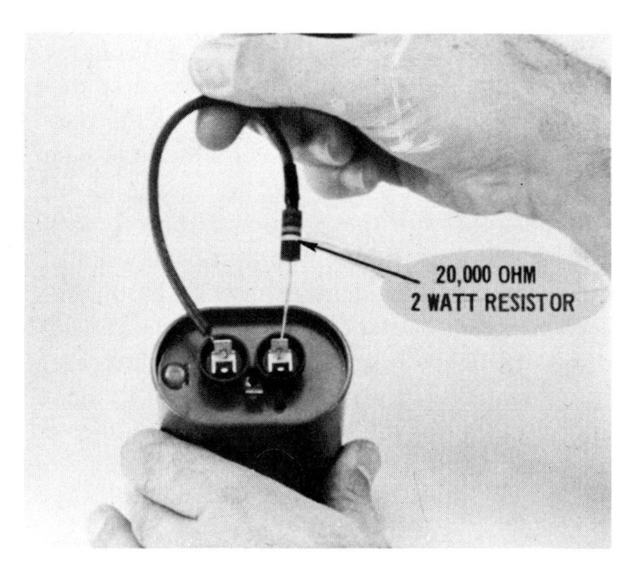

Figure 4-10. Proper method of discharging a run capacitor.

If a short should occur between the plate and the can, it would blow a fuse of a properly grounded appliance. In the event that the coil was incorrectly wired to the unmarked terminal and a short through the capacitor shell occurred, the capacitor would still act like a capacitor. A fuse would not blow, and a live circuit would be completed to ground through the chassis. The phase-winding circuit would also be grounded at the capacitor shell. The motor would not operate, and the winding would overheat. The situation would become more serious if the chassis were not properly grounded.

A capacitor, as mentioned in *Basics of Electric Appliance Servicing*, may hold a charge of electricity for a long time. Do not handle a capacitor until you know it has been discharged. The safest way of discharging a capacitor is with a 20,000-Ω, 2-W resistor (Fig. 4-10). This allows the discharging action to take place slowly and under control. Do not discharge a capacitor by shorting across its terminals. The uncontrolled flow of current from one side of the capacitor to the other may damage or weaken the plates or the thin paper separating them. Some capacitors are fused internally. If the fuse blows, the capacitor is useless.

Run capacitors are manufactured in many sizes and shapes. Microfarad ratings will range from 2 up to 5 μF. When replacing capacitors, always use a replacement of the same microfarad rating as the original unless an alternate is approved by the manufacturer in the service manual. Generally, most makers will permit a replacement of a capacitor with one having a *slightly* higher microfarad rating, but never lower.

The voltage rating of the run capacitor on a 240-V unit will be much higher than line voltage. This one is rated at 366 to 430 V. There is a reason for this. During a running cycle there is a generator action between the motor windings, and the voltage of the phase winding will rise above line voltage. The capacitor must be able to handle this higher voltage. Surprisingly, the less load on the motor, the higher this voltage will become. On a cool day, therefore, a capacitor on the verge of blowing will most likely fail. Thus, a replacement capacitor may have a higher working voltage rating, but never lower.

Dual capacitors are found on some units. It is two capacitors in one body with three terminal connections marked HERM (for *hermetic*), C (for *common*), and FAN. The PSC compressor is connected to the HERM terminal. The fan motor lead is attached to FAN. The L2 line to both motors is attached to C. One capacitor unit serves both motors. A few models contain a four-terminal capacitor which is really two separate capacitors in the same body.

Under normal conditions, the PSC compressor and run capacitor will start and operate with no problem. Difficulties may arise under abnormal conditions such as marginal voltage or unusual load conditions. If a starting problem is encountered, a start capacitor and a voltage or potential relay may be added. Check the parts list for part numbers of components needed for a particular model.

When properly wired into the compressor, the start capacitor is parallel with the run capacitor. The potential or voltage relay coil is parallel to the start winding and is sensing the voltage across the winding. The relay contacts are normally closed. When the motor starts, the voltage drops slightly. As the motor gains speed, the voltage in the phase winding rises above normal. At about 80 percent of full speed, the relay coil becomes strong enough to open the contacts and remove the start capacitor from the circuit. The start capacitor uses a dry electrolytic material and can stay in the circuit only a very short time.

The relay may be tested by momentarily connecting an insulated jumper wire between the compressor terminals with the switch set on the COLDEST position. When the compressor starts, remove the jumper wire. If the compressor continues to run with the jumper wire removed, the relay is defective and must be replaced. If the compressor fails to start, or runs for only a brief interval, the relay is probably satisfactory and the trouble is caused by other defective electric components.

Testing capacitors and compressors. There are several methods of checking the efficiency of a capacitor; here are three of the most popular.

Test Method 1. A capacitor analyzer is possibly the easiest and most exact method of checking the capacitance and efficiency of a capacitor. As detailed in *Basics of Electric Appliance Servicing*, it will diagnose shorted, open, or inefficient capacitor operation. An analyzer will also indicate the microfarad rating of the capacitor as well as its power factor percentage. As a rule, a run capacitor with a power factor greater than 15 percent or a start capacitor greater than 5 percent should be discarded. It is consuming electricity, meaning that heat is being generated somewhere internally. Early failure is probable.

Test Method 2. A volt-ohmmeter can be used to check a suspected defective capacitor. (Do not use the ohm scale on an ampere probe meter since you may receive a false indication.) But an ohmmeter test of a capacitor will indicate only if it is shorted or open. Always discharge the capacitor with a 20,000-Ω, 2-W resistor before using an ohmmeter.

To test, touch the meter probes to the capacitor terminals while watching the dial. A shorted capacitor will cause the dial to register near zero. An open capacitor will not register on the meter. A normal capacitor will cause the dial to swing quickly toward zero and then fall back to a resistance reading of 100,000 Ω or better. It can be read only once. For additional readings the capacitor must be discharged again or the meter probes reversed.

Test Method 3. You can build an inexpensive device to test capacitors. This tester can be used to test for a short between the plates and the can on the running capacitors or an open circuit inside the capacitor. You can also establish the approximate microfarad rating of a capacitor. The following parts are required to build this test device.

1. Two flush-mount receptacles that will accommodate a standard-base light bulb
2. Two 200-W incandescent light bulbs
3. A snap-on volt ammeter
4. One flush-mount receptacle that will accommodate a standard-base fuse
5. One standard-base quick-acting fuse plug not to exceed 10 A
6. Two test probes
7. Approximately 10 ft of No. 16 wire
8. A service cord of No. 16 wire equipped with 120-V plug

Mount the two light bulbs and fuse receptacles onto a base and install the light bulbs and the fuse. Form the coil of wire by wrapping the wire around a $1\frac{1}{2}$-in cylinder six full turns. Remove the wire from the cylinder and tape the wires with electrical tape in a couple of places to keep the coil together. To wire and complete the tester, check the circuitry shown in Fig. 4-11.

To operate, plug the tester into a 120-V power source and touch the test probes together. The bulbs will light to full brilliance if the tester is in proper working condition. Then clamp a snap-on ammeter into coil A and momentarily touch the test leads to the capacitor terminals as outlined below and note the ampere reading.

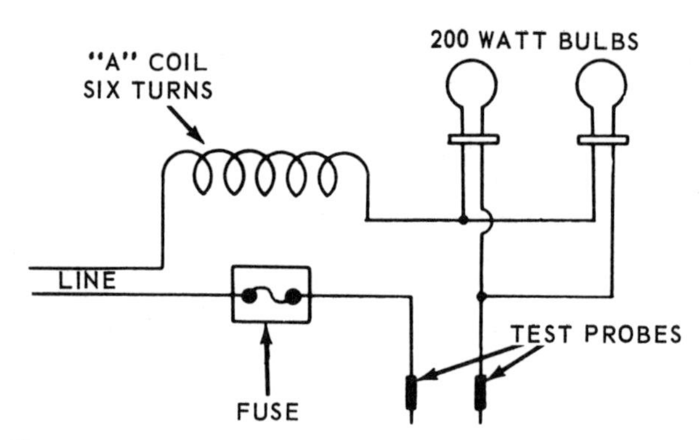

Figure 4-11. Capacitor tester.

1. Two-terminal capacitors.
   a. Take the ampere reading between the two terminals.
2. Three-terminal running capacitors.
   a. Take the ampere reading between the terminals marked C and HERM for the high-microfarad factor.
   b. Take the ampere reading between terminals marked C and FAN for the low-microfarad factor.
3. Four-terminal running capacitors.
   a. Take the ampere reading between terminals marked HERM for the high-microfarad factor.
   b. Take the ampere reading between terminals marked FAN for the low-microfarad factor.

Caution: Before handling the capacitor as outlined below, discharge the capacitor through a 10,000-Ω, 1-W resistor placed across the terminals for 1 min. Do not discharge any capacitor by means of a direct short as this will cause dielectric breakdown, and in case of an internal fused capacitor the fuse may blow, necessitating the replacement of the capacitor.

Once you have noted the ampere reading of the capacitor under test, you can determine the microfarad capacitance by reference to the chart on p. 139.

If the ammeter does not show an ampere reading, the capacitor under test is open, and the capacitor must be replaced.

If both test bulbs light to full brilliance when the test probes are touched to the capacitor terminals, the capacitor is internally shorted and must be replaced.

To test a compressor, remove the wiring from the three compressor terminals. Continuity between any one of the three terminals and the case indicates that the compressor is grounded, or if the continuity shows an open circuit between any two terminals, the compressor is defective and the complete chassis must be delivered or shipped to the nearest authorized factory service station for replacement.

Before going to the work and expense of changing starting components, however, it is good policy to check out the voltage supply. Is there a voltage drop at the receptacle due to incorrect wiring? This can be solved by the owner's electrician. Is the voltage supply at the building service box correct? If it is too low, the owner should request correction by the utility company. Voltage test must be made with the unit operating.

Air conditioners operating on 240- and 208-V circuits are equipped with service cords and plugs designed specifically for the amperage rating of a particular unit. The correct wiring and fusing information is included with the operating and installation instructions and should be complied with along with all applicable codes. Taking a running voltage and wattage check with a standard volt-watt meter is impossible without some sort of adapters. Robinair adapters were developed for this use. One adapter is equipped with alligator clips and a male plug to connect any service cord to the meter. The second adapter has a female plug to receive the meter cord and has flexible blades which will fit any one of the three receptacles. Since this meter does not have a ground-wire connection, a separate wire must be used to ground the third prong of the appliance cord.

Ampere factor	Microfarad rating
Starting capacitor	
4.9– 5.2	21/25
5.9– 6.2	25/26
10.6–11.0	38/46
12.0–12.4	50/56
14.8–15.2	72/87
14.8–15.2	86/103
15.0–15.4	88/108
16.0–16.4	108/120
16.8–17.2	121/146
16.9–17.3	135/155
18.4–18.8	161/193
Running capacitor	
1.1– 1.5	5
1.5– 1.9	8
3.3– 3.7	15
4.4– 4.8	20
5.9– 6.2	25
6.9– 7.2	30
7.9– 8.2	35

Certain power utilities reduce their voltages in periods of excessive load conditions. Low voltages will cause problems in starting and operating motors, particularly in air conditioners. This brings up the subject of minimum operating voltage. Electric motors have a tolerance factor of ± 10 percent of their rated voltages.

Examples of tolerances are listed below.

Rated voltage	*Minimum voltage at start*
120	108
240	216
208	187.2
240/208	197.6

This is not the voltage during a normal running cycle; it is the voltage at the instant of start. How do we read it? When the unit is running, turn the compressor off and immediately turn it back on. This will cause an overload condition for 5 to 10 s. During this time of high-current draw, read the voltage. If the voltage reads lower than 10 percent below the nameplate rating, the voltage supply needs correction. The motor should not be run under this condition. Notice that on the units with 240/208 dual voltage rating the minimum minus voltage tolerance for 208-V application is only 5 percent or 197.6 V.

It is important to keep in mind that the analysis of any air-conditioner system would not be complete without a check on the amount of current being drawn by the compressor. The amount of load or work being done by the compressor is in direct relation to the amount of amperes or current being consumed. The performance chart given in the service manual shows the normal amount of current the compressor should draw, and allowing for local variations in temperatures and humidity, the service technician can utilize the current data as an aid to diagnose certain troubles. For example, abnormally high ampere draw would indicate trouble in one or more of the following areas.

1. Recirculation of condenser air
2. Obstruction of condenser air
3. Leaking or partially shorted running capacitor
4. Partially shorted or grounded compressor winding
5. Overcharged system

On the other hand, abnormally low ampere draw would indicate possible trouble in one or more of the following areas.

1. Clogged intake filter
2. Dirty evaporator air
3. Recirculation of evaporator air
4. System partially discharged—refrigerant leak
5. Partially restricted capillary
6. Partially restricted strainer
7. Undercharged system
8. Inefficient compressor

Occasionally a new air conditioner may fail to start because of the "tightness" of the moving parts of the compressor. *Caution*: Do not declare the compressor defective. The compressor may cycle on the overload protector for 10 to 15 min before starting. Such compressors will usually require a brief run-in period after which they will start satisfactorily.

A temporary expediency to start a "stiff" compressor is outlined below and may be employed by a service technician for the initial starting or breaking-in of the compressor.

1. Use a starting capacitor with a capacity of approximately 150 μF and 250 V; connect a pair of lead wires 12 to 15 in long to the terminals of the capacitor.
2. Expose the terminals of the running capacitor of the air conditioner.
3. Plug the service cord into the wall receptacle and set the switch to the cooling position.
4. Momentarily contact the starting capacitor wires to C and HERM on the running capacitor terminals. The compressor should start instantly. *Caution*: Do not leave the starting capacitor connected for more than 1 or 2 s.
5. Once the compressor is started, allow the air conditioner to run several hours for the break-in period. After this, the compressor should start satisfactorily without the aid of the starting capacitor.

Evaporator. The evaporator is a finned coil which not only cools the air but also removes its moisture before returning it to the room. Room air is circulated through the fins and over the coils, and water vapor in this air condenses out onto the cold surfaces as a liquid and drips down into a drain pan under the evaporator. The collecting water must then be removed from the room side to the outside of the conditioned area. To do this, the room air conditioner must slope down toward the condenser; a $\frac{1}{4}$-in drop is enough. If a level is used, one-half bubble is a rule of thumb. The condensate will flow to the outside by gravity.

There are two systems of condensate removal: the wet base and the dry base. In the latter, two hoses carry the water from the evaporator pan to a similar pan beneath the condenser. The water will not normally come in contact with the steel baseplate. In fact, in most models using a dry base, the water flows into a sump area beneath the condenser fan blade slinger ring. The moisture is picked up by the slinger and sprayed into the condenser. The addition of moisture to the air passing through the condenser will increase the efficiency of the unit. Moist air will remove a greater amount of heat from the coil than dry air.

The slinger ring may not be able to handle

all the water when the humidity is very high. The cut-out in the condenser pan will allow the overflow to enter the baseplate and flow on to an overflow drain. When condensate and humidity levels become too high, a pinging sound caused by droplets hitting the fan blade may be heard by the user. This is to be considered normal.

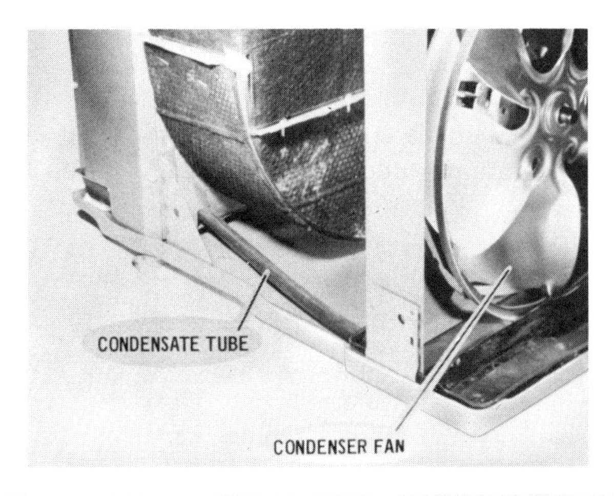

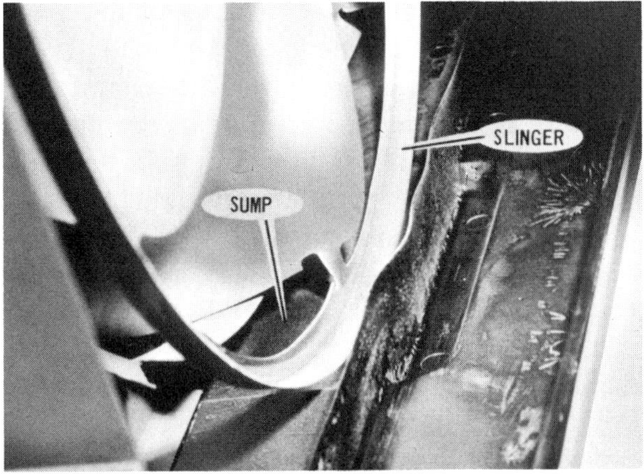

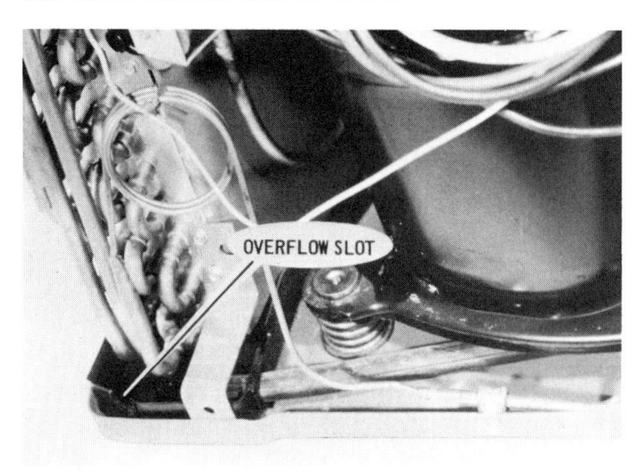

Figure 4-12. Parts of dry-base unit.

Some models have a drain tube projecting from the unit base for draining excess water. A length of tubing may be attached to the outlet to carry the water to an area where it is not objectionable. Other makes, especially late models with slide-out chassis, have a drain hole in the baseplate near the back edge. When the unit is inserted into the cabinet, the baseplate outlet will line up with a drain fitting in the cabinet rail. A hose attached to the fitting may be used to control the drain water.

The condensate of wet-base models flows across the steel baseplate to a slinger ring sump. A grille-type insert is placed beneath the fan on some units to protect the fan motor from water being splashed up from the fan air currents. Some models do not have a baseplate drain. If the humidity is so high that the slinger ring cannot handle the water, it will accumulate until deep enough to overflow at the shipping bolt holes. If the dripping occurs and is objectionable, a controlled drain may be constructed according to instructions in the service manual. For example, a drain pan with a hose connector may be placed below the baseplate. The drain pan must be made locally and be sized to fit the chassis. It may be suspended from the unit rails of the cabinet on slide-out chassis models. The pan must be creased for positive drainage. A drain connector is then soldered to the crease point for attachment of a regular garden-hose fitting.

The drain pan for a wraparound cabinet is similar to that used on the slide-out chassis except it is attached by the same screws securing the cabinet to the baseplate.

Water dripping from the front of the unit is usually the result of an improper installation. The air conditioner must be installed with a slight slope toward the outside. This slope will allow evaporator condensate to flow toward the rear of the air conditioner and be disposed of as described previously. Dripping water may also be the result of blocked water channels in

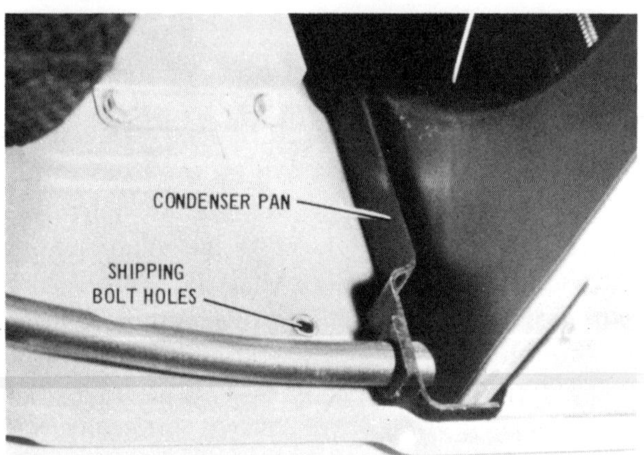

Figure 4-13. Wet-base condensing unit.

the bulkhead. This condition can be easily corrected by taking the chassis from the window and removing any obstructions from the drain hose which passes from the drain pan through the upright bulkhead located behind the evaporator coil. These obstructions may be in the form of foreign material or algae. A thick yellowish jellylike substance in the base pan could be a form of algae.

To clean the base pan, move the chassis from the building. Flush the base pan with water by using a hose with a high-pressure nozzle. If algae are suspected, a reoccurrence may be prevented by adding an algaecide. Several types of algaecides are on the market to prevent algae from forming. *Caution*: Never use copper sulfate as an algaecide.

Condensation on the decorative front may occur if the insulation on the rear of the front panel is broken, dislocated, or missing. In such cases, the insulation must be repaired or replaced. But under extremely humid conditions, condensate may form on portions of the discharge louvers of the front panel. This condition may occur even though the insulation is in good condition. Such sweating will usually disappear after the air conditioner runs for a few hours or long enough to dehumidify the room air.

COOLING CAPACITY TEST

A cooling capacity or performance test enables the service technician to accurately determine whether the air conditioning unit is operating satisfactorily. When making capacity tests, be sure the filter is clean, the evaporator and condenser fins are not dirty or bent, the damper doors are closed, and the window curtains or drapes are not obstructing the airflow.

To check the capacity of a room air-conditioning unit, it is necessary to take wet-bulb (WB) temperatures, dry-bulb (DB) temperatures, and a total wattage reading and to compare these readings with the capacity check table usually given in the service manual. (A typical capacity check table is given below, but remember that this is only an example and should not be used in any service work.) Operate the unit on FULL COOL for about 15 min before taking readings. Connect the wattmeter into the circuit; see the illustration on p. 146 for details.

1. *Wet-bulb temperature.* The simplest method for service application for wet-bulb measurements is by placing a fairly accurate thermometer, supplemented with a water-soaked cloth tube or sock over the bulb, in the airstream being measured. Subtract the evaporator outlet temperature from the evaporator inlet temperature given in the service manual.

Table 4-1. Typical capacity check table

Unit model	Air inlet temperature to condenser dry bulb, °F	Min. temperature drop through evaporator wet bulb, °F	Total watts (approximate)	
			230 V	208 V
Model A	80	15.5	3,080–3,400	3,120–3,460
Model B	90	14.0	3,230–3,570	3,270–3,620
Model C	100	12.5	3,380–3,740	3,420–3,780
Model D	110	11.0	·3,520–3,900	3,470–3,950

a. Inside, evaporator inlet wet-bulb temperatures should be taken from $\frac{1}{4}$ to $\frac{1}{2}$ in away from the decorative front grille along the lower half of the grille face.

b. Inside, evaporator outlet wet-bulb temperatures should be taken where the cooled air comes out of the adjustable grilles. Take temperature across grille and use average figure. The temperature from these vanes will vary.

The sling psychrometer (Fig. 4-14) is also used to obtain the wet-bulb temperature in determining the percent relative humidity. To obtain the wet-bulb temperature, operate the sling psychrometer as follows: Saturate the wick (only once during the procedure of obtaining wet-bulb readings) with clean water slightly below the room temperature. The psychrometer reading should be acquired 5 to 6 ft in front of the unit and approximately 4 ft off the floor. Direct the discharge louver so the cold air will not hit the sling psychrometer.

2. *Dry-bulb temperatures.* The dry-bulb temperatures should be taken from $\frac{1}{4}$ to $\frac{1}{2}$ in away from the condenser air inlet louvers (where the outside air enters the louvers). Do not allow the thermometer bulb to touch the metal louvers. *Note*: This thermometer can be attached with a side clip type clothespin by raising the lower window sash and reaching out over the unit. If it is not possible to raise the lower sash, this dry-bulb temperature can be taken outdoors on the same side of the building where the condenser is located.
3. *Wattage.* For the capacity check, connect the wattmeter into the circuit before starting the unit. Allow the unit to run at least 15 min before taking readings. Reading should be within the minimum and maximum limits given in the service manual's capacity test tables.

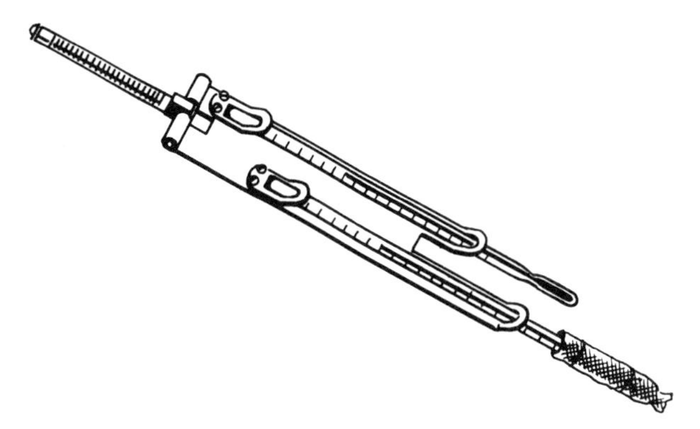

Figure 4-14. Standard sling psychrometer.

If the capacity test tables are not available, a simple and reliable cooling performance test of an air conditioner may be made by checking the temperature of the airstream which enters the evaporator through the front panel and the temperature of the outlet air which is discharged through the discharge louvers, as follows:

1. Suspend a thermometer in the stream of air entering the evaporator coil.
2. Suspend another thermometer in the stream of cold air that is discharged from the evaporator.
3. Operate the air conditioner for 15 min with controls set for maximum cooling. Close the fresh-air vents (if used) and put the thermostat on the coldest setting.

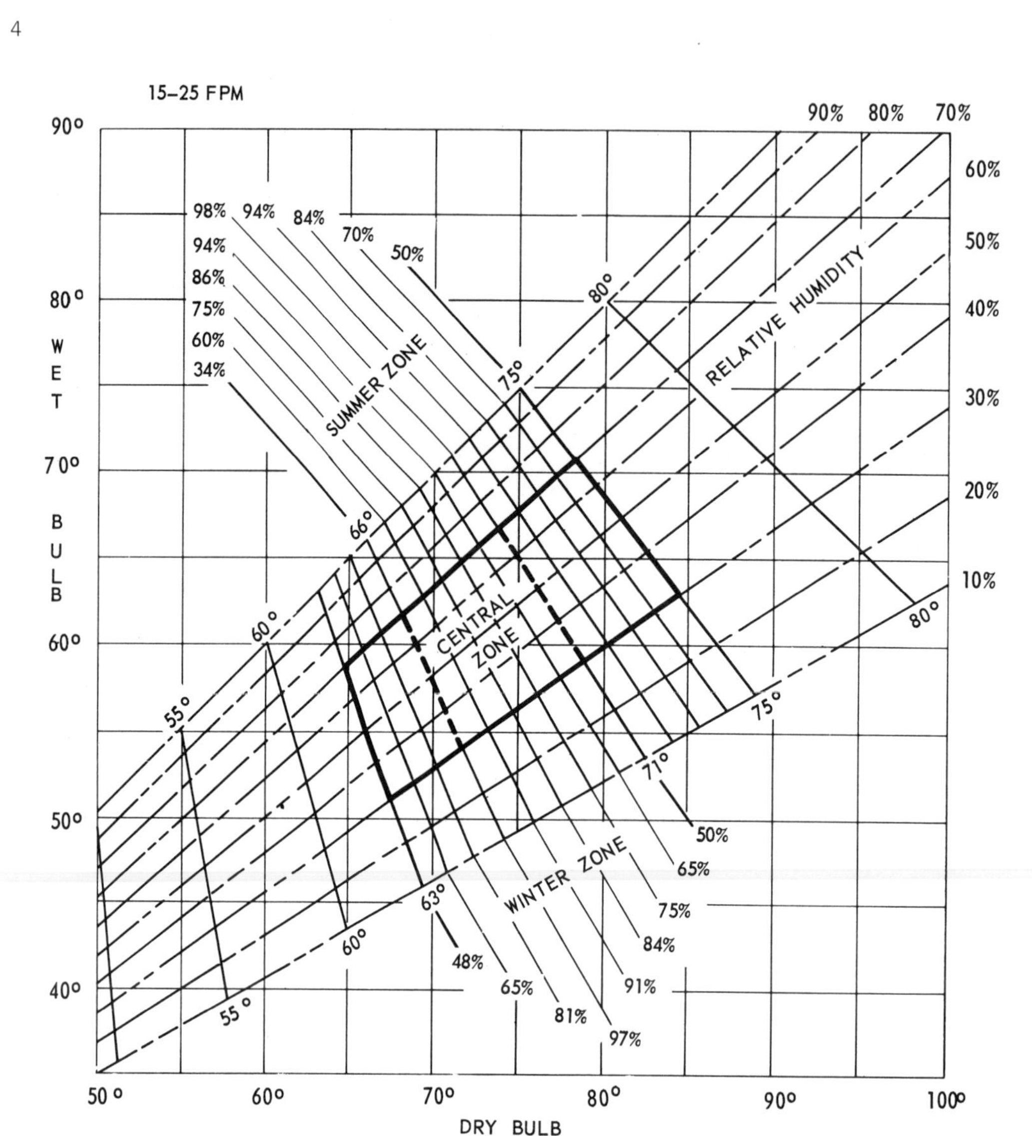

Figure 4-15. Typical comfort chart.

4. Obtain the temperature differential (TD) by subtracting inlet air temperature from discharge air temperature.
5. Compare the temperature differential (TD) of the unit under test with the TD shown in the applicable performance chart in the service manual.

Temperature differentials shown on most performance charts are measured with an ambient room temperature of 80°F and 40 percent relative humidity. Any variations from these conditions will affect the temperature differential. An air conditioner which produces a temperature difference of 15 to 20°F under average conditions is considered to be operating properly. In extremely humid weather, a temperature differential of 10°F is considered adequate because of the loss of cooling capacity in dehumidifying the air.

Filter care. The most common loss of cooling and ventilating efficiency is caused by a dirty or clogged filter. Filters must be inspected periodically as detailed in the consumer's instruction booklet and thoroughly cleaned when dirty.

To inspect a filter, hold it up to the light.

If the light is not clearly visible through the filter mesh, clean the filter. In cleaning extremely dirty filters, the bulk of the dirt may be removed with a vacuum cleaner. Filters should then be held under a stream of warm water, clean side up, to remove the dirt embedded in the filter mesh. (Bleach solutions or dry detergents should never be used to clean filters.) Shake the excess water from the filters and allow them to dry. When replacing filters, be sure the thermostat sensing tube is in its proper position.

Never operate an air conditioner without a filter. Dirt will enter the evaporator coil, clogging it and reducing cooling capacity.

Electrostatic air filter. Some air conditioners feature an electrostatic air filter frequently called a *generator*. This electrostatic filter is a unique application of the Van de Graaff principle of generating static electricity. The static electricity is transmitted to the generator, thus allowing it to pluck the very fine particles of dust and dirt from the air passing through the filter. The amount of foreign material in the air and the frequency with which the filter is cleaned will determine how often the generator must be cleaned. Each time the filter is cleaned, the generator should be inspected for excessive accumulation of surface dust. If the generator surface appears to be coated over with surface dust, the generator should be removed and cleaned.

A clean, lintless cloth dampened with rubbing alcohol provides a very satisfactory agent for cleaning the generator. Only the outside surface of the generator body and the generator mounting bracket should be cleaned.

PERFORMANCE CHECKING PROCEDURES

The performance of an air conditioner can be checked by using the performance charts found in the service manual. There are far too many to be included in this book but their usage will be demonstrated with the following example using a hypothetical unit and conditions. The following check procedure and the example chart (Fig. 4-16) will be based on these conditions and temperatures.

EER (energy efficiency ratio). There has been a lot of talk about EER and high-efficiency air conditioners. Most of this stems from the fact that a shortage of electrical power has become a real national problem. Higher-efficiency air conditioners could possibly help ease the power crises during the peak air conditioner season. But regardless of the power situation, when it gets hot customers want comfort.

Understanding the efficiency of an air conditioner need not be confusing. Just as automobile gas mileage is figured by dividing the miles driven by the amount of gasoline (gallons) used (250 mi divided by 20 gal = 12.5 mi/gal), so the efficiency of an air conditioner is expressed in Btu per watt. In other words, 1 W of electricity will remove a given number of Btu of heat from the space to be air conditioned.

8,000 Btu divided by 860 W = 9.3 Btu of cooling per watt

In this case, Btu and miles are the net result of using watts of electricity or gallons of gasoline. In the case of the automobile, it is generally accepted that the customer is better satisfied with 20 mi/gal than 10 mi/gal of gasoline. The same is true with an 8,000 Btu air conditioner. A 9.3 Btu/W system will cost less to operate than a 5.9 Btu/W air conditioner.

Much has been written about air conditioner efficiency and many different terms have been used to describe it. Terms such as energy efficiency ratio (EER), watt-wise, high efficiency, and many others. What all of these terms refer to is the Btu of cooling per watt of electricity used. And just as with automobile gas mileage, the higher the Btu per watt the more efficient the air conditioner. Most manufacturers describe air conditioners with an EER of 7.5 Btu/watt or greater as high-efficiency air conditioners.

Why are high-efficiency air conditioners more efficient than their standard counterparts?

Performance Checking Procedures

The following check procedure and the example chart will be based on these conditions and temperatures.

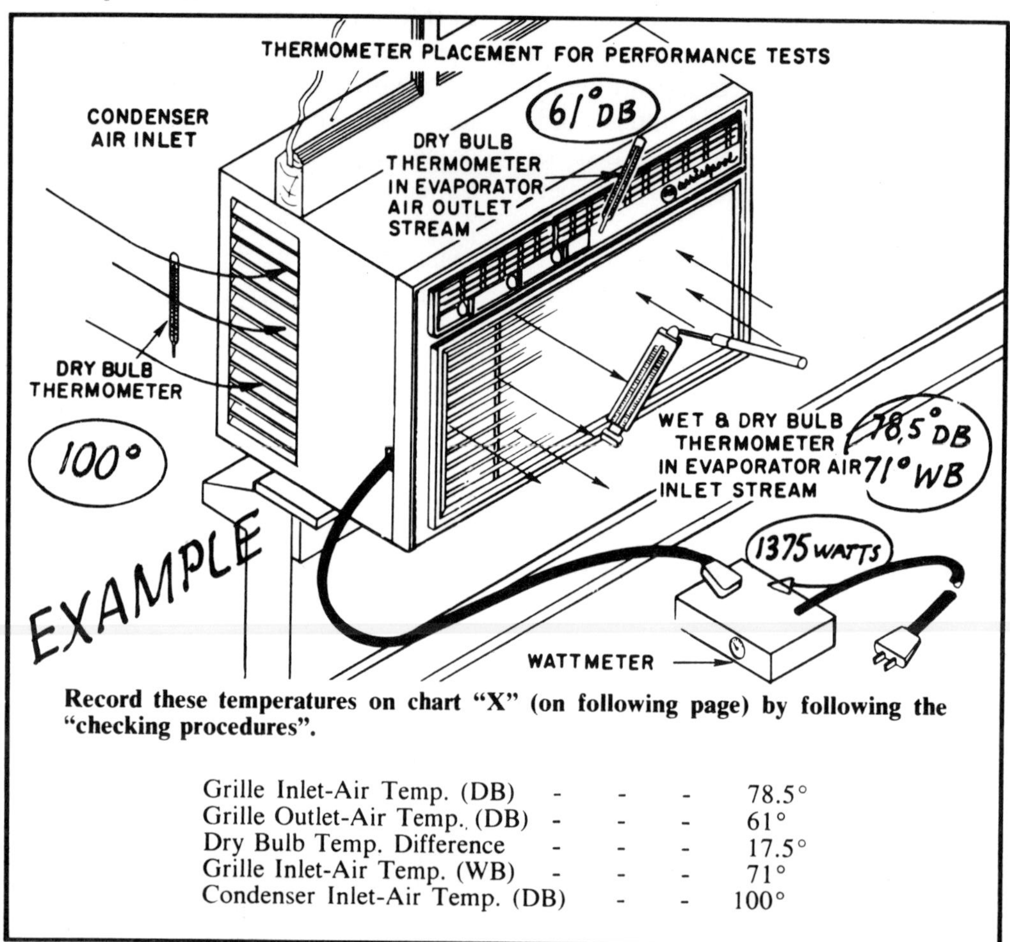

Record these temperatures on chart "X" (on following page) by following the "checking procedures".

Grille Inlet-Air Temp. (DB) - - -	78.5°
Grille Outlet-Air Temp. (DB) - - -	61°
Dry Bulb Temp. Difference - - -	17.5°
Grille Inlet-Air Temp. (WB) - - -	71°
Condenser Inlet-Air Temp. (DB) - -	100°

CHECKING PROCEDURES

The method of checking an air conditioner to determine if it is providing the amount of cooling for which it was designed is given below. The PERFORMANCE TABLES and wattage data (listed on the model specification pages) have been obtained from the representative production models.

To Make Performance Check:

1. Use a clean (and correct) filter.
2. Be sure front fits properly and air is directed upward (to prevent recirculation).
3. Close AIR EXCHANGE (or exhaust) door.
4. Close any doors or openings to the room so that no outside air is allowed to enter the room and no cooled room air is allowed to escape.
5. Insert wattmeter into receptacle and air conditioner cord into wattmeter and operate unit for one-half hour before temperature readings.
6. Take and record dry-bulb temperature reading in the condenser air inlet stream. (Do not allow thermometer to contact any parts. Shield it from the direct rays of the sun.)
7. Take and record the wet bulb and dry bulb sling psychrometer readings as near as possible to the air inlet portion of the front grille. (Remember, you want the readings to reflect the conditions of the air going onto the evaporator coil.)
8. Take and record a dry bulb temperature reading of the air coming out of the discharge grille. (Place thermometer in area of greatest air discharge velocity.)

Figure 4-16. Typical performance checking procedure.

NOTE: Readings in 6, 7 and 8 should be taken as nearly simultaneously as possible.

9. Subtract air discharge dry bulb reading (taken in 7) from the **dry bulb** front grille inlet reading (taken in 6) and record this temperature **difference.**
10. Refer to PERFORMANCE CHART for model involved.
11. Locate temperature difference (found in 9) in lefthand vertical line (line "A") and the wet bulb room temperature along the horizontal line (line "B") at the bottom of the chart.
12. Extend these two points horizontally and vertically until they intersect and note the location with respect to the performance line.
13. If the condenser air inlet temperature is other than 95°, allow for it by using the compensation chart (found on each performance chart) and mentally relocate the performance line near the point where the lines intersected.
14. **If the unit's performance point falls below the new performance line (after compensation) by more than 1 1/2°, the unit is not performing properly and the trouble should be located and corrected.**

NOTE: The charts are based on regular production units operated at line voltage as specified on the serial plate.

15. The wattage reading taken should be compared to the wattage shown on the specifications for the model involved. Compensation should be made for outside temperature increase or decrease from the 95° wattage reading shown in the model specifications. The amount to compensate is shown on each chart.

PERFORMANCE CHART "X"

(Example)

TO USE CHART:

1. Record inlet and outlet air dry bulb temp. . . . .
2. Subtract outlet temp. from inlet temp. . . . . .
3. Record inlet air wet bulb temp. . . . .
4. Locate temp. difference (2) on line "A" . . . . .
5. Locate inlet air wet bulb temp. on line "B" . . .
6. Extend points to intersection and mark.
7. Compensate for proper condenser inlet air temp. per Chart "C". Performance should not be lower than 1 1/2° below corrected performance line.
8. Compensate wattage rating per chart "C" below . . . .

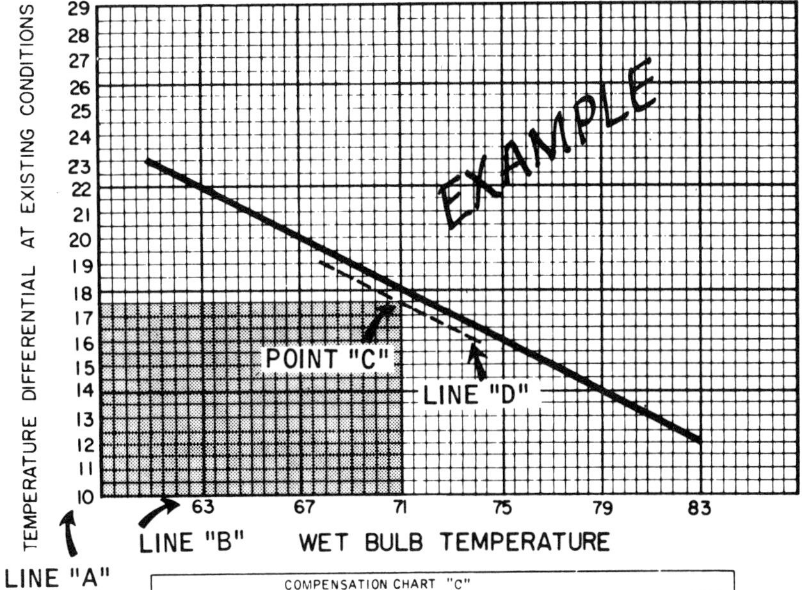

COMPENSATION CHART "C"					
If Cond. Inlet Air Temp. is:	Correct Performance Line By:	Select column nearest to wattage shown in specifications for model involved.			
		850	1300	2100	2600
30	Adding 1 1/2°	675 ± 50	1050 ± 75	1700 ± 100	2100 ± 125
35	Adding 1°				
90	Adding 1/2°				
95	No Correction	850 ± 50	1300 ± 75	2150 ± 100	2600 + 125
100	Deducting 1/2°				
105	Deducting 1°				
110	Deducting 1 1/2°	1000 ± 50	1500 ± 75	2400 ± 100	3100 ± 125

Figure 4-17. Cooling hours per season.

Understanding the technical approach to higher-efficiency air conditioning may not seem important or useful, but consider it similar to the question customers sometimes ask, "How does an air conditioner work?"

In terms of a real customer benefit, the higher an air conditioner's operating efficiency, the lower its operating cost. Many of today's customers shop for the initial or installed cost and do not concern themselves with efficiency or what the unit costs to operate. Possibly this is because the initial dollar outlay is something they understand and Btu, watts, amps, hours of operation, etc., they find difficult to relate to dollars. For example, if a room air conditioner ran an average of 2,000 h per season, a standard 8,000 Btu air conditioner could cost more in added operating costs per year compared to an air conditioner having higher operating efficiency. It is possible that in a little over two years, savings in operating costs could pay for the high-efficiency premium. As an example, to determine the cost savings comparison for two 8,000 Btu air conditioners, first determine the number of cooling hours per season for your region using the map in Fig. 4-17. (For our example we will use 2,000 h.)

Model M, Standard

1,380 W
× 2,000 Cooling hours/season
2,760,000 W/Season

Model C, High efficiency

860 W
× 2,000 Cooling hours/season
1,720,000 W/season

To determine kilowatt hours, divide by 1,000 W. (1000 W = 1 kW.)

2,760,000 W divided by 1,000 W = 2,760 kW

1,720,000 W divided by 1,000 W = 1,720 kW

The cost savings of the high-efficiency model compared to the standard model is found by subtracting the difference in kilowatts used and multiplying that figure by your local residential electrical rate. (For this example we will use 2.3¢/kW.)

$$\begin{array}{r} 2{,}760 \text{ kW} \\ -1{,}720 \text{ kW} \\ \hline 1{,}040 \text{ kW} \end{array}$$

1,040 kW × 2.3¢*/kW = $23.92 cost savings

Three factors in room-unit design affect efficiency or how much power an air conditioner uses for every Btu of cooling: (1) refrigerant cycle; (2) airflow capacity; and (3) electrical efficiency. While the refrigerant is alternately vaporizing and condensing, the temperature changes, from a high of about 130°F in the condenser to about 40°F in the evaporator—a difference of 90°. Temperature is directly related to pressure. To reduce this temperature difference to, say 70°F, would make the cycle more efficient. Less pressure would make the cycle more efficient. Less pressure would be needed to pump the refrigerant, hence less power would be needed for the compressor. To reduce the temperature difference, engineers have to design condensers and evaporators with bigger surface areas. Result: bigger, more expensive coils, but more efficiency.

Airflow capacity is another area that can contribute to higher efficiency. More free airflow through the cooling (evaporator) and heating (condenser) surfaces means the power it takes to drive the fan motor can be lower. With larger coil areas, there is more airflow.

Last, higher-efficiency motors, which need more copper and iron (and hence are heavier, larger, more expensive) and are less costly to operate, are used. Therefore, the main way to improve the efficiency is to use a bigger condenser coil and blow lots of air through it. The consumer will probably note a quieter operation with the high-efficiency air conditioner because of a bigger throat (air discharge) area resulting in lower-velocity airflow from the air conditioner. As you can see, the consumer benefits are many in owning a high-efficiency air conditioner.

* Based on the national average residential cost per kilowatt hour.

SEALED REFRIGERANT SYSTEM

The refrigerating system is similar, except for size, to that of a household refrigerator. However, system repairs should not be attempted before checking the unit electrically and for performance. In fact, there are only three main reasons for entering the sealed system of a cooling-only air conditioner: an incorrect refrigerant charge, a restriction, or an inoperative compressor.

When the temperature difference is appreciably less than it should be, remove the front panel and filter to check the evaporator coil for cooling. The coil should be uniformly cool from bottom to top. If the evaporator is not cool, or is cool only part way up, the system may have lost part of or all its refrigerant.

An additional check for a shortage of refrigerant may be made by stopping the flow of air through the evaporator by placing a piece of cardboard over the entire evaporator coil and allowing the compressor to run for 15 or 20 min. A slight, uniform film of frost should form on the entire surface of the evaporator if the system is adequately charged. If it does not, the refrigerant charge is too low. But this same frost-back will occur if a unit is operated with a dirty filter, inoperative fan, or in too-cool an ambient temperature. There just is not enough heat load to vaporize all the refrigerant before it gets back to the compressor. After a short running time the compressor will kick out on overload.

A wattage reading taken during operation and used with a performance chart will also indicate a low-charge possibility. Such a chart as given on p. 147 is usually contained in the service manual.

Table 4-2. **Failure disposition chart.**

Failure	Disposition							
	Evacuate and recharge	Replace compressor	Replace capillary	Replace condenser	Replace evaporator	Repair by brazing	Replace process strainer	Replace process tube
A. Compressor								
1. Noisy	×	×					×	×
2. Stuck	×	×					×	×
3. Inoperative	×	×					×	×
4. Burned out	×	×					×	×
5. Leak	×	×					×	×
B. Capillary								
1. Broken	×		×				×	×
2. Restricted	×		×				×	×
C. Condenser								
1. Internal leak	×			×			×	×
2. Leak at ends	×					×	×	×
D. Evaporator								
1. Internal leak	×				×		×	×
2. Leak at ends	×					×	×	×
E. Refrigerant leak								
1. Copper to copper	×					×	×	×
2. Copper to steel	×					×	×	×

When a need for system repair is determined, make the following checks.

1. *Undercharged system.* Disconnect fan motor leads from selector switch, and start the unit on HIGH COOL. A slight, uniform film of frost should form on the *entire* evaporator within about 10 min. If it does not, the system is undercharged. An undercharged system must be leak-tested, purged, the leak repaired, and the system evacuated and recharged with the proper amount of refrigerant. Always leak-test the system after recharging.
2. *Overcharged system.* Operate the unit on HIGH COOL for at least 30 min in a room temperature of 75 to 85°F. The top of the compressor should be warm to the touch. If it is cool or cold, the system is overcharged. An overcharged system must be evacuated and recharged. Purging off the excess refrigerant is not practical, as there is no way of determining the amount to purge.
3. *Pressure equalizing (system unloading).* If an attempt is made to start the unit too soon after it has stopped, the compressor may fail to start, and it will cycle on the overload protector. This is caused by the combination of high head pressure and low suction pressure. Allow the pressures to equalize through the capillary tube. Pressure equalizing (system unloading) will usually take from 2 to 5 min.

Refrigerant leaks. An undercharge of refrigerant is usually caused by a leak in the system. Such leaks must be located and repaired before evacuating and recharging, as simply adding refrigerant or recharging will not permanently correct the problem and may lead to a compressor burnout. *Note*: Do not replace a component because the system is undercharged, unless you find a nonrepairable leak within the component.

If you suspect a leak, attempt to find it before opening the system. Leaks, particularly small ones, are easier to find if the surrounding air is not contaminated with refrigerant from the system. Leaks are also more easily found if the hermetic system is pressurized to at least 75 lb/in^2. If necessary, attach a line-piercing valve to the compressor process tube and add enough refrigerant for testing.

Most leaks can be found with a halide torch (see Leak Detectors in Chap. 1). However, to pinpoint a very small leak, it may be necessary to use an electronic leak detector, or the bubble method. *Note*: Be sure the system has a positive pressure before using the bubble method of leak testing. A vacuum within the system could draw in moisture and other contaminants.

To test for refrigerant leaks in the hermetic system with a halide torch, remove the end of the leak-detector suction hose over all joints and parts which contain refrigerant. For the most accurate diagnosis, have the flame just high enough to keep the copper element plate in the leak-detector head red hot. In pure air, the flame will be light blue in color. As the suction hose sucks in air containing refrigerant, the color of the flame will change depending on the size of the leak.

1. Small amounts of Freon leakage will cause the blue flame to turn green.
2. Large amounts of Freon leakage will cause the blue flame to flare purple.

Do not breathe the fumes from this torch when the section tube is exposed to large amounts of refrigerant.

After servicing a system, always leak-test the entire system, especially new joints, before final recharging. Clean off any soldering flux, if used, from the joints before leak-testing, as flux can seal off pinhole leaks that would show up later.

Component replacement. When replacing components, all copper-to-copper joints must be made with silver solder or Sil-Fos. Copper-to-steel joints must be made with silver solder only. Be careful not to damage adjacent parts when using a torch on soldered joints. If necessary, use a sheet of asbestos as a heat shield, and wrap a wet cloth around the tubing to reduce heat transfer.

Replacement compressors are frequently shipped with oil and a holding charge of dry nitrogen or refrigerant, and the refrigerant lines are sealed with rubber plugs. Mechanical disassembly will vary with the model of air conditioner when a component is to be replaced. (Some have separate outer shells.) Careful examination will disclose the easiest method of gaining access to a given component.

To replace a typical compressor, you should proceed as follows.

1. Disconnect electric leads from the compressor.
2. Cut the process tube on the compressor to purge refrigerant charge.
3. Unsolder discharge and suction lines at the compressor. *Note*: On some models the discharge and/or suction lines are not easily accessible at the compressor. On these models the lines should be unsoldered at the condenser and/or evaporator, and unsoldered from the compressor after it is removed from the chassis.
4. Remove the clips or bolts holding the compressor to the base. Remove the compressor from the chassis.
5. Mount the replacement compressor in the chassis.
6. Attach service valve to the process tube of the new compressor, leaving room for pinchoff after charging.
7. Clean and connect the discharge and suction lines. Silver-solder all joints.
8. Evacuate and recharge. Connect electric leads and check for proper operation. Pinch off and silver-solder the process tube. (See sections on Evacuating and Recharging, below.)
9. Seal stubs on the defective compressor with rubber plugs taken from replacement compressor. A small amount of oil on plugs will aid in their insertion into stubs.

To replace a typical evaporator, condenser, or the capillary tube, you would proceed as follows.

1. Remove the chassis parts necessary to gain access to the component to be replaced.

2. Cut the process tube on the compressor to purge the refrigerant charge.
3. Unsolder the two connections for the component to be replaced.
4. Clean, connect, and silver-solder the two joints.
5. Attach service valve to process tube of the compressor, leaving room for a pinchoff (additional tubing may be required).
6. Evacuate. (See sections on Evacuating and Recharging, below.)
7. Recharge (check specifications for correct amount).
8. Leak-check and test for proper operation.
9. Pinchoff and silver-solder the process tube.
10. Reassemble the chassis parts previously removed.

Evacuating system. The operating characteristics of most air-conditioner systems, and the high pressures and temperatures, make complete evacuation a must. Very small quantities of moisture left in the system will chemically combine with the refrigerant to form acids. Once the formation of acid starts, it will continue at an ever-increasing rate, as it is a self-feeding process. The acids will cause sludge to form and attack the copper plating in the compressor, and eventually destroy the motor-winding insulation, resulting in a compressor motor burnout. Because of the high operating temperatures, moisture will not freeze out at the capillary tube outlet; thus no advance warning is given of its presence.

The use of a high-vacuum pump and an electronic micron vacuum gauge is recommended as the most positive method of moisture removal from a system. With this high-vacuum equipment, the final vacuum should be 500 μm or less. Obtaining this vacuum will ensure a proper job of moisture removal and a system free of leaks. Where high-vacuum equipment is not available, and a rough vacuum pump is to be used for evacuation, the system should be partially charged, the compressor operated for a few minutes, and the system then evacuated for 20 to 30 min. This procedure should be repeated two more times before final charging.

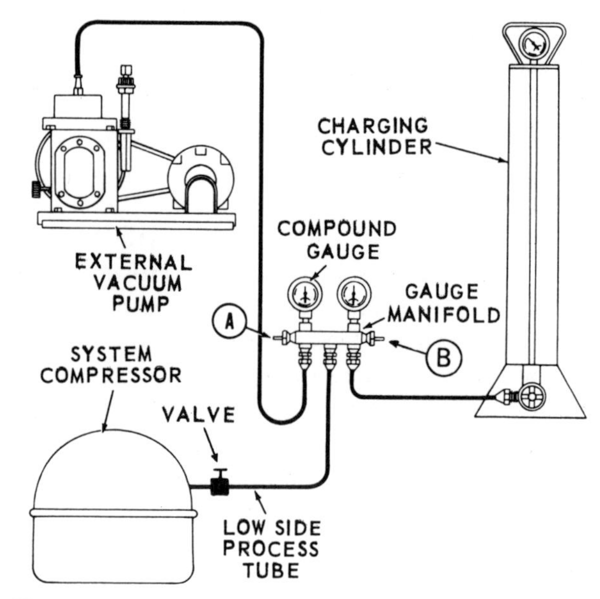

Figure 4-18. Evacuating and recharging a system.

A typical evacuation procedure is as follows:

1. Cut the compressor process tube and install a service valve, leaving sufficient room for a pinchoff. If necessary, silver-solder a new process tube to the compressor.
2. Connect the vacuum pump to the service valve, and leave the valve closed, Figure 4-18 illustrates a typical evacuation-charging hookup. However, any preferred hookup may be used that is capable of performing the required functions.
3. Start the vacuum pump and open its discharge valve. Then, slowly open its suction valve and the service valve on the compressor process tube. *Caution*: If high-vacuum equipment is being used, just crack the vacuum-pump suction valve for the first minute or two, and then open it slowly. This will prevent compressor oil from foaming and being drawn into the vacuum pump.
4. Operate the vacuum pump for 20 to 30 min, or until a vacuum of 600 μm is obtained. Close the vacuum pump suction valve and observe the vacuum gauge for a few minutes. A rise in pressure would indicate a possible

leak in the system or that moisture is still present.

5. Close manifold valve A and stop the vacuum pump. Connect a calibrated charging cylinder to the manifold and purge the line. Open manifold valve B and partially charge the system to a pressure of 35 to 40 lb/in^2 gauge. Leak-test the low side. Run the compressor for a few minutes. Leak-test the high side.
6. Purge the partial charge from the system, and evacuate for 20 to 30 min, or to a vacuum of 500 μm or less. If a round vacuum pump is being used, charge, operate the compressor, purge, and evacuate. Repeat a total of three times. Close manifold valve A and stop the vacuum pump. The system is now ready for final charging.

Recharging system To operate at full capacity, air conditioners must be charged to within an accuracy of $\frac{1}{4}$ oz of refrigerant. This accuracy can be achieved with a calibrated charging cylinder, which is accurate regardless of ambient temperature. *Do not operate the compressor while charging*, as the refrigerant enters the system as a liquid, and damage to the compressor could result. Wait at least 5 min after charging the system before starting the compressor.

The recharge usually goes as follows.

1. Connect a calibrated charging cylinder to the manifold and purge the line. Leave the charging cylinder valve open. After the refrigerant has stabilized within the cylinder, observe the cylinder pressure gauge and rotate the cylinder sleeve to the correct setting for pressure and type of refrigerant being used.
2. Open manifold valve B and add correct refrigerant charge, as shown in specifications in the service manual. If the refrigerant bubbles in the charging cylinder, close valve B and momentarily invert the cylinder, then continue charging. *Caution*: If it is necessary to raise the internal pressure of the charging cylinder *do not use a torch*, but place the cylinder in a container of warm water (not over 125°F). Heating the cylinder with a flame can build up pressures sufficient to cause the cylinder to explode.
3. When the system is correctly charged, close manifold valve B, the service valve on the compressor, and the cylinder valve. Check the unit for proper operation. Pinch off the compressor tube. Remove the service valve and silver-solder the end of the process tube (closed).

In many models, the sealed system may be entered for checking pressure without discharging the system. Most makers have some type of piercing valve or similar device which may be installed and left on the system without jeopardizing the warranty. [Remember that the warranty of most manufacturers is automatically canceled when the sealed refrigerant system is opened and serviced by an unauthorized service agent. A sealed refrigeration system repair may be identified as any type of service which requires opening the hermetic system (consisting of the compressor, condenser, evaporator, and interconnecting tubing) for the purpose of repairing a leak, adding refrigerant, removing an obstruction, or replacing a component part. With this type of warranty, the entire chassis must be shipped or delivered to the nearest authorized repair station for in-warranty service.]

Where some type of access valve is permitted or suggested by the manufacturer, it is usually installed on either the process stub or the suction line. Select a location away from other solder joints or adjacent tubing, and where it cannot touch the outer cabinet when the unit and the cabinet are reassembled.

Once the piercing valve or similar access device is installed as directed by its maker, the valve's cap may be removed, and a purging or manifold hose containing a core depressor is attached. When discharging a unit, use a purge hose to release the refrigerant to the outdoors. Letting it discharge into the room may contaminate the air.

AIR-CONDITIONER CONTROLS

The controls for the room air conditioner are usually isolated from other components in a control box on the front of the chassis.

Thermostat

The thermostat controls the compressor. It senses the temperature of the room air being drawn into the evaporator. Temperature selection is made by turning the dial knob toward a warmer or cooler setting. The user finds the point which satisfies his or her needs and adjusts it later if those needs change. Most controls use the words WARMER, COOLER, and NORMAL as reference points, but some makers use numbers to indicate temperature settings.

The majority of cooling thermostats are equipped with a vapor-filled temperature-sensing tube. This capillary-sized tube is placed in front of the evaporator and senses the temperature of the room air being drawn in for cooling. The vapor-filled sensing tube is affected by the coldest point along its entire length including the bellows. Whichever point is coldest will take over control. The thermostat body is therefore located in the control box away from the cold airstream. The thermostat sensing tube must not touch the evaporator or the cold-air outlet. Incidentally, the switch of a gas-type thermostat is of single-pole, single-throw design.

The thermostat cut-in and cut-out differential on most models is $4\frac{1}{2}$°F $\pm$ $1\frac{1}{2}$°. With a differential this close, it is possible for a cycled-off unit to cycle back on too soon (called *short cycle*) if a door is opened or another source of heat is felt by the thermostat. To alleviate the short-cycle possibility, some models .are equipped with a temperature anticipator. During a cooling cycle, the anticipator control is assuming a temperature somewhere between the cold temperature of the coil and that of the room air. It will be more sensitive to air temperature and will cause the thermostat to open a little sooner than it normally would. This will compensate for overshooting temperatures because of residual cooling coils. Excessive fluctuation of room air temperature is reduced.

A second type of room thermostat, which also acts as a single-pole, single-throw switch, may be found on some models. It uses a bimetal temperature-sensing device in place of the gas-filled sensing tube. The bimetal is placed in a small air duct. Room air is pulled through the duct passing over the bimetal. When the room air temperature drops to the bimetal cut-out point, the thermostat opens and stops the unit. Some models will experience evaporator icing in cooler ambient temperatures. Replacing the bimetal thermostat with a gas-filled type may be necessary.

To test either type of thermostat, set it at its coldest position and the switch for cooling, and connect the service cord to the power outlet. With the bare ends of a short piece of No. 12 insulated wire, shunt across the electric terminals of the thermostat. If the compressor runs with the jumper wire attached, but will not run without it, the thermostat is defective and must be replaced.

Anti-Ice Control

The design temperature of a room air conditioner is 95°F. Operation in a room between 65 and 70°F will cause evaporator icing. There just is not enough work for the air conditioner to do. The coil becomes too efficient, drops to a freezing temperature, and ices up. But icing of up to 70 percent of the coil is considered normal on all models under low ambient conditions with a near-maximum cooling-thermostat setting.

While excess icing of the coil is usually due to operation in too-cool an ambient temperature, it could also be caused by a dirty filter, a dirty coil, or an inoperative fan. Check out these possibilities before going further.

The switch action in the anti-ice or de-ice control is a single-pull, single-throw type. As a rule, the pressure from the bellows will hold the switch contact closed until the sensing element of the control sets a temperature of 28°F. The controls will then open the circuit to the compressor. The fan motor will continue to run as the coil de-ices and reaches a temperature of 58°F; then the control closes the circuit to the compressor, and the air conditioner continues to cool. Generally, the anti-ice or de-ice control feeler tube (capillary) is located at the center top edge of the evaporator.

To test the anti-ice or de-ice control, prepare a solution of ice, water, and salt. Use a mercury-filled thermometer to determine the solution temperature. Place the feeler bulb into the solution. When the thermometer reads 28°F ± 1°, the switch contacts should open. Warm the water to 58.5°F ± 1°, and the switch contacts should close. If the control is not defective, rinse off the solution with clear water and remount into the original position.

Operation of the air conditioner with the cold discharge air deflected back into the return airstream entering the intake grille will result in the recirculation of chilled air. This condition may reduce the amount of heat normally delivered to the evaporator coil by the evaporator air, causing the internal coil temperature to fall to 32°F or lower. With the coil temperature running below freezing temperature, the moisture in the air collects onto the surfaces of the fins and quickly freezes into ice, thus blocking or impeding the airflow. To remedy this situation, all that is required is the adjustment of all discharge air louvers or relocation of any furniture which might deflect the air back into the intake opening.

In some cases, especially with air conditioners installed in large apartment houses, freeze-ups occur when the air conditioner is operated during periods with outside temperatures dropping below 70 to 75°F at night. Should this occur, suggest that the thermostat be set on a lower position to allow the air conditioner to cycle more frequently. Also, during such weather the user should be instructed to operate the air conditioner on HIGH COOL or a similar position only.

Some units are now being equipped with an expansion valve which is used in the refrigerating system to control the flow of refrigerant into the evaporator and subsequently to the compressor. The valve is adjusted to the system to obtain the most favorable low-side pressure. This pressure is closely maintained in the range of operating temperatures to which the air conditioner is expected to be exposed. The principal advantage of the valve when compared with a capillary tube is that high temperatures are maintained in the evaporator when low operating temperatures exist; thus coil freeze-up does not occur under normal operating conditions.

Selector Switch

The selector switch is the ON-OFF switch of the room air conditioner. It also controls fan speeds and compressor operation. It may be a rotary or a push-button switch, depending on the particular model, but both types of switches perform the same functions. Internal contacts complete circuits from L1 to the operating components. The solid terminal points in the switch indicate which circuits are made by any one of the selections. On LOW COOL, for instance, L1 is connected to the compressor by terminal 2 and to low fan speed by terminal 3. The selector switch symbol may also be shown in this form (Fig. 4-20) on wiring diagrams for some models.

To test a selector switch, insert the service cord into the power outlet and set the thermo-

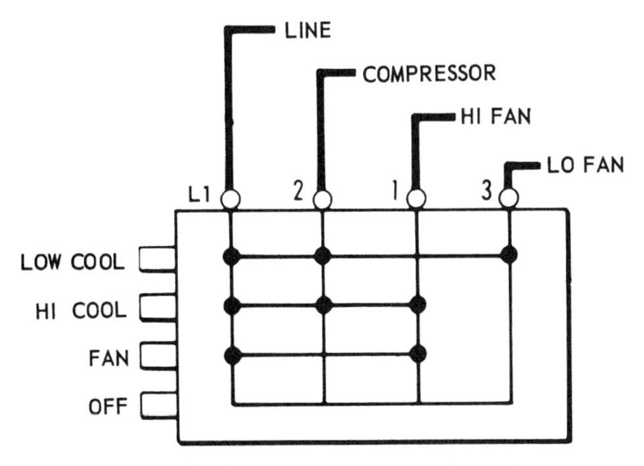

Figure 4-19. Push-button selector switch diagram.

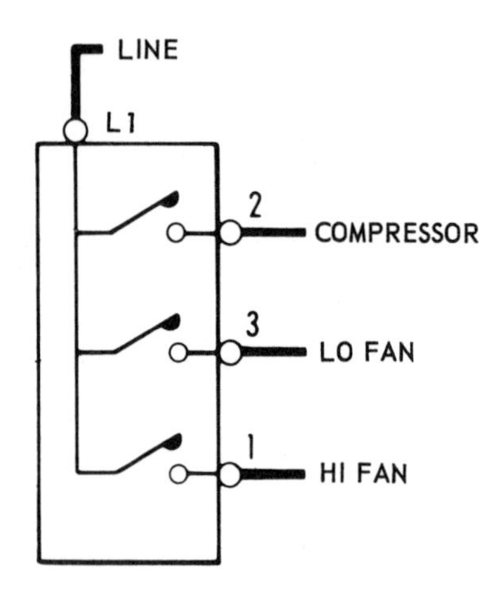

Figure 4-20. Another form of selector switch symbol.

stat on the coldest position. Review the schematic wiring diagram and engage the selector switch into all the various positions. If the fan motor or compressor fails to operate in any one of the particular selector switch positions, use an insulated jumper wire and shunt across the particular selector switch element by connecting the wire from the "hot" side of the selector switch to the terminal which feeds the high- or low-speed circuit of the fan motor, or to the circuit feeding the thermostat (or the overload) in case the compressor fails to run.

If the fan motor or compressor runs with the jumper wire connected, but will not with the wire removed, the selector switch is defective and must be replaced. If the compressor does not run, check the continuity of the overload.

In normal use the overload contacts are closed. The overload protects the compressor by opening the circuit when excessive heat or current conditions prevail. If there is no external overload mounted in the compressor terminal compartment, the overload is located in the compressor motor windings and must be tested as a part of the compressor motor. *Caution*: Do not perform a continuity test on the motor windings or condemn the compressor motor until the compressor has been out of electric circuit for at least 20 min. This will allow the compressor to cool down and the internal overload to reset. The overload is not replaceable. To test for external overload, remove it from the terminal box of the compressor, disconnect the wires, and test for continuity between all terminals; if the overload is defective, it must be replaced. Current will flow between all terminals of a good overload.

Ventilation Control

The ventilation or air-control knob is usually marked AIR, EXHAUST, or AIR CHANGER. It operates wires or chains to open and close dampers in the bulkhead between the evaporator and the condensing unit.

When the knob is turned to EXHAUST, a damper is opened on the pressure side of the evaporator blower. Some of the cooled air to the room is diverted through the damper to the condenser fan area and expelled to the outside.

Turning the ventilation control knob to FRESH AIR opens a damper in the suction or negative-pressure area of the evaporator blower. Outside air is drawn in from the compressor side of the unit, mixes with the cooled air from the evaporator, and is expelled into the room. When maximum cooling is desired, the FRESH AIR damper must be closed; otherwise the cooled air from the evaporator would be diluted by the incoming warm air.

INSTALLATION PROCEDURE

An installation and operating instruction booklet is included with each air conditioner. An installation kit is also available which contains all parts necessary to install any specific model. In fact, kits are available for most types of windows—double-hung, casement, etc.—and for through-the-wall installations. Therefore, when called upon to make a room air installation, be sure to follow the manufacturer's instructions to the letter. In addition, keep these installation pointers in mind.

1. Whenever possible, the air conditioner

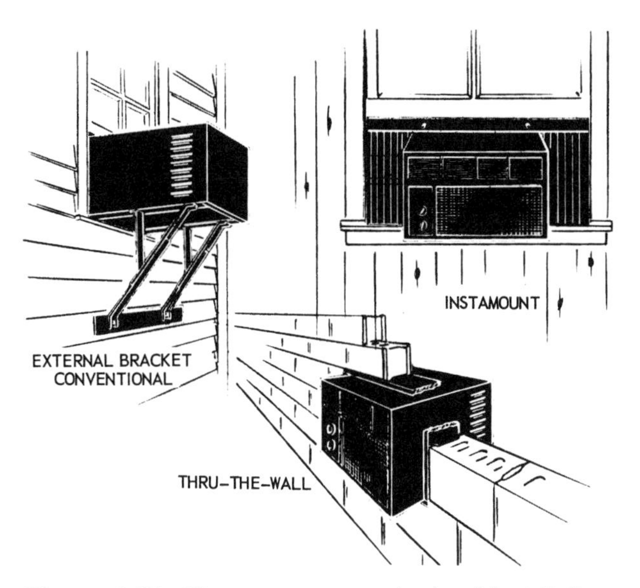

Figure 4-21. Three common methods of installation: (left) standard window mount; (top) outside mount; and (bottom) through-the-wall installation.

should be located in a window on the shady side of the house. If this is not possible and the unit is exposed to the sun, an improvised shade should be provided such as an awning. The awning should be open-ended or equipped with air vents to allow a flow of air to the unit.

2. When making a through-the-wall installation, particularly of a conventional unit, be sure the side louvers of the outer cabinet are not blocked up or restricted by a brick or cinder block veneer, or by the wall itself. This would prevent sufficient air being available for cooling the condenser. The efficiency of the unit would be greatly reduced.
3. The air conditioner should slope toward the outside about $\frac{1}{4}$ in/ft or the length of a bubble in a level. This ensures the flow of condensate water from the evaporator back to the slinger ring on the condenser fan blade or to the drain hose. Drippage of excess condensate outside the building during highly humid weather is sometimes a problem. If the slinger ring cannot handle the water, a $\frac{3}{8}$-in hose may be installed to direct the condensate to a remote location.
4. Voltage furnished at the air conditioner is another important factor. It must be within 10 percent of the voltage requirement stated on the data plate. (Loose wiring connections, plugs, or fuses can cause low voltage.) Fuses should be the time-delay type and of the correct size as indicated on the data plate. Electric supply lines to the receptacle should be at least No. 14 gauge wire on 15-A models; 20-A models should be supplied through at least No. 12 gauge wire. In all cases observe local codes. Extension cords should not be used unless constructed of the adequate-sized wire and a ground wire. *Note*: Do *not* change or alter the plug on the supply cord attached to the air conditioner.
5. When lower-than-normal wattage is found

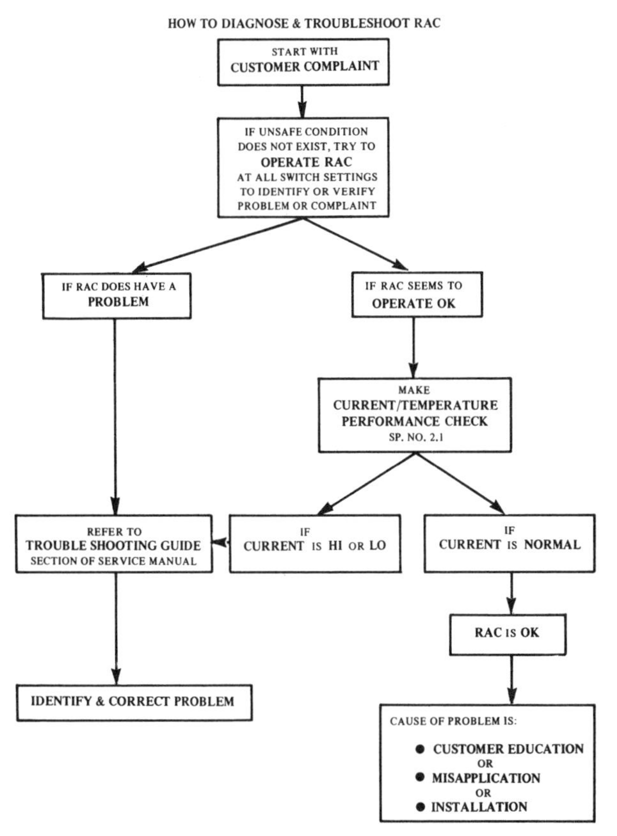

Figure 4-22. How to diagnose and troubleshoot a room air conditioner (RAC).

during a performance test, one of the reasons could be a low charge of refrigerant. To check refrigerant charge easily, block air entry to the evaporator with a piece of cardboard. Operate on full cooling at least 5 min. The suction line should frost back to the compressor dome. If it does not, check for refrigerant leak.

TROUBLE DIAGNOSIS

The purpose of the following troubleshooting chart is to cover and review some of the more common troubles encountered in a room air conditioner, their probable causes, and possible solutions.

Problem: The air conditioner unit does not run.

Possible cause	*Solution*
1. No power at unit.	1. Check for current at wall receptacle. Replace blown fuse with correct size time-delay fuse. Check service cord connection at receptacle and at switch of air conditioner. (Have customer contact electrician or power company if no power is available to fuse box.)
2. Low voltage.	2. Raise voltage if more than 10 percent below nameplate rating. Improve wiring and/or use transformer.
3. Defective selector switch.	3. Turn switch to COOL and check for current at load side of switch. If there is current going into switch but no continuity through it, replace switch.
4. Incorrect or defective wiring within unit.	4. Check correctness of wiring and connections. Check operation of switches, all thermostats, relays, overload, compressor, fan motor, capacitors, transformer, etc. If any are defective, replace. Tighten any loose connections.

Problem: Fan motor runs but the compressor will not run.

Possible cause	*Solution*
1. Incorrect power supply.	1. Check for proper voltage or abnormal load conditions.
2. Overload open or defective.	2. Check for overheated compressor and check overload.
3. Inoperative thermostat.	3. Turn thermostat to its coldest position. If room temperature is above 75°F but compressor does not attempt to start, short across terminals of thermostat. If compressor starts when thermostat terminals are shorted, replace thermostat.
4. Loose or defective wiring connections.	4. Check wiring and connections. Take necessary corrective actions.
5. Defective run capacitor.	5. Check run capacitor. Replace if defective.
6. Defective switch.	6. Check for current at compressor terminals of switch. Replace switch if defective.
7. Defective compressor relay.	7. Check and replace if necessary.

8. Defective starting capacitor if used (open circuited, short circuited, loss of capacity).	8. Check starting capacitor. Replace if faulty.
9. Thermostat set too low.	9. Adjust control to proper setting.
10. Incorrect wiring or defective component.	10. Check correctness of wiring and connections. Check operation of switches, all thermostats, relays, overload, open or shorted starting capacitor, shorted run capacitor. If any are found defective, replace.

Problem: Compressor starts and runs but fan motor does not run.

Possible cause	*Solution*
1. Loose or defective wiring in fan circuit.	1. Check fan circuit wiring from load side of switch to fan motor. Tighten connections or replace defective wiring.
2. Shorted, grounded, or open fan capacitor.	2. Check fan capacitor and replace if necessary.
3. Open circuit in fan motor.	3. Check continuity of motor windings. Replace motor if windings are open or grounded.
4. Air-handling fans binding on shrouds or baseplate.	4. Inspect fans for clearance. Adjust fan on shaft or fan motor on fan motor mounts. Also check for dirt buildup on fins, or bent fins. Clean or straighten fin if necessary.
5. Incorrect wiring.	5. Check correctness of wiring and connections.
6. Defective reactor or incorrect wiring.	6. Check reactor and wiring.
7. Defective controlling relay (remote-control models).	7. Check relay and wiring.
8. Partial short circuit in compressor motor.	8. Check compressor winding with ohmmeter.
9. Faulty switch.	9. Check and replace if defective.
10. Bearings on fan motor sized.	10. Oil fan bearings. Replace fan motor, if bearings are worn.

Problem: Fan runs, the compressor starts and runs, but the compressor occasionally stops (on overload).

Possible cause	*Solution*
1. Low voltage.	1. Low voltage due to overload circuits within the building or throughout local power system (due to varying power demands, this condition might exist only at certain times during the day). Consult with local power company. Run a separate line to the unit.
2. High voltage.	2. High voltage due to fluctuations in local power system—usually occurs at

2. (*cont'd.*)

Possible cause	Solution
	various low load periods of the day. Consult local power engineers.
3. Abnormal location.	3. Extreme heat on outer cabinet of unit during the hottest part of the day. Shade outer cabinet with a ventilated awning.

Problem: Unit gives electrical shock.

Possible cause	*Solution*
1. Grounded electrical circuit.	1. Check and eliminate ground.
2. Faulty capacitor.	2. Check capacitors for ground. Replace if one side of capacitor is grounded.

Problem: No cooling, but compressor and fan motors operating.

Possible cause	*Solution*
1. Airflow restriction.	1. Dirty filter. Air passages obstructed.
2. Compressor not pumping.	2. Check for low wattage, restriction, loss of refrigerant, low-capacity compressor.
3. Complete clog in refrigerant system.	3. Replace clogged component.

Problem: Insufficient cooling, but compressor and fan motors operating.

Possible cause	*Solution*
1. Thermostat set too high.	1. Instruct customer on proper setting.
2. Dirty air filter.	2. Clean or replace.
3. Restricted air to condenser.	3. Clean condenser and evaporator fins. Straighten any bent ones. Check condenser for fresh-air intake and re-circulation of hot exhaust air.
4. Air conditioner undersize for area to be cooled.	4. Make heat load survey. Check that room openings are closed. Check for added heat load since installation.
5. Low volume of air (in cubic feet per minute) passing over evaporator and condenser.	5. Fan motor not running up to prescribed speed. Blower wheel or slinger ring fan slipping on shaft. Evaporator air duct obstructed. Check for wrong motor blower wheel for blade. Check for bind of fan blade, blower wheel, or motor shaft. Lubricate fan motor, if necessary. Low voltage: adjust top on reactor (if present).
6. Refrigerating system defective. Compressor runs but unit does not cool.	6. Low or lost refrigerant charge. Restricted capillary tube or filter screen, inefficient compressor.
7. Compressor not pumping at full capacity.	7. Check for low wattage, low voltage, low capacity compressor, restriction, loss of refrigerant.
8. Operating 60-Hz unit on 50-Hz current.	8. Advise customer unit is operating normally under these conditions if unit is delivering only 5/6 of 60-Hz operating output.
9. Low voltage.	9. Check power supply.

Possible cause	Solution
10. Improper seals: insulating seals out of place or missing. Doors or windows open excessively.	10. Readjust or replace seals. Instruct customer to minimize window and door openings.
11. Improper use or excessive load.	11. If load is excessive or installation is incorrect or inadequate, advise customer. Instruct on proper operation of thermostat and other controls.
12. Undercharged of refrigerant.	12. Purge and recharge unit.
13. Ventilate door open.	13. Close ventilate control, repair linkage, if binding.
14. Loose insulation restricting airflow.	14. Reattach with rubber cement.
15. Excessive frost or ice buildup on evaporator.	15. See Evaporator Frosts Over, p. 163.
16. Unit standing too long without being run.	16. If a room air conditioner is allowed to stand for an extended length of time without being run, it is possible for the Freon to become absorbed in the oil. If this should happen, there will be no cooling until the necessary working pressures have been established. This process of getting Freon out of the oil may take several hours of continuous running. Therefore, before rejecting a unit on the complaint that it will not cool, this running-in time must be allowed for.

Problem: Compressor stops and starts: running time too short (short cycling).

Possible cause	*Solution*
1. Thermostat differential.	1. Check for proper location of sensing tube or bulb in evaporator intake airstream. Tube or bulb must not touch evaporator. Check for recirculation of evaporator discharge air.
2. Inoperative or slow condenser fan.	2. Check operation of condenser fan. Correct rubbing on base pan or shrouds. Replace if warped.
3. Dirty condenser fins. Restricted airflow over condenser.	3. Check and clean condenser. Check for recirculation of hot discharge air. Check for adequate supply of fresh intake air to condenser.
4. Defective or incorrect overload.	4. Check if overload has tripped. Check for correctly rated overload and for overheated compressor. If compressor is not overheated and amperage is not excessive, but overload trips out, replace overload.
5. Attempting to start unit too soon after shutoff.	5. Instruct customer of need for 3-min wait for pressure equalization.

Possible cause	Solution
6. Incorrect voltage.	6. Check for proper voltage.
7. Thermostat set too warm.	7. Instruct customer.
8. Defective thermostat or control.	8. Check temperature thermostat and de-icer control.

Problem: House circuit fuses blowing.

Possible cause	*Solution*
1. Incorrect size fuse.	1. Check amperage on unit's specification plate; use correct size time-delay fuse.
2. Air-conditioner circuit overloaded.	2. Check circuit load and have necessary correction made. Check for extension cord with too-small wire size.
3. Compressor stops and attempts to start too soon.	3. See Compressor Stops and Starts, p. 161.
4. Wiring shorted or grounded.	4. Check wiring and make necessary correction.
5. Shorted, grounded, or open compressor run capacitor.	5. Check run capacitor and replace if defective.
6. Stuck or grounded compressor.	6. Check compressor for ground. Wire compressor directly.
7. Compressor has starting difficulty.	7. Check for low voltage.
8. Turning air conditioner off and back on too soon.	8. Instruct customer of necessity of 3-min delay for pressure equalization.
9. Improper voltage.	9. Check power supply, connections, all wiring. Improve wiring and/or use transformer. Advise customer to contact electrician or power company if house wiring is improper or inadequate.

Problem: Compressor cycles on its overload.

Possible cause	*Solution*
1. Incorrect voltage.	1. If above or below 10 percent of rated voltage, consult electrician.
2. Starting relay.	2. Be sure start capacitor is cut out of circuit. Replace relay if contacts are remaining closed.
3. Fan speed.	3. If low (at high setting) and oiling does not correct, change motor.
4. Fan not running.	4. If defective, replace motor.
5. Compressor not working properly.	5. Check compressor, start and run capacitor, and relay. Replace if inoperative. If unit does not have a start capacitor or relay, install kit where necessary.
6. Dirt buildup on condenser fins or bent fins.	6. Clean or straighten fins.
7. Clog or partial clog in refrigerant system.	7. Replace clogged component and recharge system.

Problem: Noisy operation.

Possible cause	*Solution*
1. Tubing vibrating or striking adjacent tubing or metal surfaces.	1. Reshape tubing slightly to eliminate vibrations or to give sufficient clearance between it and adjacent parts.
2. Air conditioner impro-	2. Check cabinet for distorted mounting.

perly or insecurely mounted in window.	Check that mounting brackets and other mounting parts are secured in place.
3. Loose or bent fan blades.	3. Tighten or replace fan.
4. Fan motor out of alignment or loose on mounting.	4. Check alignment. Check setscrews in fan hubs. Check fan motor mounting. Tighten if loose, and replace grommets if worn. Check for proper positioning of fans on motor shafts and clearance of fans with shrouds and base pan.
5. Compressor could be overloaded because of high ambient temperature or airflow restrictions. It may be loose internally.	5. If compressor makes internal noise, replace.
6. Fan hitting condenser or evaporator coils, shrouds, or orifice plate.	6. Adjust fan position on motor shaft; shim between fan motor and bracket assembly.
7. Fan motor bearings dry or defective.	7. Oil or replace fan motor.
8. Loose decorative front or outer cover.	8. Check and tighten as necessary.
9. Loose electrical components.	9. Check and tighten as necessary.
10. Refrigerant absorbed in compressor oil after extended shutdown.	10. Noise will disappear after unit runs awhile.

Problem: Evaporator frosts over.

Possible cause	*Solution*
1. Restricted airflow over evaporator.	1. Dirty air filter. Air passage obstructed. Dirty blower wheel. Blower wheel loose on shaft. Dirty evaporator coil. Evaporator fins bent.
2. Outside temperature too low (below 75°F).	2. Check for air conditioner being operated when outside temperature has dropped rapidly to below 75°F. Instruct customer.
3. Thermostat defective or set too low.	3. Check thermostat and instruct customer on proper setting.
4. Defective refrigeration system.	4. Check for refrigerant leak and low charge. Check for partial restriction.
5. Fan motor too slow. Improper reactor tap. Low voltage. Wrong motor or blower wheel.	5. Customer using low fan speed; instruct customer. Adjust reactor. Check for binding of fan blade, blower wheel, or motor shaft.
6. Defective unit.	6. Check for restriction or loss of refrigerant.
7. Improperly located anti-ice or temperature anticipator.	7. Approximately 70 percent frost covering on evaporator is normal. Relocate anticipator or anti-ice device to cooler portion of evaporator.
8. Undercharge of refrigerant.	8. Purge and recharge unit.

Problem: Unit drips water.

Possible cause	*Solution*
1. Cabinet not properly tilted.	1. Cabinet should tilt $\frac{1}{8}$ to $\frac{1}{4}$ in toward outside.
2. Condensate drain from	2. Clean condensate port in partition assembly.

Possible cause	Solution
2. (*cont'd.*) evaporator to condenser area obstructed.	
3. Condenser slinger ring fan not adjusted properly.	3. Slinger ring should have approximately $\frac{1}{16}$-in clearance from base pan. Greater clearance educes water pickup.
4. Extreme humidity.	4. Advise customer of possible sweating formations under abnormal conditions. Check and improve all possible seals. Instruct customer to minimize door opening.
5. Inadequate seal.	5. Check and improve all sealed areas including gaskets around window duct.

Problem: Controls hard to operate.

Possible cause	*Solution*
1. Binding on front.	1. Reposition switch dial plate.
2. Exhaust damper wire binding.	2. Free wire inside cable. Replace if unable to free.

ALL-SEASON MODELS

Up to this point the subject has been room air conditioners for cooling only. With the addition of a few components the air conditioner may also be a heater. Additional push buttons on the selector switch control the heating cycles.

There are three methods of heat output: (1) by reverse cycle; (2) by resistance heaters; and (3) by a combination of reverse cycle and resistance heat.

Reverse-cycle models. Reverse-cycle heating is accomplished by reversing the functions of the evaporator and the condenser. If you stand in front of the condenser of an operating air conditioner, you will feel a blast of warm air. Theoretically, if you turned the air conditioner around so that the evaporator was outdoors and the condenser indoors, the warm air would be entering the room. Reverse-cycle heating-cooling conditioners reverse the gas flow through the evaporator and the condenser. They are often referred to as *heat pumps.* The evaporator becomes the condenser, and the condenser becomes the evaporator. The heat picked up from outdoors plus the heat of compression is used to warm the room. The selector switch of the heating-cooling air conditioner must be set on HEAT for the reverse cycle, and the thermostat must be set at least 4°F warmer than the room air. A solenoid-operated reversing valve is actuated when heating is called for. It automatically changes the refrigerant flow in the system. The solenoid coil is easily removed for service. The valve body is part of the sealed system.

With the selector switch and the thermostat calling for cooling, the solenoid of the reversing valve is not energized. Refrigerant vapor from the evaporator passes through the diverter of the sliding valve and on through the suction line to the compressor. The space between the sliding valve diaphragm and the left end of the main valve body assumes suction pressure through the open port of the pilot valve. The high-side pressure of the compressor discharge passes through the port in the main valve body left open by the sliding valve and on to the condenser. Its high pressure against the back of the left-end diaphragm forces the sliding valve to the left. The air conditioner is cooling the room air.

When heating is called for, the solenoid is energized, shifting low-side pressure to the right diaphragm. The high-side pressure forces the sliding valve to the right. The suction line is now pulling from the condenser through the sliding valve diverter. The high-side pressure is being pumped into the evaporator. The refrigerant flow through the condenser and the evaporator has been reversed. The room air is now

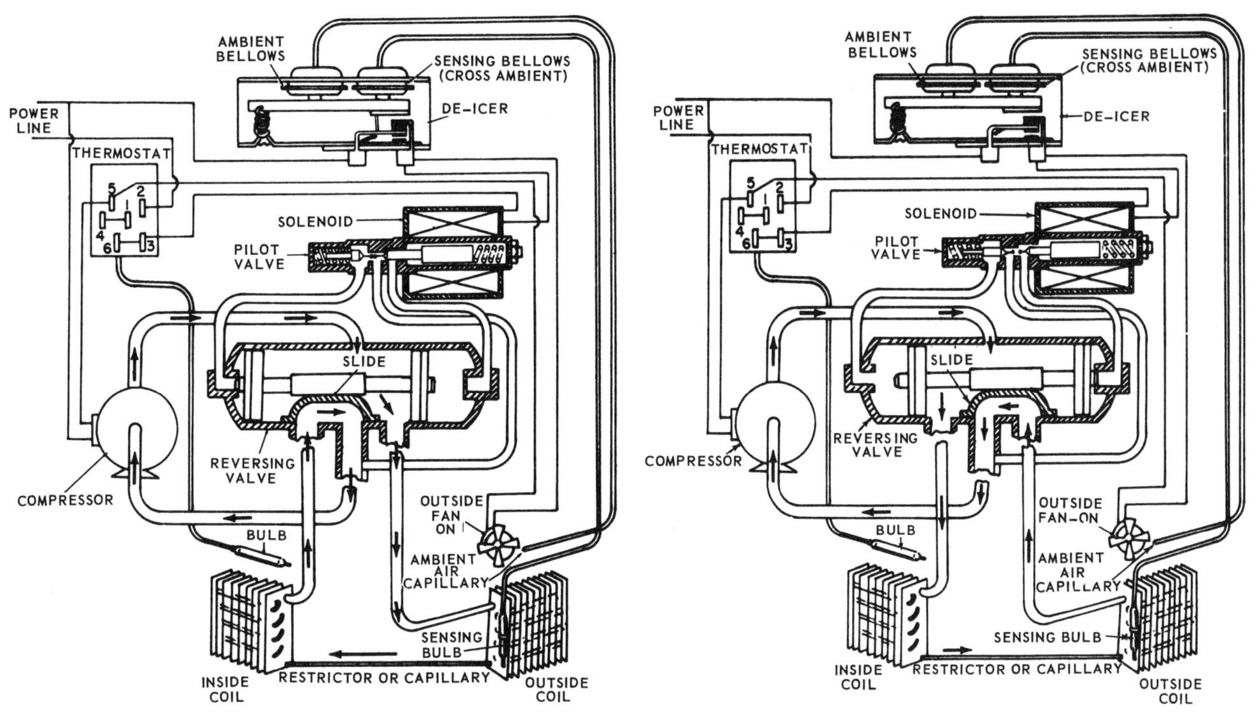

Figure 4-23. Reverse-cycle models: (top) cooling cycle and (bottom) heating cycle.

being warmed by heat absorbed from the outside air. Notice the two temperature-sensing elements located near the condenser. One is sensing the coil temperature, the other is sensing the temperature of the outside air.

Actually, these two sensing elements are usually part of a defrost control located in the control box. The outdoor evaporator has the same operating temperature limitations as room-cooling evaporators. In the absence of a sufficient heat load to completely vaporize the refrigerant, frosting and icing of the coil will result. The frost and ice must be defrosted quickly, and the heating cycle resumed. This defrost control with its two sensing elements senses the frost accumulation, then initiates and terminates a defrost cycle.

One of the sensing elements is mounted in the condenser airstream. It is sensing the outside air temperature. The second element is inserted into a sleeve or well on the side of the condenser coil and senses the temperature of the coil. Good contact of the bulb in the well is essential. When the coil ices up during the heating cycle, and the temperature of the *sensing bulb* becomes about 16°F *colder than the air-sensing tube*, the control initiates a defrost cycle.

When the defrost control calls for defrosting (Fig. 4-24), it stops the fan motor and de-energizes the reversing-valve solenoid. The compressor continues to run, pumping hot gas into the condenser. The fan is not running, so defrosting takes place quickly. When the coil-sensing bulb and coil temperature rise to 60°F, the defrost control switches back to the heating cycle. When the refrigerant flow through the evaporator and the condenser reverses, the flow through the cap tube also changes direction. A strainer is therefore found at each end of the cap tube to protect it from foreign materials.

The reverse-cycle-heating electric circuits are energized in Fig. 4-25A. The thermostat has been turned to at least 4°F warmer than room temperature. The selector switch HIGH HEAT button has been depressed. The compressor and fan motors are running, and the reversing

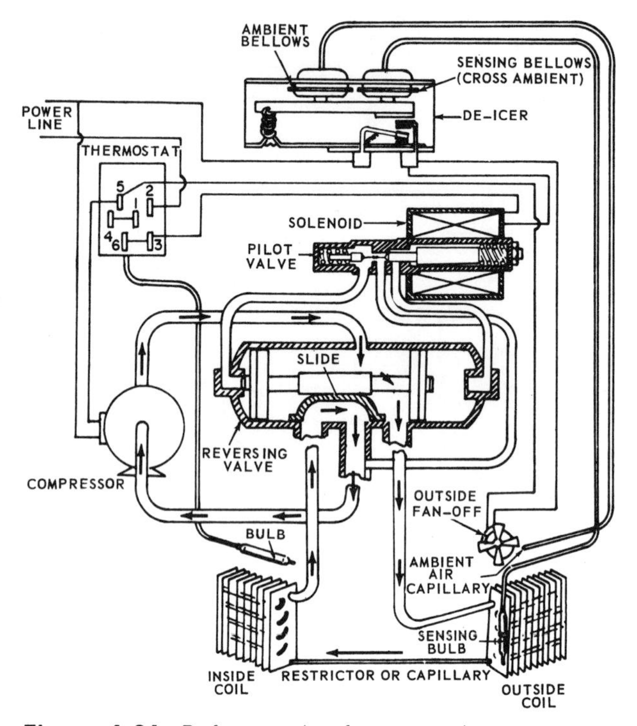

Figure 4-24. Defrost cycle—fan stopped.

solenoid energized. The evaporator is heating the room, and the condenser is absorbing heat from outdoors.

When the condenser coil ices up during a heating cycle, the defrost switch opens, de-energizing the fan and solenoid circuits (Fig. 4-25B). The compressor continues to run, pumping hot gas to the condenser to defrost it quickly. When the condenser warms up to 60°F, the de-icer will close, reopening the solenoid and restarting the fan motor. The conditioner resumes the heating cycle until the inside thermostat is satisfied or the coil ices up again.

Resistance-heater models. The second type of heating-cooling air conditioner uses a resistance heater attached to a frame. It heats like a toaster coil and is usually mounted between the blower wheel and the evaporator (Fig. 4-26). The unit and the heating element are protected by two safety devices—a fusible link and a high-limit switch in series with the heating element. The fusible link, when under a 15-A load, will usually open at about 235°F. The limit switch opens at about 175°F. It is mounted between the blower wheel and the evaporator of a standard cooling unit.

When the selector switch and the temperature control of a resistance-heat model are set for heating only, the heating coil and the fan circuits are energized. The unit will cycle on the temperature control.

Reverse-cycle and resistance-heat models. The third heating method is a combination of reverse-cycle and resistance-heat. An additional control, an outside thermostat, is used on this unit. This control will call for reverse-cycle heat until the outside temperature drops to

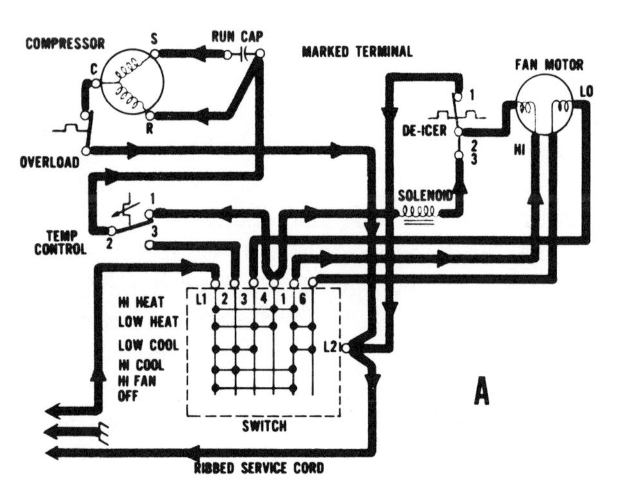

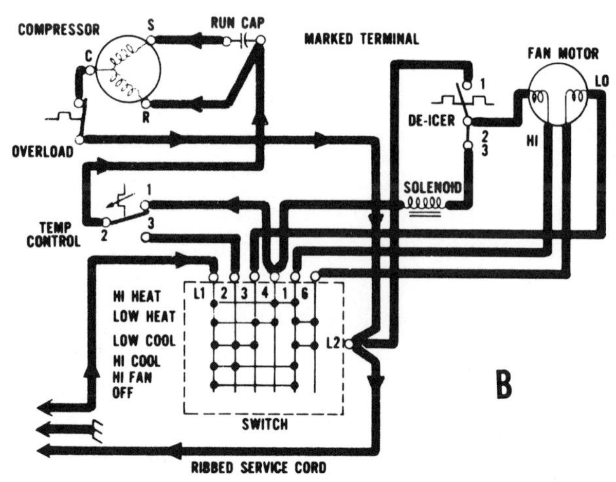

Figure 4-25. The reverse-cycle-heating electrical circuit (A) in HIGH HEAT position and (B) in DEFROST.

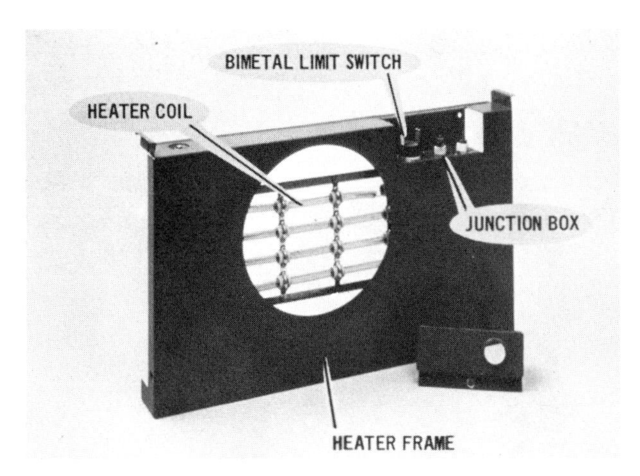

Figure 4-26. Resistance heater.

45°F when it will break the solenoid and compressor circuits and complete the resistance-heater circuit (Fig. 4-27). If the outdoor temperature rises above 45°F, the outside thermostat will return the unit to reverse-cycle heating. Note that the compressor and heater circuits are *never* energized simultaneously. This should never be attempted. The outside thermostat is located in the control box and its sensing tube is in the condenser airstream. When it senses an outdoor air temperature of 45°F, it cancels the reverse-cycle heating and energizes the resistance-heater coil.

Some combination reverse-cycle and resistance-heating models use a relay contactor

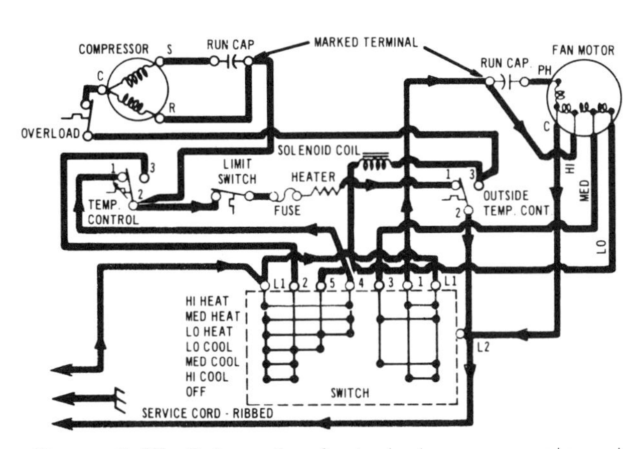

Figure 4-27. Schematic of a typical reverse-cycle and resistance-heated model.

to energize the resistance heater. The contactor is in the control box near the outside thermostat and the terminal board. It is used to eliminate arcing and burning of thermostat contacts when making and breaking the heater circuit. The relay coil is energized by the outside-temperature control.

The relationship of the contactor and the outside-temperature control is shown in Fig. 4-28. The outside temperature has dropped to below 45°F, and the outside-temperature control has closed a circuit from the terminal board through the relay coil. The relay contacts snap closed and complete the circuits to and from

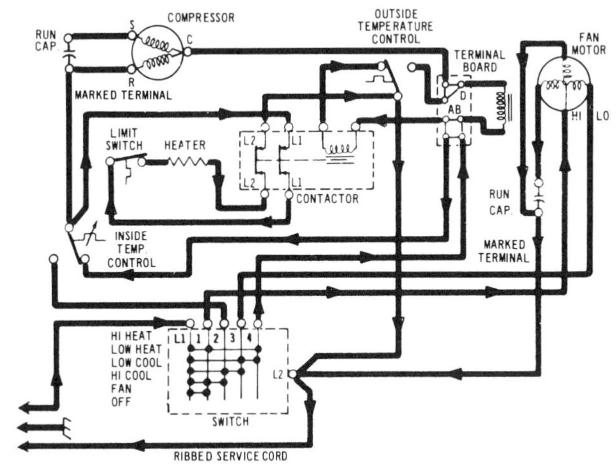

Figure 4-28. Schematic of a typical reverse-cycle and resistance-heated model with a contact relay.

the heater. When the temperature rises above 45°F, the temperature control will snap back to de-energize the coil, restart the compressor, and open the reversing solenoid valve. Follow the arrows on the diagram.

COOLING-DEHUMIDIFYING SYSTEM

The conventional cooling air conditioner provides temperature- and moisture-reducing capa-

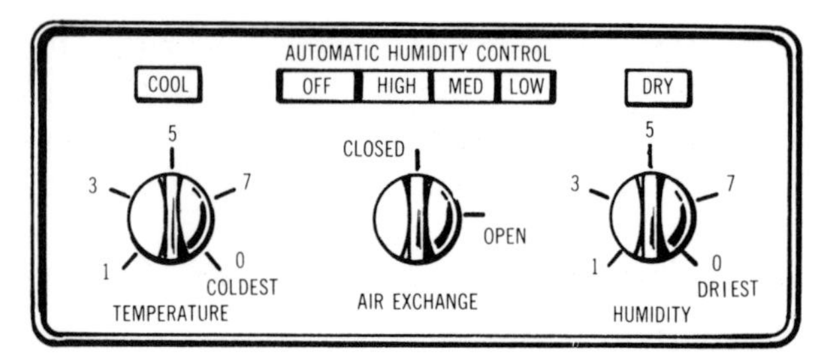

Figure 4-29. Typical control panel–cool/dry models.

city down to thermostat settings, but when the comfort temperature has been reached the air conditioner cycles and operates again only when the room temperature rises. This type of air conditioning does not provide for the removal of excessive moisture that may still be present in room air especially during mild weather. If additional moisture reduction is desired, the room must be overcooled. The automatic cooling-drying air conditioner provides a solution to this problem. This model provides cooling capacity to satisfy the comfort temperature and when this temperature is achieved, will continue to operate, removing additional moisture without a further reduction in room temperature, all automatically. Two electrical controls are used: a humidistat and a relay to control the unit during the drying cycle.

The humidistat is basically the same as used on dehumidifiers (see p. 173). It reacts to changes in relative humidity (air moisture content) to start and stop the unit. Both adjustable and nonadjustable types are used. In the adjustable type, the knob adjusts the control's operating range from 20 to 80 percent relative humidity (Fig. 4-29). The dial numbers are not intended to be calibrated to control closer than 10 percent. This is close enough for reasonable customer guidance in using the product and should be remembered when checking the control's calibration. Checking should be done with a psychrometer. For this reason, no adjustment or recalibration should be attempted in the field.

In the typical cooling-drying model shown in Fig. 4-30, the humidity control knob changes the control setting through a right-angle bead-chain drive. The chain drive pulleys are notched to receive the chain links, thus providing a positive drive. The self-lubricating plastic idler arm changes the direction of the chain drive. Loosening the two humidistat mounting screws allows repositioning of the control to take up or let out on the chain tension should adjustment

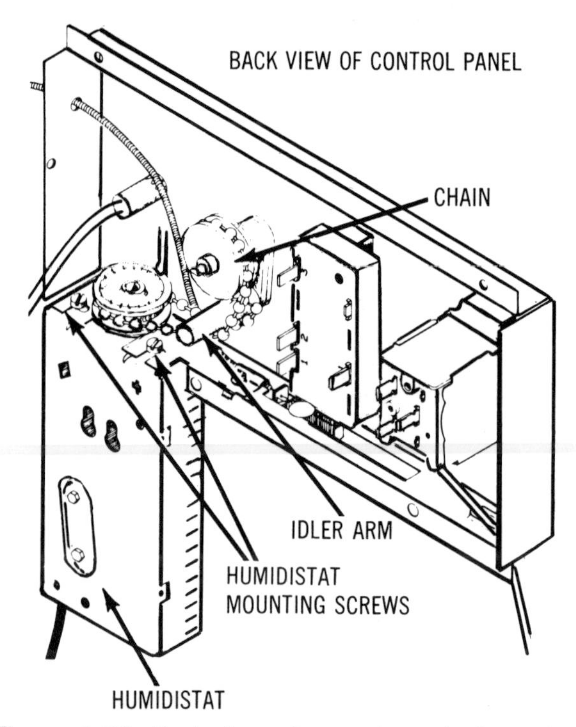

Figure 4-30. Back view of control panel of a typical cooling-drying model.

become necessary. This relay is used only because the humidistat terminals cannot carry the heavy compressor current. The coil is in series with the humidistat switch. When the humidistat switch contacts are closed, the coil magnetically closes the single-pole–single-throw switch contacts on the relay. These contacts are of the heavy current-carrying type needed for the compressor current draw. No adjustment is required or should be attempted on this relay.

Cooling cycle. As shown in Fig. 4-31, the thermostat must be set at least 4°F colder than

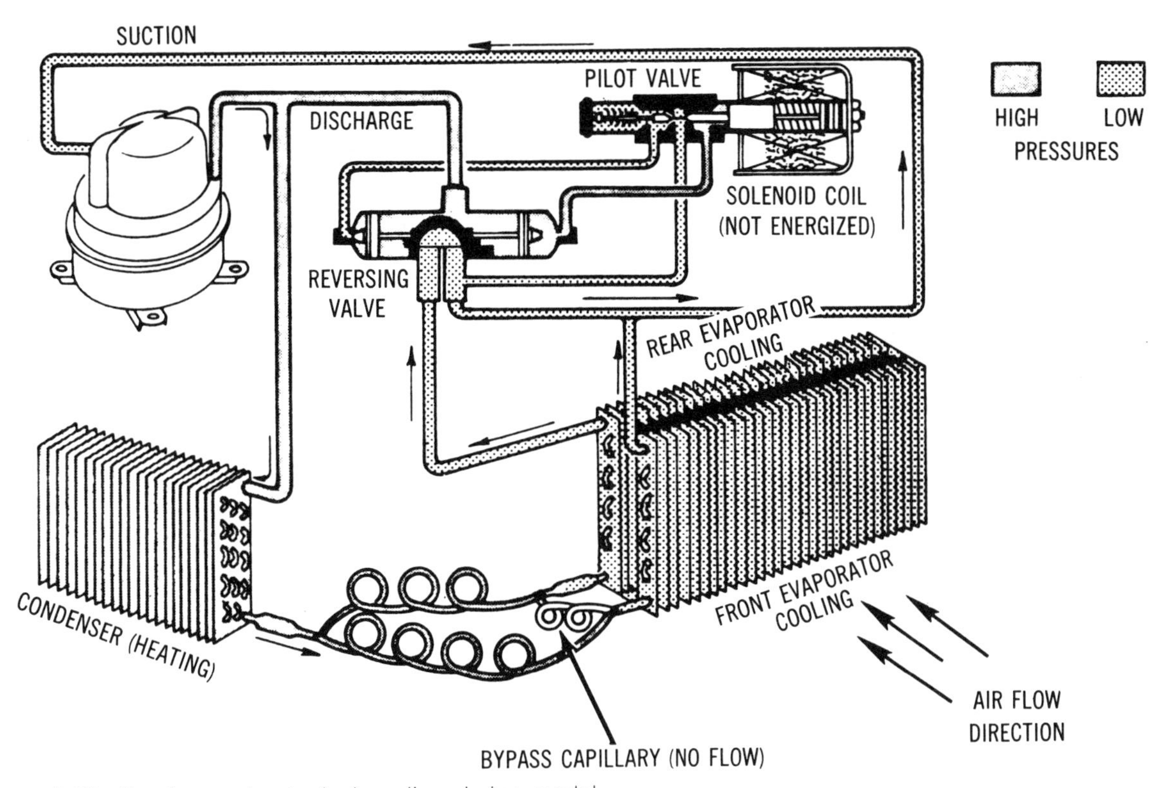

Figure 4-31. Cooling cycle—typical cooling-drying model.

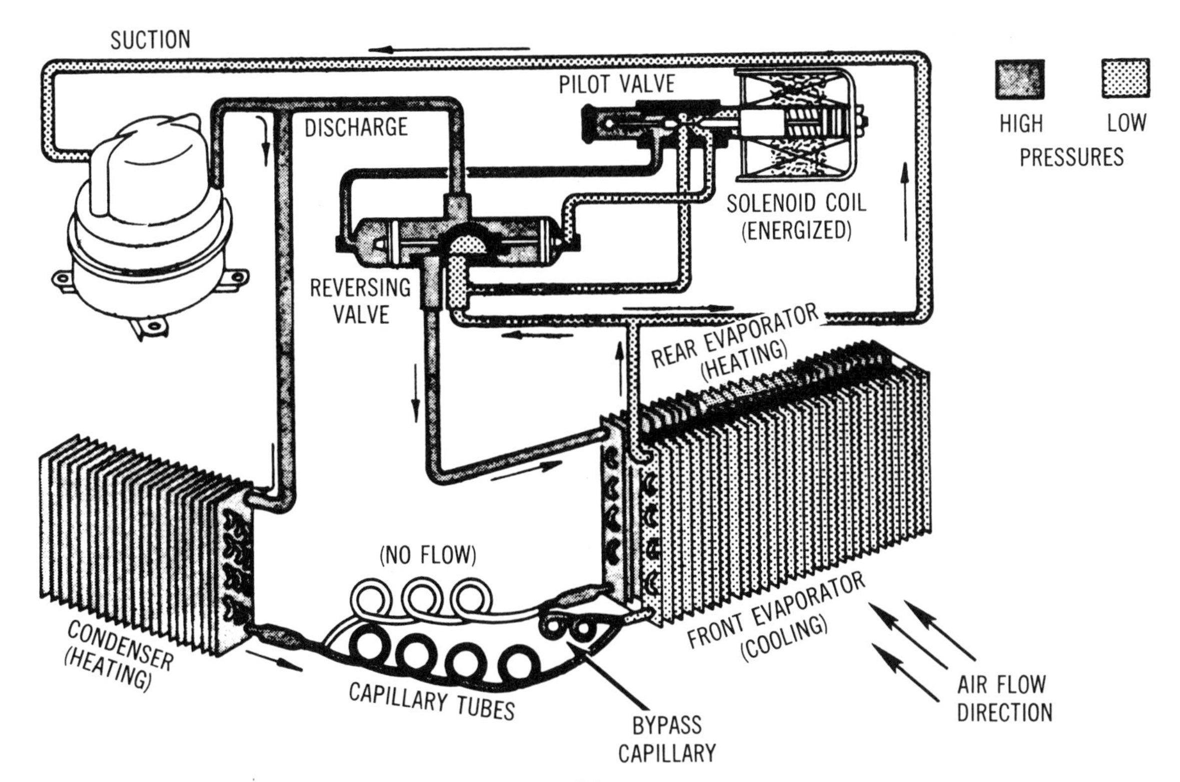

Figure 4-32. Drying cycle—typical cooling-drying model.

room temperature or, if the unit is in operation, the room temperature must rise 4°F above the temperature setting to begin a cooling cycle. When either occurs, the compressor begins to pump refrigerant. The solenoid coil is de-energized, blocking the flow of refrigerant through the reversing valve. Note that one of the reversing valve ports is blocked off. The refrigerant, therefore, passes through the T connection provided in the discharge line between the compressor and the reversing valve.

The hot gas flows to the condenser to transfer its heat. The liquid refrigerant then passes through the two capillary tubes to the evaporator. The evaporator is actually two sets of coils or two evaporators: one front and one rear. Each capillary tube feeds a separate evaporator. The bypass capillary does not function in the cooling cycle. Heat is absorbed by the refrigerant in the two evaporators, reducing the air temperature. The refrigerant from the front evaporator passes to the suction line through a T connection while the refrigerant from the rear evaporator bypasses through the reversing valve to the suction line. The low-pressure gas completes its cycle by returning to the compressor. The cycle continues until the temperature is reduced at least $3\frac{1}{2}$°F.

Drying cycle. As shown in Fig. 4-32, the cooling cycle has completed its function by reducing the air temperature to the thermostat setting. The drying cycle now begins its function—to remove excessive moisture without further temperature reduction.

When the thermostat is satisfied by the cooling cycle, the thermostat switches from its "temperature-control" to "humidity-control" position. This energizes the solenoid coil on the reversing valve and supplies current to the humidistat. If the relative humidity of the room air is below that which is required by the humidistat setting, the unit will shut off (fan continues to run). However, if the relative humidity of the room air is above the humidistat setting, the compressor will continue to run and the reversing valve solenoid is energized. The hot discharge gas from the compressor has two paths to follow: part through the T connection in the discharge line to the condenser and part through the reversing valve. That portion of the hot gas which passes through the T toward the condenser is cooled and liquefied in the condenser. This high-pressure liquid passes through the capillary feeding the front evaporator. The refrigerant does not flow through the capillary feeding the rear evaporator because of the equally high pressure at both ends. The next paragraph will explain the reason for equal pressures.

That part of the gas which passes through the reversing valve flows to the rear evaporator where it transfers its heat. This high-pressure condensed liquid then flows through the bypass capillary tube to the front evaporator. This explains why the pressures are equally high at both ends of the capillary connecting the condenser and rear evaporator, so that there is no flow. In effect, two refrigerant-condensing circuits are established, with both feeding the front evaporator.

The room air being circulated across the two evaporators contacts the cold coil (front evaporator) first. This coil reduces the temperature of the air below its dew point and thus reduces its moisture content. The cooled, drier air then passes over the warm coil (rear evaporator) where it is reheated and returned to the room. The drying cycle will continue until the heat gain in the room has raised its temperature 4°F. If this occurs, the unit will automatically return to a cooling cycle. If the temperature does not rise 4°F, on the other hand, the unit will operate until the relative humidity of the room air has been reduced to the setting of the humidistat.

Trouble Diagnosis

Most of the troubles for year-round room air conditioners can be serviced as described earlier in the chapter. Table 4-3 will serve as a review of these problems as well as covering the other functions of all-season models.

Table 4-3. All-season air conditioner trouble chart.

Directions: Locate complaint in this column below. Follow complaint line right to 1. Follow line up from 1 to test method letter and test subject. Find same letter below for test method and values. Proceed in sequence from 1 to 2 and on until trouble is found.

Directions	Electrical system													
	Rotary switch	Thermostat	Volts-watts-amps	Overload	Capacitor-compressor	Compressor motor	Capacitor-fan motor*	Fan motor	Solenoid coil*	Reversing valve*	De-icer control*	Heater thermostat*	Electric heater*	Fusible link*
Test-method letter	A	B	C	D	E	F	E	G	A	H	B	B	A	V
Complaint														
Compressor and fan motor will not start.	2		3											
Compressor will not start. Fan motor operates.	4	2	5	6	7	8								
Compressor starts and stops.		1	2	12		13	4	3						
Partial or no cooling, compressor operating.			12				2	1						
Noise						†		†	†	†				
Heating phase only														
Partial or no heating.	4	2					6	5	22	23	21	20	18	19

Directions	Refrigeration system									Condenser air								
	Refrigerant shortage	Refrigerant overcharge	Noncondensibles	Restrictions	Capillary	Strainer	Inefficient compressor	Frosted inside coil	Frosted outside coil	Low ambient	High ambient	Obstructed outer coil	Inoperative drain valve*	Inoperative fan motor	Slow fan motor	Fan motor reversed	Poor location	Faulty installation
Test-method letter	I	J	J	K	K	K	L	V	V	N	N	V	R	V	V	V	V	V
Complaint																		
Compressor and fan motor will not start.																		
Compressor will not start. Fan motor operates.																		
Compressor starts and stops.		15	16								6	5		3	7	8	9	10
Partial or no cooling, compressor operating.	13	17	18	14	15		16	5			7	6		1	8	9	10	11
Noise																	3	4
Heating phase only																		
Partial or no heating.	24	27	28		25	26	29		12	11		13	13	5	14	15	16	17

Directions	Evaporator air						General			
	Room too large	Obstructed filter	Obstructed inner coil	Grilles misdirected	Poor location	Air leaks	Thermostat setting	Switch setting	Sound level	Customer education
Test-method letter	M	V	V	V	V	V	V	V	O	P
Complaint										
Compressor and fan motor will not start.								1		
Compressor will not start. Fan motor operates.							1	3		
Compressor starts and stops.				14	11		1			
Partial or no cooling, compressor operating.	21	3	3	4	19	20				
Noise									1	2
Heating phase only										
Partial or no heating.	30	7	7	8	9	10	1	3		

	Test method	*Normal value*
A.	Check continuity.	
B.	Check continuity and cut-in.	See tables below.
C.	Check volts-watts-amps.	120 V. 208 V, 10% temporary deviation. 240 V, allowable. See manual for watts/amps.
D.	Check continuity.	
E.	Check continuity and microfarad capacity. See page 30/31 in manual for tests.	See parts list for microfarad values.
F.	Check with test cord and wattmeter.	See manual for watts/amps.
G.	Check continuity, amps, and speed.	See manual for amps/rpm.
H.	Check opening/closing audibly while moving control from heat to cool and back.	
I.	Check evaporator temperature, watts/amps and for leaks. Make cooling/heating test.	See manual for watts/amps and cooling performance test data.
J.	Check watts/amps and make cooling/heating tests.	See manual for watts/amps and cooling performance test data.
K.	Check watts/amps and evaporator temperature.	See manual for watts/amps.
L.	Check watts/amps.	See manual for watts/amps.
M.	Make cooling/heating tests.	See manual for data.
N.	Check room and condenser air.	See owners' service manuals for recommendations.
O.	Check for abnormal sound.	
P.	Explain usage and operating characteristics to owner.	
V.	Visual inspection.	

* On models so equipped.
† Check functional items for abnormal sound.

Dehumidifiers and humidifiers

CHAPTER

5

Humidity is the term applied to the moisture in the atmosphere. Humidity and heat, separately or in combination, have a decided influence on our daily existence. Nature provides both of these in the air that we breathe, but in variable quantities that greatly affect our health and comfort. Too much humidity causes discomfort by preventing the body from getting rid of perspiration at a normal rate and can contribute to respiratory disorders. It also promotes the growth of microbes such as yeast, mold, and bacteria. On the other hand, a lack of humidity is often undesirable because of the effect on health and comfort. Although low humidity is effective in the cure or relief of respiratory disorders, it can contribute to ill health and discomfort by causing the skin and the .mucous membranes of the nose and throat to become dry.

DEHUMIDIFIERS

Like air conditioners, humidifiers and dehumidifiers are of two types: central (usually built into the heating system) and room (free-standing) units. Also like air conditioners, the central systems are, as a rule, installed and serviced by air-conditioning-heating contractors, and are not covered in this book. On the other hand, room-size humidifier and dehumidifier units fall in the realm of the appliance service technician.

Operation of a Dehumidifier

An electrical room dehumidifier operates on the same refrigeration principle that was previously discussed for refrigerators, freezers, and room air conditioners. That is, moisture-laden air is generally drawn into the rear of the dehumidifier and over the cold evaporator coils by the fan. As this air is cooled, some of the moisture is condensed out of it and is deposited on the coils in the same manner as moisture collects on a glass of ice water on a humid day. The drops of water which form on the coils run down and fall into the water container or through the hose to a drain.

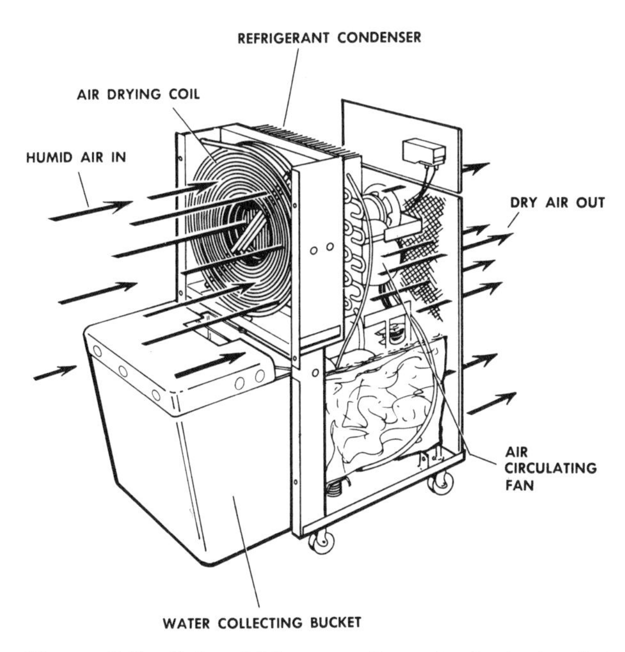

Figure 5-1. Dehumidifier operation of a typical unit.

The cooled air is then passed over the warm condenser coils which raise its temperature. Since the air now contains less moisture and is warmer, its relative humidity is much lower. The warm, drier air is expelled through the front of the dehumidifier and mixes with the room air. As this process continues, the relative humidity of the air in the entire room is lowered, and the room and its contents are made drier.

Humidistat. Most dehumidifiers employ a humidistat rather than a control thermostat to maintain their operation. Many dehumidifiers are now using humidistats that contain a human-hair element (Fig. 5-2) which stretches when the humidity is high. Other humidistats use a nylon strip in place of the human hair. In either case in a typical humidistat of this kind, high humidity causes the hair or nylon strip to become long enough to relieve pressure on the pivot bracket. When the pressure on the pivot bracket is relieved enough, the plunger of the electrical switch will move upward because its pressure is then greater than the pressure of the pivot bracket. This allows the switch to be in the closed position, which then allows the compressor to run. As the air is dried during the operation of the compressor, the hair or nylon strip will contract and force the pivot bracket against the plunger of the electrical switch, forcing the plunger downward and into an open position, which will break the compressor circuit.

Since operation of a dehumidifier varies to such an extent from one application to another, no specific relative humidity values can be indicated on the automatic control. However, for many applications the dehumidifier will correct the dampness condition for which it was purchased with the control at the DRY setting. For a more severe dampness condition, some dehumidifiers must be operated with the control closer to the EXTRA DRY setting. Of course, some dehumidifier controls have numbers rather than specific settings. (The higher numbers

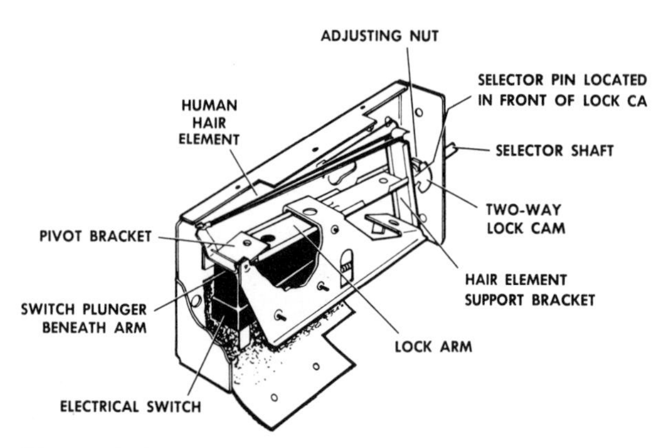

Figure 5-2. Cutaway of a "hair-type" humidistat. No attempt should be made to repair this humidistat.

usually mean more dehumidification.) But remember that none of these settings produces a specific relative humidity condition.

Refrigerant cycle. A dehumidifier usually is designed to maintain a predetermined temperature in the evaporator coils. The proper charge, length and size of capillary tube, compressor design coil surface, and fan motor and blade maintain a balanced system, which keeps the relative humidity under control. The refrigerant flow from the condenser to the evaporator coils is controlled by a capillary tube permanently bonded to the suction line in heat-exchanger relationship. A strainer, located ahead of the capillary tube, prevents the entry of foreign material, which might clog the tube. The cycle of operation of the refrigeration system for the dehumidifier is as follows. Hot, high-pressure refrigerant vapor is forced by the compressor through the discharge tube to the top of the condenser. In passing down through the condenser, the vapor gives up some of its heat to the room air and condenses to liquid. (Room air is drawn through the condenser by the fan motor located within the dehumidifier.) The liquid refrigerant passes out the condenser bottom and through a strainer and the capillary tube, which is bonded to the suction line to form the refrigerant heat exchanger. Here the warm liquid gives up some of its heat to the cold vapor in the suction tube. The liquid refrigerant passes from the capillary tube into the larger tube which makes up the evaporator coils. In removing heat from the air, there is some evaporation so that a combination of liquid and low-pressure gas flows out the evaporator coil. Low-pressure gas then flows through the refrigerant heat exchanger and suction tube down to the compressor before being compressed into hot high-pressure vapor and made ready for another cycle.

The refrigeration component parts in a dehumidifier are designed to maintain a high suction pressure which in turn provides the evaporator coil with liquid maintained at a high enough temperature that the water collected on the evaporator coils will not freeze under normal ambient temperatures.

Electric circuit. When the service cord is plugged in (and the relative humidity of the room rises to the cut-on setting of the humidistat, thus closing the contacts in the humidistat switch), the circuit is completed through the compressor motor running winding and the relay coil. Like other refrigeration units, the motor cannot start on the running winding alone and draws a heavy current. This current passing through the relay coil creates a strong magnetic field which lifts the relay plunger, closing the relay contacts, and completing the circuit through the starting winding. Because of the

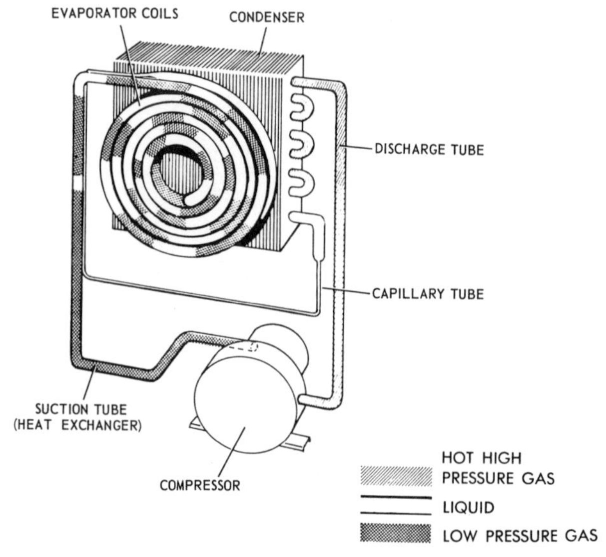

Figure 5-3. Dehumidifier refrigerant cycle.

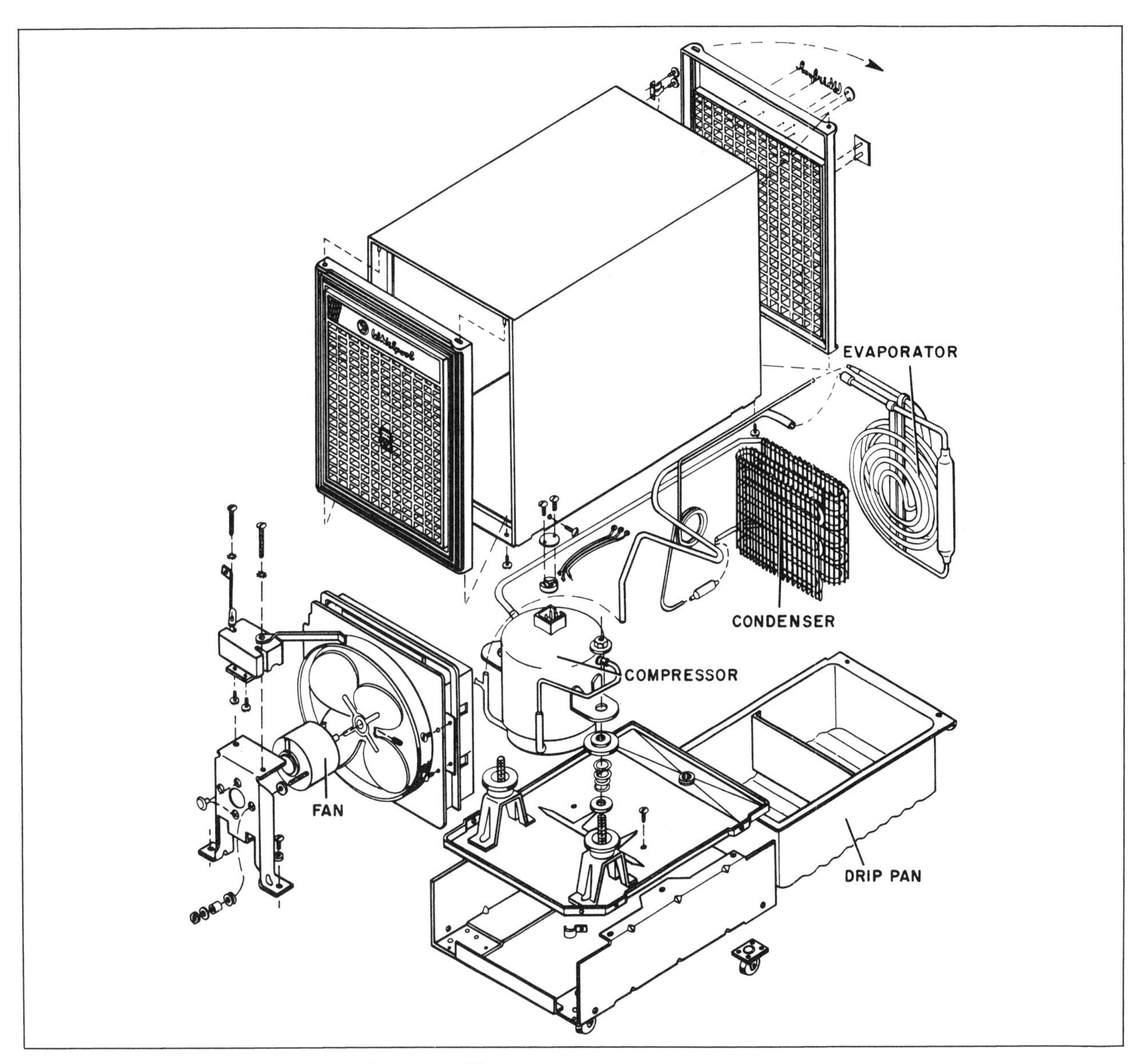

Figure 5-4. Major parts of a typical dehumidifier.

phase angle between the starting and running windings, the motor has sufficient torque to start. As the motor comes up to speed, the current drawn decreases, thus decreasing the strength of the magnetic field in the relay until the weight of the relay plunger causes it to drop, opening the relay contacts and the circuit to the starting winding. The motor continues to run on the main winding only. (When the relative humidity of the room is lowered to the cut-off setting of the humidistat, the humidistat switch contacts open the circuit and the motor stops.)

The fan motor is usually a two-phase motor with its running winding in parallel with that of the compressor motor and its auxiliary winding in parallel with the starting winding of the compressor motor. The fan motor running winding is energized along with the compressor motor running winding. When the compressor motor

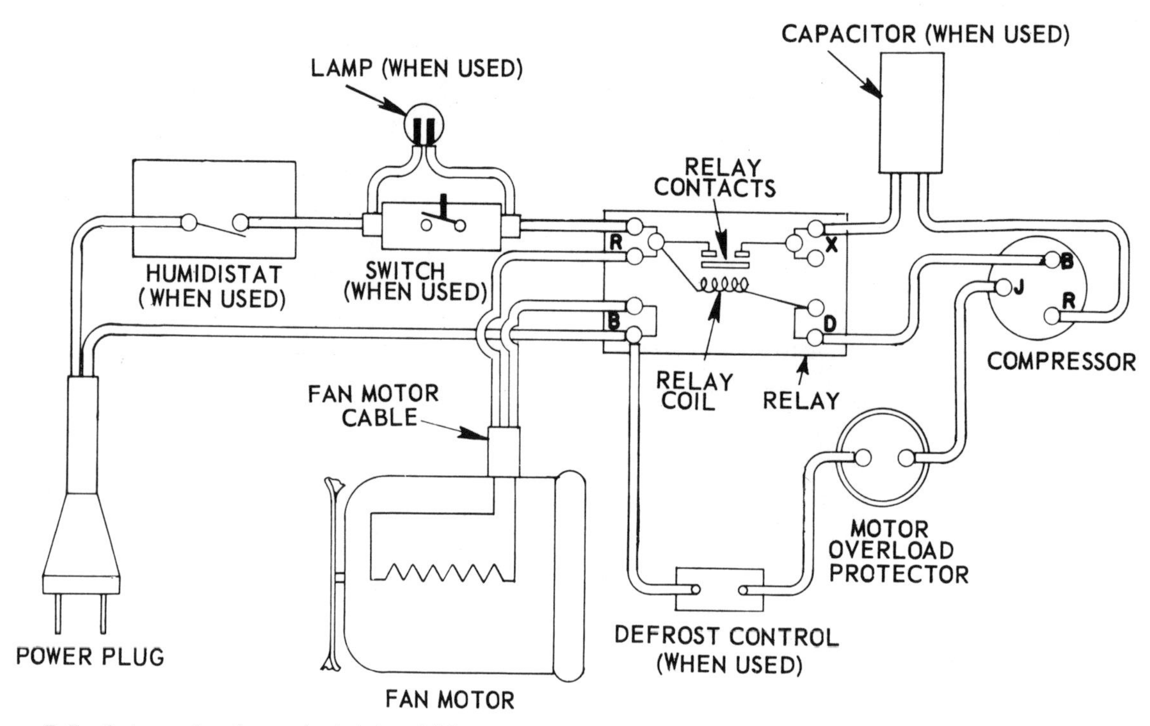

Figure 5-5. Schematic of a typical dehumidifier.

starts and the relay opens, voltage is induced in the compressor motor starting winding. This induced voltage energizes the auxiliary winding of the fan motor and, because of the phase angle between the two windings, provides the necessary torque to start and run the motor.

Current flowing through the compressor motor windings also flows through a small heating element in the motor thermostat, an automatic reset circuit breaker, which is mounted on the compressor housing. The motor thermostat will open the circuit if the current flow through the motor windings is excessive or if the compressor motor overheats, thus protecting the motor from damage. It closes the circuit automatically after a short time. If the abnormal condition continues, the thermostat will continue to cycle, opening and closing the circuit.

Locating and Maintaining a Dehumidifier

As a service technician, you must be able to advise your customer on the best locations for a dehumidifier and how it should be maintained. This advice can help reduce servicing problems.

For best results, the dehumidifier should be located near the center of the area being dehumidified. However, when it is desirable to locate the dehumidifier elsewhere so as to utilize a hose connection from the collecting tray to a drain, the dehumidifier may be located some distance from the center of the area. In this case, the dehumidifier should be located so that the airflow from the front of the dehumidifier is along a wall rather than either toward or away from a nearby wall. Remember that good circulation of the air is essential for any part of the area requiring dehumidification.

A dehumidifier is not too effective in the living quarters of a home when a number of people are present and entering and leaving the home regularly. Also, since the doors and windows must be kept closed at all times when it is in use, the dehumidifier will substantially raise the temperature of the air in the room where it is operated. For these reasons, it may not produce more comfortable conditions in the living area and may not be too effective.

It is important that air circulation from the dehumidifier be maintained to all areas requiring dehumidification. Hence, the dehumidifier can maintain very satisfactory conditions throughout the major part of a basement including the outside of a basement closet, but conditions inside a basement closet can be controlled only if the dehumidifier is placed inside the closet or if the closet door is kept open at all times to allow air circulation in and out of the closet. The dryness in a closet cannot be maintained merely because the air is dry around the outside of the closet. The dry air must have free passage into and out of the closet, or the dehumidifier must be located inside the closet.

The dehumidifier always heats the surrounding air. The heat removed from the water during condensation at the cooling coil in addition to the heat equivalent of the power consumption is utilized to raise the temperature of the surrounding air. This method of heating the air to reduce the relative humidity is used to gain the greatest efficiency for the dehumidifier.

Most dehumidifiers employ a water-collecting pan and bucket to gather the water from the evaporator coils. Many models provide a hose connection which can be fastened to the house's drain system. In this way, the water has a continuous run. Units providing a bucket outlet for water usually contain a pressure switch which under normal operating conditions (bucket not full) is closed, and the compressor circuit is also closed. But when the bucket has been filled, an indicator light circuit is closed, and the compressor circuit is opened by the pressure switch. When this occurs, of course, the compressor will not operate.

When the dehumidifier is first put into operation, it will remove relatively large amounts of moisture until the relative humidity in the area being dried is reduced to the value where moisture damage will not occur. After this, the amount of moisture removed from the air will be considerably less than that removed when the dehumidifier was first put into operation. This reduction in the amount of moisture being removed indicates that the dehumidifier is doing its job and that it has reduced the relative humidity to a safe value. The performance of the dehumidifier should be judged by the elimination of the dampness and dampness odors rather than by the amount of moisture being removed and deposited in the bucket.

The mechanical type of dehumidifier described in this chapter ordinarily will not operate satisfactorily below 65°F. At this temperature it becomes necessary to operate the evaporator coil below freezing temperatures in order to reduce the relative humidity to a reasonable value. Although the dehumidifier can be operated with ice forming on the evaporator coil, it becomes necessary to defrost this coil at least once an hour to maintain satisfactory performance.

For the most economical operation, the evaporator and condenser should be cleaned with a vacuum cleaner whenever dust or lint accumulates. An inspection through the rear opening will reveal when this is necessary.

The evaporator and condenser may be cleaned from the rear. Occasionally the evaporator coils, drip tray, and the inside of the water container should be wiped with a damp cloth. *Caution*: Do not touch or attempt to clean the human hair or nylon membrane on the humidistat.

In some cases, there will be a collection of fungus attached to the bottom of the dehumidifying coil after an extended period of operation. Generally, this is a mixture of fungus and airborne dust. The fungus is an airborne spore that alights and grows on the evaporator coil; since it is airborne and peculiar to the particular location involved, there is nothing the manufacturer can do about it. This condition is worse in some areas than others. Where the fungus is present, it tends to collect the airborne dust and aggravate the situation. The best way to clean this material from the evaporator coil is by the use of a soft brush, such as a corn-buttering brush, and an adequate amount of clean water. As the material is loosened from the evaporator coil with the brush, it should be flushed away with clean water. It may be necessary to use a test-tube brush or a pipe cleaner to clean out the drain of the collecting tray.

Trouble Diagnosis for Dehumidifiers

Here are the problems a technician usually finds when servicing a room dehumidifier.

Problem: Unit does not run (no water collected, compressor does not operate).

Possible cause	*Solution*
1. No power.	1. If the unit does not operate, check the power supply with a voltmeter or other appliance. If no power is available, check for a blown fuse. Advise customer to call the power company or notify an electrician. If power is available, check for broken wires inside the service cord.
2. Poor plug contact at outlet.	2. See that the prongs make good contact in the wall outlet.
3. Low voltage.	3. Check the voltage at the outlet to which the dehumidifier is connected while the appliance is in operation. Dehumidifiers are designed for operation on 120-V, 60-Hz current with an allowable variation of ± 10 percent. Use a good voltmeter to make the voltage test. If the voltage is low, see that the circuit is not overloaded with other appliances. Avoid the use of long extension cords. If required, use No. 16 gauge wire or heavier.
4. Humidistat knob at OFF position.	4. Turn knob to DRY or similar position.
5. Humidistat inoperative.	5. Check humidistat. With the service cord plugged in and knob turned clockwise as far as possible, place insulated jumper wire across terminals of humidistat switch. If compressor starts, replace the humidistat.
6. Low relative humidity.	6. Humidistat switch contacts may not close if relative humidity of room is low (20 to 30 percent) even with knob turned clockwise as far as possible.
7. Relay inoperative.	7. Check relay. Remove the cap from the motor-compressor connections on top of the motor-compressor and use the test starting cord illustrated on p. 31. If the motor-compressor operates with the starting cord hookup, the relay is defective and will have to be replaced.
8. Open circuit or ground in electric connections.	8. Check electric circuit. Repair defect.
9. Compressor motor thermostat.	9. Check motor thermostat. Replace.
10. Open windings, compressor motor.	10. Check compressor. Replace unit or dehumidifier.
11. Pressure switch (when used).	11. Under normal operating conditions (bucket not full) the switch is closed to the compressor and open to

indicator light circuit (see Locating and Maintaining a Dehumidifier, p. 176). If switch is defective, replace.

12. De-icer control (when used). — 12. This is a bimetal control clamped to the suction line and is designed to open the compressor electric circuit when ice develops on the evaporator (approximately 80 percent blocked). This control is used on the higher-capacity units because of their ability to reduce evaporator temperatures quickly when air restriction (usually icing) is encountered.

Problem: Unit runs, but there is no performance (no water or insufficient water collected —compressor operates).

Possible cause	*Solution*
1. Abnormal use.	1. Room dehumidifiers are designed to operate efficiently under certain conditions. They perform most efficiently at a temperature of about 90°F and relative humidity in excess of 70 percent and will perform satisfactorily well below these conditions. However, when the unit operates at a temperature below 65°F and a relative humidity of 60 percent, the air is too cool and dry. Advise the customer that the unit is not designed for these conditions, as dehumidification is not necessary, and to turn off the unit.
2. Location.	2. The room dehumidifier will effectively reduce the humidity in an enclosure up to 10,000 ft^3. For larger rooms, additional dehumidifiers may be required to accomplish the desired results.
3. Air circulation.	3. The grilles on the cabinet shell must be free of all obstructions to allow for unhampered air circulation. The grilles should not be covered, and furniture and equipment should not be placed closer than 6 in.
4. Evaporator and/or condenser clogged with dust and dirt.	4. Clean evaporator and/or condenser.
5. Fan motor not operating.	5. Check faulty wiring first. Disconnect the fan leads and check for continuity. If no continuity is indicated, replace fan.
6. Frost or ice on evaporator.	6. See Frost or Ice Forms on Evaporator Coils, p. 180.
7. Humidistat knob at high-humidity position.	7. Turn knob to dryer position.
8. Humidistat out of calibration.	8. Replace humidistat.

9. Compressor cycling on thermostat.	9. See Compressor Cycles on Motor Thermostat below.
10. Refrigerant leak, broken compressor valves, or restricted tubing.	10. If discharge line and condenser are cool and wattage low, the cause may be loss of refrigerant or broken compressor valves. Pull plug. If compressor coasts to a stop, valves probably are broken. If discharge line is hot, wattage high, and compressor noise level high—possibly compressor cycling on thermostat—cause may be obstruction in liquid line, either foreign material or ice (latter is unusual since normal evaporator temperature is above 32°F). May also be kinked tube. Remove kink if possible. Shut off unit to allow ice to melt or move obstruction. If no results, replace unit or dehumidifier.
11. Low-capacity compressor.	11. Replace compressor or dehumidifier.

Problem: Compressor cycles on motor thermostat.

Possible cause	*Solution*
1. High or low line voltage.	1. Check line voltage. Must be between 105 and 130 V. If intermittent, provide new supply. If steady, install transformer.
2. Poor air circulation.	2. Relocate dehumidifier where it will have free and unobstructed airflow.
3. Condenser clogged with dust or dirt.	3. Clean condenser.
4. Fan motor not operating.	4. Check fan. Repair or replace.
5. Relay stuck closed.	5. Check relay. Repair or replace.
6. Short circuit or ground in electric circuit.	6. Check electric circuit. Repair.
7. Unit pressures not equalized.	7. Allow 2 or 3 min for pressures to equalize before starting compressor.
8. Thermostat out of calibration.	8. Replace thermostat.
9. Unit liquid line restricted.	9. Check for damaged or bent tubing. Straighten tubing if possible, or replace unit or dehumidifier.
10. Stuck compressor.	10. Check compressor. Replace unit or dehumidifier.

Problem: Frost or ice forms on evaporator coils.

Possible cause	*Solution*
1. Compressor just starting to run.	1. Normal. Will disappear after 10 to 15 min.
2. Temperature or humidity of room low.	2. Normal. Turn off dehumidifier or turn humidistat knob to a less dry position. Operation not needed.
3. Partial restriction in liquid line.	3. If tubing is kinked, straighten if possible, or replace unit or dehumidifier.
4. Unit short of refrigerant.	4. Repair leak. Recharge.

Problem: Noisy operation.

Possible cause	*Solution*
1. Caster rattle.	1. Move dehumidifier so weight is evenly distributed on all

	casters.
2. Evaporator coils rattle.	2. Position coils correctly in three evaporator clamps. Tighten clamps. Position evaporator spacer rod between two banks of coils.
3. Grille rattle.	3. Remove cabinet. Squeeze cabinet channels slightly for tighter fit on grille.
4. Tubing rattle.	4. Bend tubing carefully to correct. Remember that no tubing should be touching any other tubing, the cabinet, compressor, or condenser. Securely tape any loose condenser wires to condenser to prevent any unnecessary vibration.
5. Fan blade hitting condenser or evaporator.	5. Adjust condenser or evaporator for clearance. Check fan motor mounting brackets. Remember that most fan motors must be oiled every 6 months when operated continuously. Oil before operating after long storage periods. To oil the fan motor, add four or five drops of detergent-free automotive oil (SAE-20) in either the oil tubes or oil holes provided in front and rear of fan housing. The fan blade may become stuck in the fan shroud or entangled in the wiring. See that the fan blade is free of obstructions.
6. Fan blade bent.	6. If fan blade is bent badly, replace blade.
7. Fan motor excess play.	7. See that fan blade is tight on motor shaft. A loose blade causes noisy operation. If there is too much end play, replace motor.
8. Compressor mounting stud washers rattle.	8. Remove hairpin cotter and flat washer. Remove and replace spring washer with concave side up or add spring washers to take up play.
9. Incorrect compressor.	9. Straighten mounting studs if bent. Carefully bend tubing to allow compressor to ride freely, with studs centered in holes in compressor feet. Replace rubber grommets if worn or spring resiliency lost.
10. Internal compressor noise.	10. Replace unit or dehumidifier.

Problem: Water drips on floor.

Possible cause	*Solution*
1. Dehumidifier not level.	1. Move to level part of floor.
2. Evaporator coils stopping up drip tray outlet.	2. Reposition coils to clear outlet.
3. Foreign material plugging drip tray outlet.	3. Clean drip tray.
4. Foreign material	4. Remove and clean drain hose.

4. (*cont'd.*) plugging drain hose.	
5. Poor drain hose connection.	5. Check connection to drip tray. If hose end distorted, cut off end and reconnect or replace hose.
6. Leak in water container.	6. Replace water container.
7. Water drips when container removed for emptying.	7. Normal. Return container as quickly as possible or provide other means of catching water while emptying.

As with all refrigerant systems—those in refrigerators, freezers, and room air conditioners—any time the line must be opened, the system must first be evacuated, and after the repairs have been made, it must be recharged. Methods of evacuating a dehumidifer's refrigerant system, checking for a leak, cutting a suction tube, recharging a system, replacing a compressor, etc., are the same as for a room air conditioner's refrigerant system (Chap. 4).

HUMIDIFIERS

During the wintertime, the air is generally many degrees colder and considerably dryer than during the summer months. For indoor comfort and health, the air must be heated and moisture must be added. Heating the air and controlling its temperature present no particular problem. Heating can be accomplished by supplying hot water or steam through heating coils or radiators, or by hot-air furnaces. However, providing humidity often represents a considerable problem, especially on installations using electric, steam, or hot-water radiators. A humidifier, which is a nonrefrigerant type appliance, remedies this problem.

While humidity can be added to the air by several methods, the water-pan type is one of the oldest and still most commonly used methods and is the one that will concern the appliance service technician. In this method, a pan or container of water is placed in the conditioning unit housing, and water is permitted to evaporate into the airstream.

There are many variations of the water-pan type of humidifier. In one of the simpler types, the dry room air is drawn into the top of the humidifier by the enclosed fan. As the air passes through the enclosure, it is washed in a water spray and then forced through the water-saturated filter where water is absorbed into the air. Clean, humidified air is then dispelled from the sides and front of the unit.

The automatic version of this design uses a humidistat (usually the same type used in a dehumidifier) to turn the unit on and off. To operate it, the user turns the humidistat knob fully clockwise to the CONTINUOUS ON position and lets the humidifier run in this fashion until the humidity has achieved the desired level. This may easily be determined by looking for condensation or "steam" on windows or by using a humidiguide to measure humidity. Humidity should then be as follows.

Outside temperature, °F	*Desired humidity, %*
−10	20
0	25
10	30
20	35

When humidity reaches the desired level, the user depresses the switch to the AUTO position and turns the humidity control knob counterclockwise just until the fan motor shuts off. (The knob should be rotated *very slowly* as the humidity sensor requires a few seconds to respond to the setting.) The speed control in some units will automatically increase or decrease the fan speed as more or less moisture is needed to maintain the humidity setting. The humidity level may be increased or decreased at any time by rotating the humidity control knob as indicated. The humidifier can usually

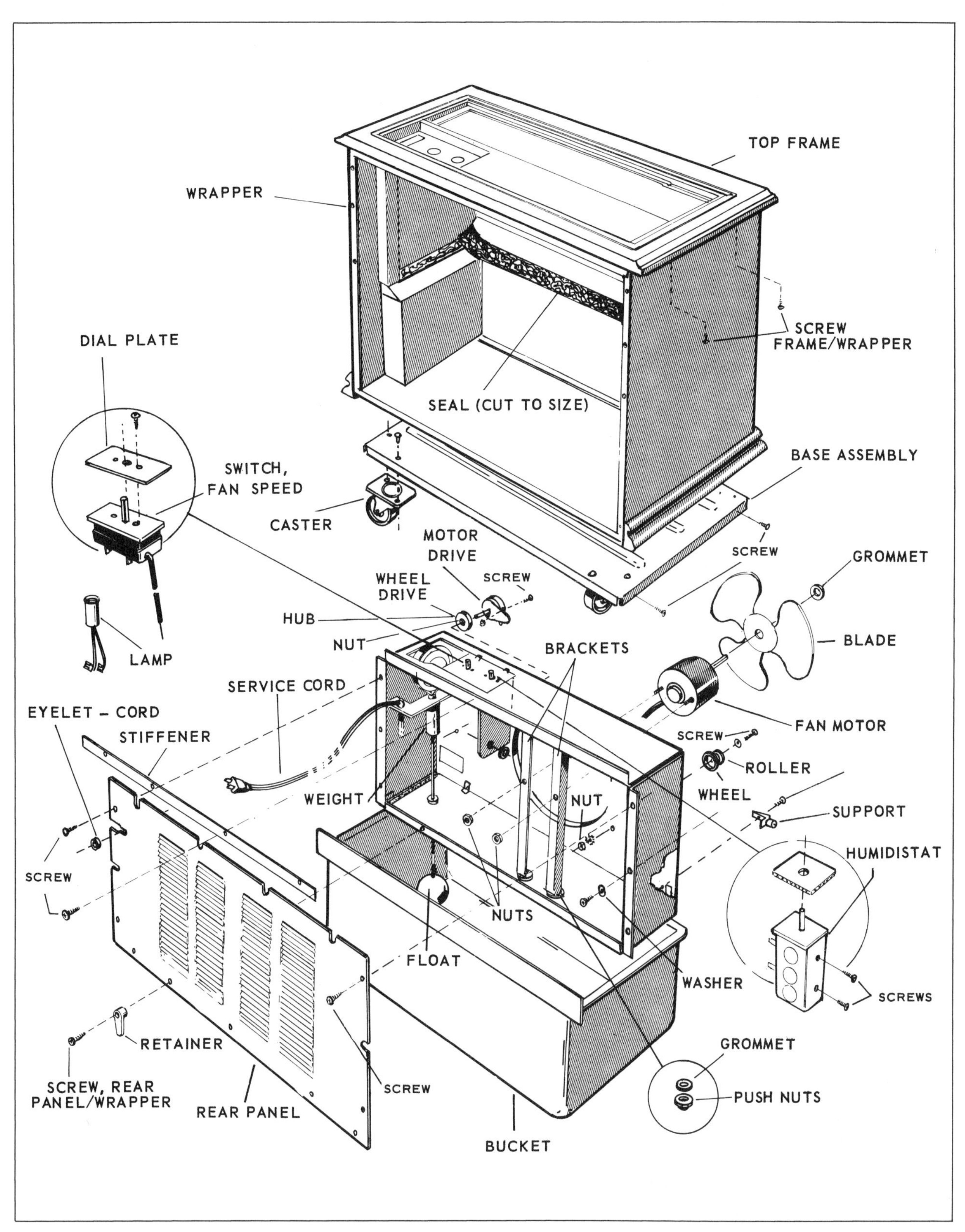

Figure 5-6. Wrapper, bucket, base, and controls of a drum-type humidifier.

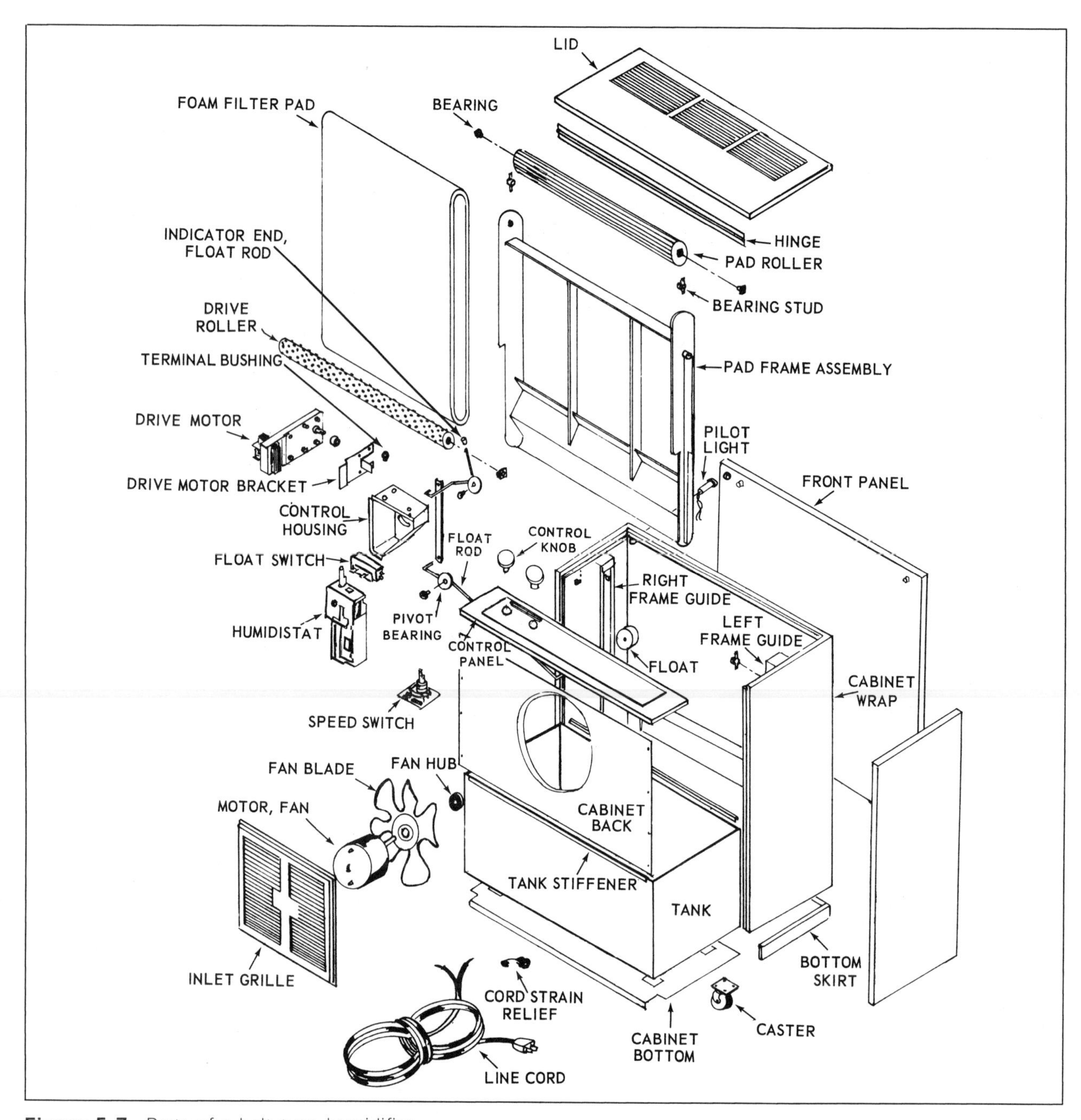

Figure 5-7. Parts of a belt-type humidifier.

be made to run continuously at full output by depressing the switch to the ON position. The humidifier may be refilled at any time to the level indicated in the instruction manual. When the water level gets too low, the REFILL light will go on and the unit will shut off and stay off. The humidifier then must be refilled to be operative. The pilot light indicates that the unit is plugged in and is operational.

In the automatic version, the air is usually drawn in through the back of the unit by the fan. The air is then forced down through the saturated filter pad, under the baffle, and up through the wet filter pad on the outer side.

The filter pad remains wet during operation by continually dipping into the water stored in the tank. The fan motor drives the roller which turns the filter pad. There are many variations of this action.

Components

As we just stated, there are many versions of the water-pan type of humidifier. But with the automatic type, the most popular, the service technician must be familiar with these major components.

1. *Automatic humidistat control.* This is usually the same as for the dehumidifier (see Humidistat the beginning of this chapter). Nonautomatic models use a simple ON-OFF switch.
2. *Float control.* This device, which includes a switch and float lever, will turn off the unit when the water tank is empty. An out-of-water signal light is usually turned on when the float control turns off the fan motor and filter pad drive motor.
3. *Circulation fan.* The fan with its motor draws room air into the unit and forces it out as humidified air. The fan motor is usually controlled by a so-called "speed" control in most automatic humidifiers.
4. *Filter pad drive motor.* This motor drives the filter pad through the water tank. In some units the fan motor and drive motor are one. Many simple humidifiers do not have a drive motor.

Cleaning

A humidifier should be cleaned when the water in the tank becomes dirty or when the pad fails to rotate. Usually the white appearance of the pad will indicate when cleaning is required. Frequency of cleaning will depend upon the hardness of the water used, the cleanliness of the air, and the amount of usage.

To clean a filter pad, most makers of humidifiers suggest that the dirty pad be soaked in water and detergent for 1 hour; for excessive lime deposit, soak in a half-and-half solution of water and vinegar. After soaking, knead gently to remove residue and rinse under running water. The pad should be replaced when the openings do not reappear after cleaning.

When a humidifier has been standing idle for some time, the pad dries out. If enough lime and other chemicals have accumulated on and in the pad, causing it to harden, it will be necessary to either wet the pad thoroughly or remove and clean it before it will rotate.

Trouble Diagnosis for Humidifiers

Because of the wide variations in humidifier design and part layout, it is difficult to give specific troubleshooting diagnosis as has been done with the other major appliances. However, we can give a general analysis of the major complaints a service technician may face when servicing a humidifier and their solutions.

Unit will not operate.

1. Check the following for continuity: float control, speed control, humidistat, cordset, connections, and motor.
2. Check for binding gears or binding motor.

Unit will not shut off when tank is empty.

1. With float control in filler position, make sure the float control switch is closed.
2. Check for intermittent operation of the float control switch.
3. Check for binds in float mechanism.
4. Check for proper adjustment of float mechanism.

Humidistat does not operate.

Under certain conditions, a humidistat may not appear to be operating. This can be the result of the humidistat setting or to humidity losses, particularly in an older home. The probable complaint is that the unit is running continuously or does not shut off. The humidistat control, as previously mentioned, usually operates from a nylon strip or human hair which expands with moisture or contracts from dryness. Check for proper operation following the procedure below.

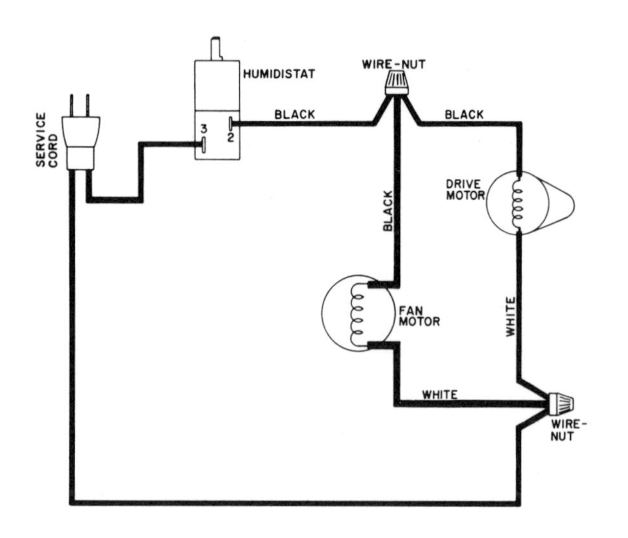

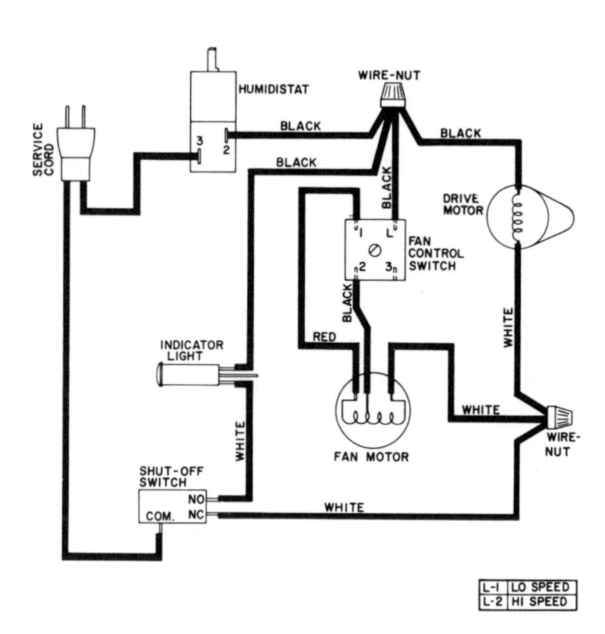

Figure 5-8. Two typical humidifier electrical schematics.

Set the speed control to the ON position. Adjust the humidistat to an ON position so that the unit just starts to operate. Breathe heavily once or twice into the louver of the humidistat chamber. The unit should shut itself off because the moisture-laden air, coming in contact with the nylon strip, causes it to expand. Blow cool dry air into the humidistat chamber by means of a small fan or a hair dryer operated at the COOL setting. The humidistat should turn back on within 2 or 3 min. This will indicate proper cycling operation.

Unit provides no moisture.

1. Check for excess lime accumulation on filter. Wash filter, tank, and drive roller.
2. Speed setting may be too low. Output is proportional to speed setting. Suggest a higher speed setting.
3. Filter drive roller does not rotate.

Noisy operation.

1. Check for proper mesh of all gears.
2. Check for rubbing of the drum or drive roller and the tank ends.
3. Check fan unit and blade.
4. Check for slippage between drive motor and drive roller.

Filter pad and drive roller do not rotate.

1. Check for proper installation of drive roller and filter pad.
2. Check for faulty drive motor.
3. Check for slippage between drive motor and drive roller.
4. Check float control switch.
5. Check electrical wiring connections.

Fan motor does not operate.

1. Check for faulty fan motor.
2. Check electrical wiring connections.
3. Check fan blade.

Water leaks appear.

1. Check tank for cracks or holes.
2. Check for proper installation of filter pad, making sure ends do not hang over causing spillage at tank edge.
3. Check for funnel position.

Odor traced to unit.

1. Check for excessive accumulation of foreign matter on filter pad or in tank. Clean if necessary.
2. Bad water; replace.

Condensation forms on windows and/or walls.

The unit is operating at too high a setting. (Reduce humidistat setting.)

Water coolers

CHAPTER

6

Although water coolers are not classified as home appliances, service technicians are increasingly being called upon to maintain them in schools, offices, factories, and stores. While service technicians seldom installs the units, they may service them on either a call or a contract service basis.

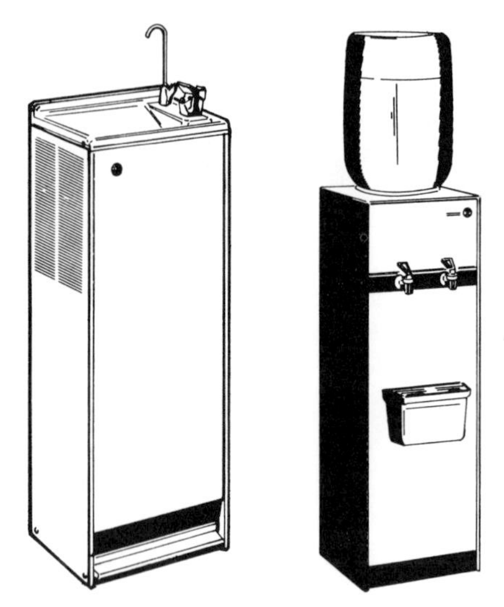

Figure 6-1. Two basic types of water coolers: pressure type (left) and bottle type (right).

There are two basic types of water coolers: (1) the so-called pressure type which employs an outside source of water, and (2) the bottle type which has a self-contained water source. Some of the newer models of both types have a hot-water outlet which permits the making of instant coffee, cocoa, tea, or soup.

OPERATION OF PRESSURE WATER COOLERS

Three basic functions are performed in the water cooler: (1) refrigeration cycle, (2) water cooling cycle, and (3) electrical circuit. Here is how these functions are achieved in a typical pressure water cooler.

Refrigeration Cycle

Low-pressure refrigerant gas (usually R-12) enters the compressor cylinder and is compressed and discharged as a hot, high-pressure gas into the condenser. Here it is cooled and condensed into a liquid by giving up heat to the air passing through the condenser. High-pressure liquid then passes through a strainer or drier into the capillary impedance tube which is bonded to the suction line to form a heat exchanger. Some heat is removed from the liquid in the heat exchanger. Because of the resistance of its small inside diameter, the impedance tube lowers the pressure of the liquid. Low-pressure liquid then enters the cooling coils wrapped around the water storage chamber, boils, and evaporates into a cold low-pressure gas. This evaporation cools the coils and removes heat from the water in the chamber. In the suction line the cold low-pressure gas removes some of the heat from the liquid in the impedance tube before it passes into the compressor shell. The cold gas passes over the compressor motor, cooling it, before entering the compressor cylinder again.

On compartment models of pressure-type water coolers, all the liquid is not evaporated in the water-cooling coils. Both low-pressure liquid and gas pass from the water-cooling coils to the restrictor valve. This is a nonadjustable, weight-loaded valve which lowers the pressure of the refrigerant passing through it approximately 25 lb/in^2. The lower-pressure liquid and gas pass into the compartment evaporator where the

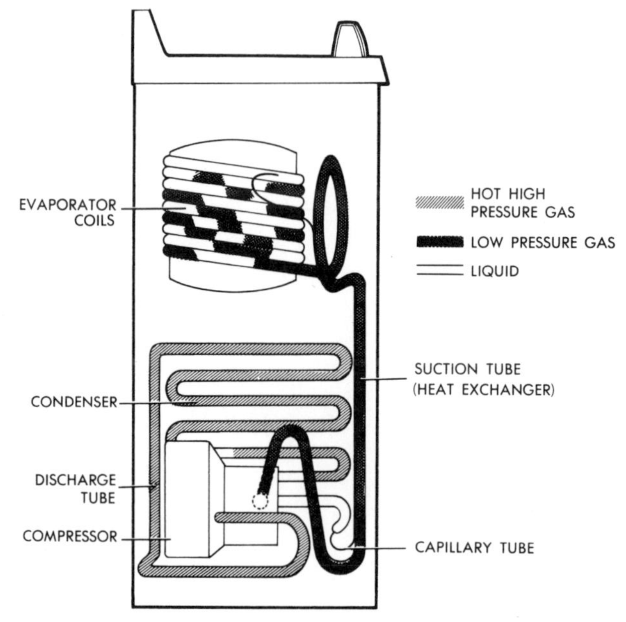

Figure 6-2. Refrigerant cycle of a typical pressure water cooler.

remaining liquid evaporates, thus removing heat from the compartment. A surge header prevents any remaining liquid from entering the suction line. The cycle is otherwise the same as on the other models.

Water-Cooling Cycle

City water enters the cooler through the inlet connection (inside cabinet), flows upward through the coiled tube of the precooler and into the top of the water-cooling chamber. It is directed down the refrigerated sides of the chamber for quick cooling. Cooled water is drawn from the bottom of the cooling chamber to the fountain valve and out the foutain orifice.

The height of the fountain (often called a *bubbler* by the trade) stream is set at the factory, and is controlled automatically by a regulator built into the fountain or bubbler valve. The regulator maintains the height of the stream relatively constant for water pressures between 20 and 100 lb/in^2 and may be adjusted if desired.

Air, which enters the cooling chamber along with the water, collects in the top of the chamber. As cooled water is drawn off, either through the bubbler orifice or water outlet fitting, it lowers the pressure at the end of the air purge tube. Full water line pressure in the chamber forces air out from the top, through the air purge tube to the point of lower pressure, from which it passes out the bubbler orifice with the water.

Waste water from the fountain flows through the drain opening in the basin top of the cooler to the precooler, where it cascades over the coiled water baffle. This baffle carries it around the precooler wall, cooling the incoming water. From the precooler the waste water flows out to the drain line.

When a side fountain is used, the water flows through the chamber drain connection, through a tube which supplies an adjustable stream height regulator, through a push-button valve, and from the valve to the bubbler.

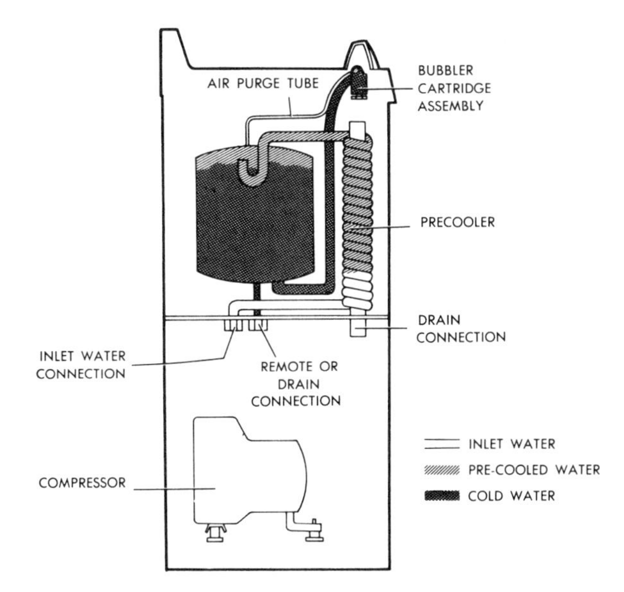

Figure 6-3. Water-cooling cycle of a typical pressure water cooler.

Electric Circuit

The principal components of the electric circuit are the compressor motor, motor thermostat, water chamber temperature control, and water cooling circuit. Most models also contain a compressor motor to aid in the cooling operation.

Compressor motor. When the temperature of the control bulb rises to a predetermined temperature, the control main contacts close, completing the circuit through the compressor motor main winding and relay coil. As a rule, the motor cannot start on the main winding alone and draws heavy current. This current passing through the relay coil creates a strong magnetic field which lifts the relay plunger, closing the relay contacts, and completing the circuit through the auxiliary winding. Because of the phase angle between the auxiliary and main windings, the motor has sufficient torque to start. As the motor comes up to speed, the current drawn decreases, thus decreasing the strength of the magnetic field in the relay until the weight of the relay plunger causes it to drop, opening the relay contacts and circuit to the auxiliary winding. The motor continues to run on the main winding only, until the temperature of the control bulb is lowered to a

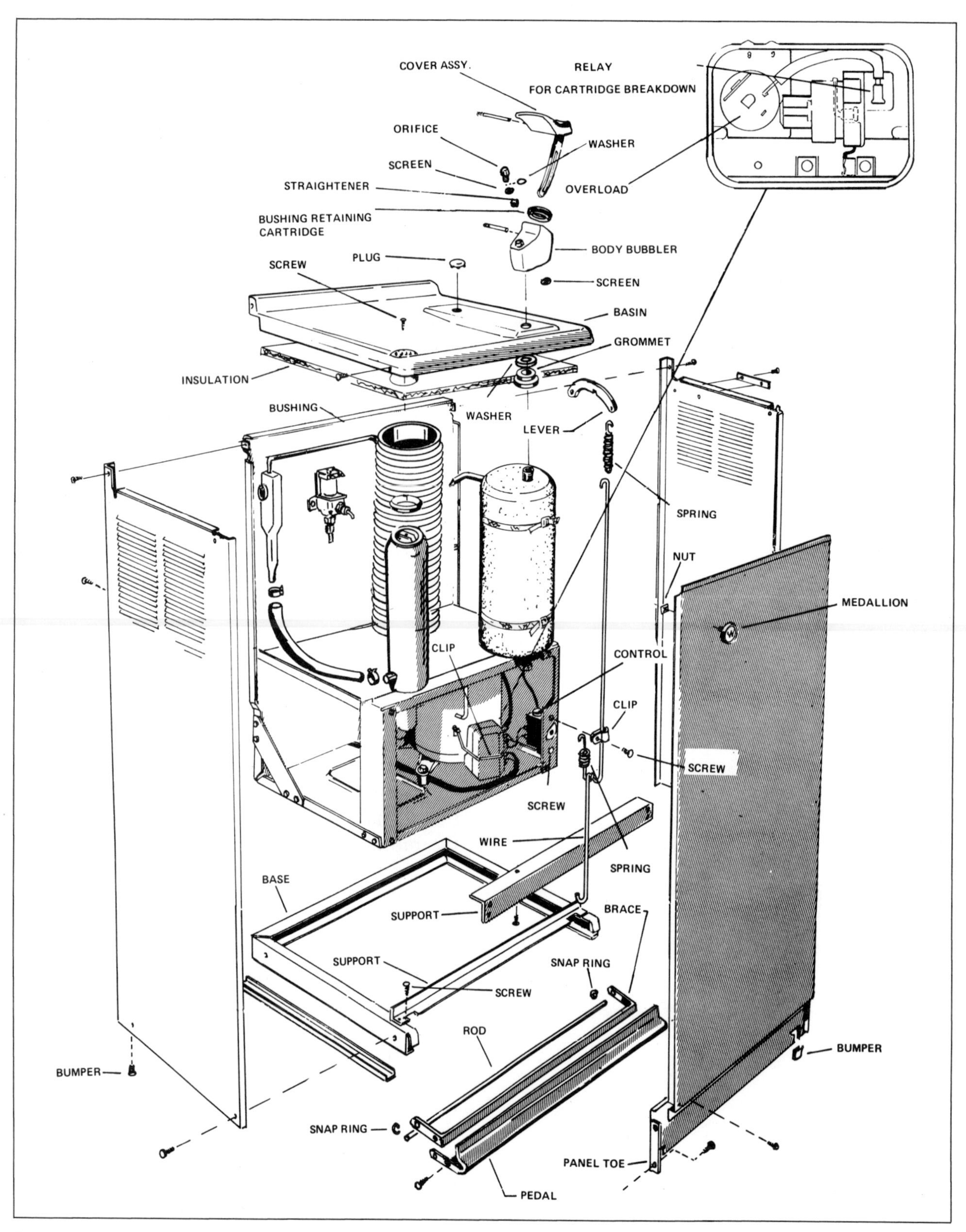

Figure 6-4. Components of a typical pressure water cooler.

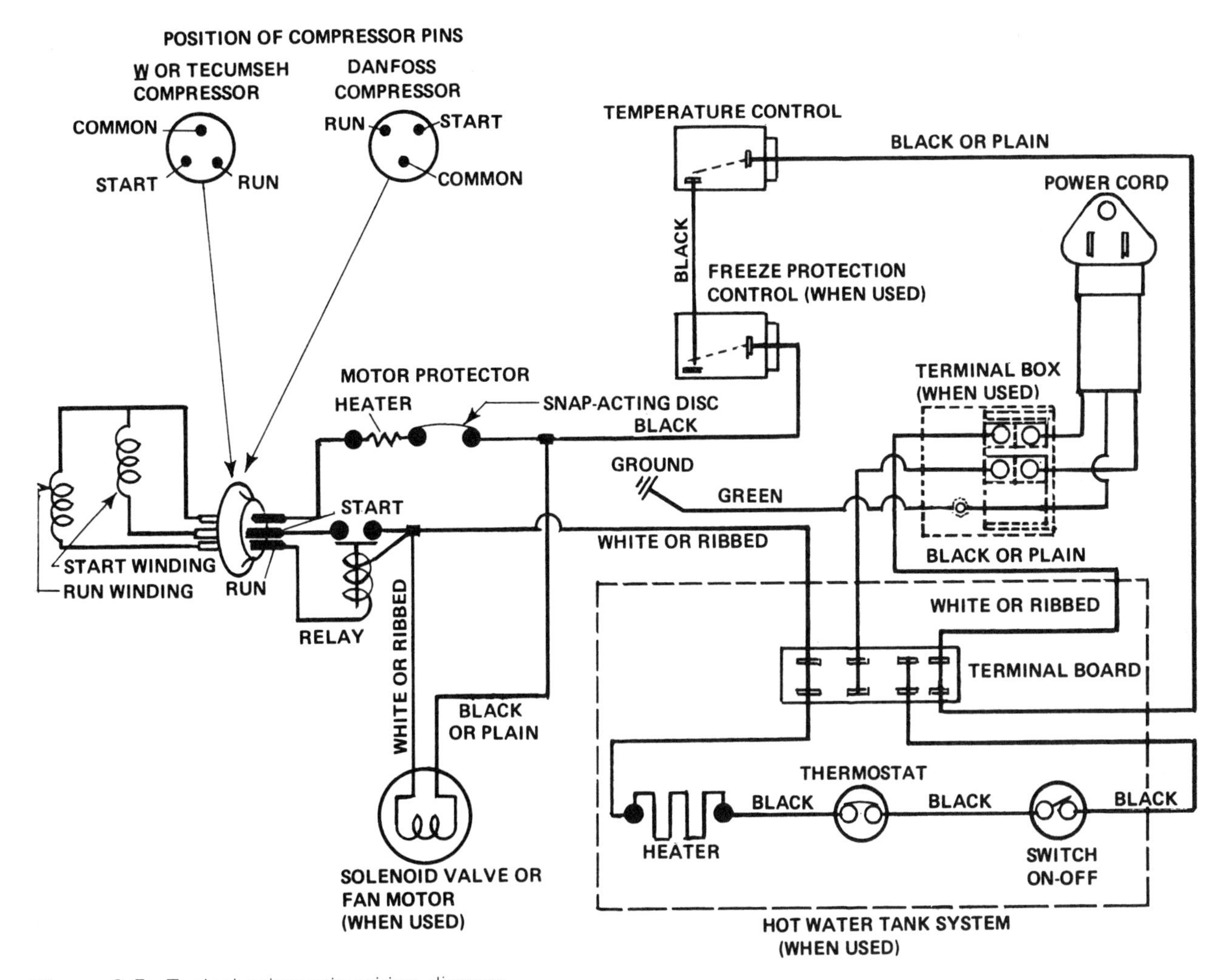

Figure 6-5. Typical schematic wiring diagram.

predetermined temperature and the control main contacts open.

Fan motor. The fan motor is usually a single-phase, two-wire, shaded-pole motor connected across line in parallel with the compressor motor main winding. It starts and runs with the compressor motor when control contacts close.

Motor thermostat. Current flowing through the compressor motor windings also flows through the small heating element in the motor thermostat, an automatic reset circuit breaker, mounted on the compressor motor housing. The thermostat will open the circuit if current flow through the motor windings is excessive or if the compressor motor overheats, thus protecting the motor from damage. It closes the circuit automatically after a short time. If an abnormal condition continues, the thermostat will continue to cycle, opening and closing the circuit.

Water chamber temperature control. The water chamber control is the main control; when the contacts are closed, current will flow through the control allowing the compressor and fan motor to operate. With most models, when the control bulb reaches approximately 40°F, the contacts will open and stop the flow of current to the compressor and compressor fan motor. Most water coolers try to maintain average water temperature of about 45°F.

Water-cooling circuit. Power is usually fed

in through the water chamber control contacts. When these contacts are in the closed position, the current will flow through the relay coil to the running windings of the compressor, then through the running windings of the compressor out through the compressor thermostat to the other side of the line. The current passing through the running windings of the compressor will create high current draw in the starting relay, causing the relay to energize and completing a circuit through the starting winding of the compressor motor. Then there will be a circuit through the control, through the relay contacts, through the starting windings of the compressor, and out through the thermostat to the other side of the line. Again simultaneously, as soon as the compressor reaches approximately three-fourths running speed, the current draw will drop, allowing the relay to de-energize and breaking the circuit to the starting windings; the compressor will, therefore, run with current passing through the running windings only. Power for the compressor fan motor circuit flows through one side of the line, through the water chamber control, through the relay coil, and through the compressor fan motor. From the compressor fan motor the current flows to the other side of the line.

Hot-water tank thermostat. On coolers with a hot-water tank, the thermostat in the tank is usually set at the factory to provide hot water at approximately 180°F. The thermostat is usually located on the tank and has some type of adjustment arrangement so that the temperature of the water may be lowered or raised. While a few models have a dial, most have an adjusting screw which turns counterclockwise to lower the water temperature and clockwise to raise it. About a one-eighth turn will usually result in approximately a 12°F change in temperature.

The actual operation and servicing problems of the hot-water tank unit are the same as for the instant hot-water dispenser described in detail in Chap. 5 of *Major Appliance Servicing*, in this series.

Cooler Maintenance

Accumulations of lint and dust on the condenser will reduce its cooling capacity and affect the economy of operation. The cooler should be kept free of such accumulations by the use of a vacuum cleaner, stiff-bristle brush, or a supply of compressed air. Before attempting any cleaning, be sure to remove the service cord plug from its outlet.

Clogged or partially clogged water lines may cause insufficient fountain or bubbler stream or no stream. To determine the location of the clog and remove it, proceed as follows.

1. Shut off water inlet valve.
2. Remove remote or chamber drain fitting plug or disconnect remote fountain line.
3. Attach a pressure gauge to this fitting (usually 3/8-in female pipe), open the water inlet valve, and observe the reading on the gauge, or make provision for catching the water, open water inlet valve (with no gauge), and observe the flow of water. If no pressure is read on the gauge or no water flows out the fitting, the water lines are plugged in the inlet line, precooler, water chamber, or outlet line from the chamber to the remote fitting. If full line pressure is read on the gauge, open the fountain or bubbler valve by push button or foot pedal. If the pressure reading drops more than 2 or 3 lb/in^2, the water lines are partially plugged in above-mentioned parts. Shut off the water inlet valve, remove the gauge, and turn the valve on again to flush the lines with full line pressure, making provision to catch the water at the remote or drain fitting. Inserting a flexible wire into the fitting may also clear the obstruction in the tube from the chamber. If still clogged, shut off the inlet valve, disconnect the inlet water line and use compressed air to blow out the lines, first at inlet fitting, then at the remote or drain fitting.
4. If full line pressure is read at the remote fitting with no pressure drop when the

fountain valve is opened, or if a full flow of water is obtained, shut off the water inlet valve, remove the gauge, and replace the remote fitting plug.

5. Remove the fountain valve assembly and open the water inlet valve, making provision to catch the water at the valve body. If no water or only a small stream of water flows from the valve body, the obstruction is in the line from the chamber to the valve body. If full line pressure does not free the obstruction, shut off the water inlet valve and use compressed air at the inlet or remote connection.
6. If a full flow of water is obtained from the valve body, the clog may be in the fountain valve assembly. Disassemble and clean the bubbler valve assembly and fountain or bubbler orifice. Water lines should be protected from foreign material by the use of an accessory line strainer.

To purge air from the cooling chamber, all that is usually necessary is to depress the push-button bubbler or fountain. This will permit a flow of water and air mixed. Place a hand or cover over the bubbler to confine the water until spurting stops.

The drain line may become clogged by accumulations of algae and slime which act as binders for dirt, verdigris, and other solid material. To loosen this vegetable matter and flush out insoluble material, proceed as follows.

1. Remove the basin top and drain baffle.
2. Disconnect the trap or drain line from the drain connection.
3. Remove as much solid foreign material as possible from the top and bottom of the drain.
4. Plug the drain outlet with a rubber stopper or wooden plug.
5. Prepare a solution of Oakite composition No. 22, 4 oz/gal of hot water, and pour it into the drain. Let the solution stand for about 30 min.
6. If the drain outlet is not accessible, plug the top of drain immediately after pouring solution to prevent entry of air and to retard the flow of solution down the drain.
7. Repeat step 5 or 6, if necessary. Then flush the drain with water; preferably pressure-flush it by hose connected to the water supply line.
8. Replace the drain baffle, basin top, and drain trap or drain line.

Heavy hard-water deposits may be removed by nitric acid (15 to 20 percent by volume) or a mixture of vinegar and water. This treatment should be followed by a thorough water rinse. To polish, apply a commercial cleanser or metal polish as directed on the container.

Trouble Diagnosis for Pressure Water Coolers

Here are the major problems, possible causes, and solutions that may arise with the pressure type of water cooler.

Problem: Drinking water is warm; compressor is not operating.

Possible cause	*Solution*
1. Power supply cut off.	1. Check power supply at outlet.
2. Poor plug contact at outlet.	2. Spread plug contacts.
3. Temperature control main contacts open.	3. Adjust control setting. Check mechanical action of control. If adjustment does not correct trouble and mechanical action is all right, replace control.
4. Open circuit or ground in electric connections.	4. Check electric circuits. Repair defect.
5. Compressor motor thermostat open.	5. Check thermostat. Replace.

Possible cause	Solution
6. Open windings, compressor motor.	6. Check compressor. Replace unit or cooler.

Problem: Drinking water is warm; compressor is operating.

Possible cause	*Solution*
1. Cooler overload: water being drawn too fast.	1. Refer to water-cooling capacity tables. Reduce load.
2. Condenser clogged with dust or lint.	2. Clean condenser.
3. Poor air circulation around condenser.	3. Allow ample space around cooler for air circulation.
4. Condenser fan motor not operating.	4. Check fan. Repair or replace.
5. High or low line voltage.	5. Check line voltage. Must be between 100 and 125 V. If intermittent, provide new supply. If steady, install transformer.
6. Starting relay stuck closed.	6. Repair or replace relay.
7. Short circuit or ground in electric circuit.	7. Check electric circuit. Repair defect.
8. Motor thermostat out of calibration.	8. Replace thermostat.
9. Defective compressor motor.	9. Check compressor. Replace unit or cooler.
10. Refrigerant line restricted.	10. Check for damaged or bent tubing. Straighten tubing or replace unit or cooler.
11. Stuck compressor.	11. Replace unit or cooler.

Problem: Drinking water is too cold.

Possible cause	*Solution*
1. Temperature control set too low.	1. Adjust control (check service manual).
2. Temperature control main contacts fused together.	2. Separate contacts. Clean with fine sandpaper or magneto file or replace contact assembly.
3. Temperature control inoperative.	3. Check mechanical action of control.
4. Control bulb not properly inserted in well.	4. Insert control bulb fully into well and seal opening.

Problem: Unit has long running time.

Possible cause	*Solution*
1. Improper voltage.	1. Check voltage. Must be 120 V $\pm$ 10 percent.
2. Refrigerant line restriction.	2. Find restriction and repair.
3. Loss of refrigerant charge.	3. Locate leak and repair. Recharge.
4. Unit overcharged.	4. Purge refrigerant charge. Evacuate and recharge.
5. Faulty compressor.	5. Replace compressor.

Problem: Fountain or bubbler stream is too high.

Possible cause	*Solution*
1. Water pressure too high.	1. Check pressure; if over 100 lb/in^2, install pressure-reducing valve.
2. Stream height regulator out	2. Adjust regulator according to service manual.

Possible cause	Solution
of adjustment.	
3. Stream height regulator inoperative.	3. Disassemble fountain valve assembly. Clean and adjust or replace parts as necessary.

Problem: Fountain stream is too low.

Possible cause	*Solution*
1. Water pressure too low.	1. Check pressure; should be at least 20 lb/in^2. Install accessory low-pressure orifice.
2. Stream height regulator out of adjustment.	2. Adjust regulator according to service manual.
3. Stream height regulator inoperative.	3. Disassemble fountain valve assembly.
4. Water lines partially clogged by foreign material.	4. Clear clog.
5. Fountain valve does not open wide.	5. Check operation of fountain valve actuating lever. Adjust fountain valve.
6. Water inlet line valve not fully open.	6. Open valve wide.

Problem: No water available from bubbler.

Possible cause	*Solution*
1. Water inlet line valve closed.	1. Open valve wide.
2. Water lines clogged by foreign material.	2. Clear clog.
3. Fountain valve does not open.	3. Check operation of fountain valve actuating lever. Adjust fountain valve according to service manual.
4. Fountain valve inoperative.	4. Disassemble fountain valve. Clean or replace parts as necessary.

Problem: Water flows continuously from fountain.

Possible cause	*Solution*
1. Fountain valve does not close.	1. Check operation of fountain valve actuating lever. Check foot-pedal rod for binding in clamp or against front panel and for hook being out of hole in actuating lever. Adjust fountain valve according to service manual.
2. Fountain valve bottom gasket defective.	2. Replace gasket.
3. Fountain valve return spring defective.	3. Replace return spring.
4. Fountain valve disk worn.	4. Replace valve disk carrier assembly.
5. Fountain valve seat defective.	5. Replace cup assembly.
6. Foreign material between fountain valve seat and disk.	6. Disassemble and clean fountain.

Problem: Fountain's water stream is erratic.

Possible cause	*Solution*
1. Regulator not controlling stream height.	1. Check regulator parts. Replace, if necessary.

2. Air in water cooling chamber or tubes.	2. Purge air.
3. No straightener in fountain orifice.	3. Replace fountain orifice.

Problem: Unit operation is noisy.

Possible cause	*Solution*
1. Tubing rattle.	1. Bend tubing carefully to eliminate noise.
2. Fan blade hitting shroud or condenser.	2. Adjust for clearance. Check fan motor mounting.
3. Fan blade bent.	3. Replace blade.
4. Fan motor excess end play.	4. Replace motor.
5. Compressor mounting stud washers rattle.	5. Remove cotter pin and flat washer. Remove and replace spring washer with concave side up, or add spring washers to take up play.
6. Incorrect compressor mounting.	6. Straighten mounting studs if bent. Carefully bend tubing to allow compressor to ride freely with studs centered in holes in compressor feet. Replace rubber grommets if worn, or springs if resiliency lost. Replace unit or cooler.
7. Internal compressor noise.	7. Straighten mounting studs if bent. Carefully bend tubing to allow compressor to ride freely with studs centered in holes in compressor feet. Replace rubber grommets if worn, or springs if resiliency lost. Replace unit or cooler.

Problem: Water appears rusty or has a metallic taste.

Possible cause	*Solution*
1. Bad water.	1. Have water source checked for bad water.

Problem: No hot water available (on coolers equipped with hot-water system).

Possible cause	*Solution*
1. No power at hot water tank.	1. If the cold-water portion of the cooler is operating, power line between input terminal and tank must be broken.
2. Faulty hot-water tank switch.	2. Check and replace if necessary.
3. Faulty hot-water tank thermostat.	3. Replace thermostat.
4. "Open" hot-water heater.	4. Replace hot-water heater unit.

More information on the diagnosis of problems that may occur to the hot water portion of the cooler can be found in Chap. 5, *Major Appliance Servicing*, in this series.

As with all refrigerant systems—those in refrigerators, freezers, room air conditioners, and humidifiers—any time the line must be opened, the system must first be evacuated, and after the repairs have been made, it must be recharged. Methods of evacuating a cooler's refrigerant system, checking for a leak, cutting a suction tube, recharging a system, replacing a compressor, etc., are the same as for a refrigerator's refrigerant system (see Chap. 2).

OPERATION OF BOTTLE WATER COOLERS

The major differences between the pressure and bottle types of water coolers are as follows.

1. A bottle of water located on the top of unit is the source of supply rather than the city water line used with the pressure type.
2. A waste-water receptacle is located in front of the unit, and it replaces the drain waste of the pressure cooler. The capacity of this receptacle is usually about 1 qt, and the receptacle lifts off for emptying. Thus with the bottle cooler no outside plumbing is needed.
3. A push-button faucet is employed to deliver the water rather than the bubbler fountain that is found in most pressure coolers. (As a rule, cold-water faucets have blue buttons, while hot-water faucets usually have red ones.)

Except for these physical differences, the refrigeration cycle, the water cooling cycle, and the electric circuit are about the same. In other words, while the parts may vary in size (the water cooling chamber is usually larger in bottle coolers) and in location (the water outlet is located in the front of the unit rather than on the top as it is with the pressure type), the cooling and delivery of the water is the same. Only the source of water is different.

Trouble Diagnosis for Bottle Water Coolers

Because of the similarity of the two types of water coolers, the trouble diagnosis information given earlier in the chapter for pressure coolers

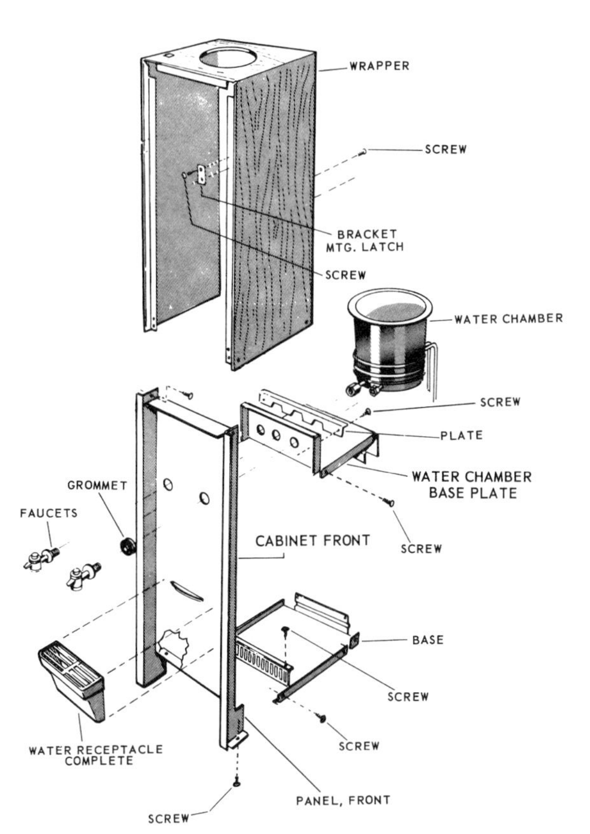

Figure 6-6. Water chamber and base plate of a typical bottle water cooler.

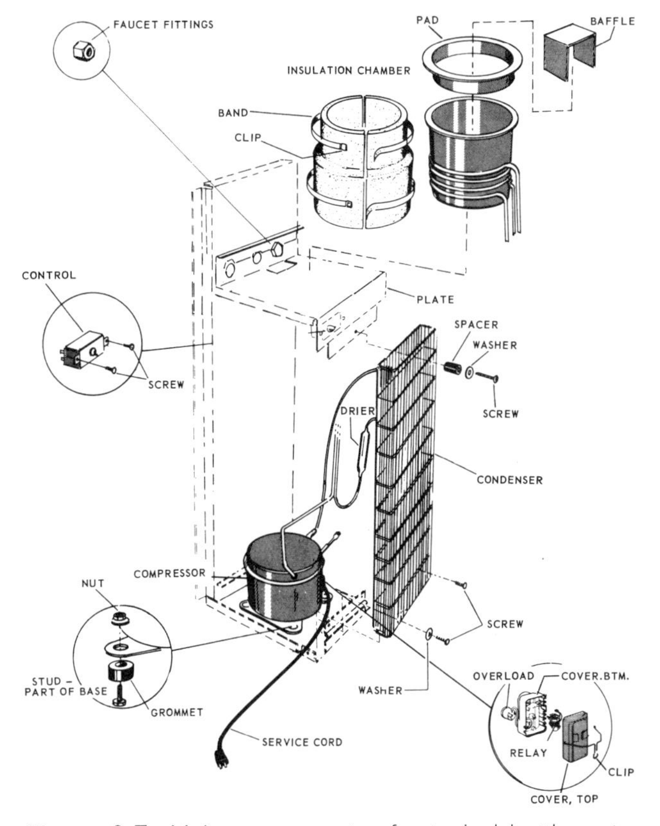

Figure 6-7. Major components of a typical bottle water cooler.

holds true for the bottle types, except most faucet troubles must be substituted for those of the fountain. Because of this and rather than just repeating the information, we have included Table 6-1, giving trouble diagnosis for bottle water coolers. (Many manufacturers use a system such as this in their service manuals to explain their trouble diagnosis procedures.) To use this chart, first locate the description of the problem on the left-hand side of the chart. This description should be a verification of the customer's description of the problem. When the problem is verified, locate the possible malfunctions (listed across the top) that can cause the problem. Check each of these by numerical sequence (No. 1 first, No. 2 second, etc.). Combining this information with the data given earlier for pressure coolers, you should be able to diagnose any problems that should occur with bottle-type coolers.

Table 6-1. Bottle water cooler trouble diagnosis chart.

These problems (below) can be caused by failure of these parts (at right). (Check parts in order shown.)	Power supply	Water temperature control	Motor protector (thermostat)	Compressor	Cooler overloaded	Condenser	Refrigerant shortage	Unit pressures not equalized	Starting relay	Liquid line restriction	Control bulb not in well	Clogged water lines	Tubing rattle	Defective faucet seat washer	Compressor mounting	Refrigerator overcharge	Bad water	Faucet	High or low voltage	Electric circuit	Broken compressor valves	Hot tank thermostat	Hot tank heater	Hot tank switch	Air lock in hot tank
	1	2	3	4	5	6	7	8	9	10	11	12	13	14	15	16	17	18	19	20	21	22	23	24	25
Drinking water warm, compressor not in operation	1	2	4	5					3											6					
Drinking water warm, compressor operating		3			1	2	4														5				
Drinking water warm, compressor operates on motor protector	9		6	7		1		2	4	8								3		5					
Drinking water too cold		1					3				2					4									
No water from cold faucet												1						2							
Continuous flow from faucet														1											
Unit runs too long		3			1	2	4									5									
Unit operation noisy				4									1		2	3									
Dirty-appearing hot water																								1	
Bad or metallic taste to water																	1								
No hot water (coolers equipped with hot-water system)	4																					1	2	3	4

Electronic air cleaners and air cleaner–humidifiers

CHAPTER

7

Until recently, the need for removing dirt from the air has been associated with convenience—improved comfort, lower cleaning bills, or general industry requirements for cleanliness. The rapidly growing problem of air pollution is changing the role of air cleaning into a "must" category. For this reason, the service technician is starting to get calls for servicing of electronic air cleaners.

AIR CLEANERS

Most electronic home air cleaners on the market today use a two-stage electrostatic precipitator, the most practical means known for collecting airborne particles and contaminants. Basically all electronic air cleaners have three functional sections: a control system to operate the electronic air cleaner (EAC); a power pack to supply high-voltage direct current; and a cell for precipitating out the dirt particles.

A two-stage electrostatic precipitator has a two-stage electronic cell. The first (ionizing) section creates a powerful electric field that places a positive charge on particles entering the cell, while the second (collecting) section has an alternating series of grounded and positive plates. The grounded plates attract and hold the positively charged (ionized) particles, while the positive plates push the positive particles toward the grounded plates.

A two-stage electrostatic precipitator operates in the following manner. Particles carried by the circulating air pass into the EAC. The particles then enter a powerful electric field (set up between a series of ionizing wires) where they are given an intense positive electric charge. When the particles reach the collecting section, they enter another electric field set up between a series of metal plates carrying alternate positive and negative charges. The positively charged particles are attracted and hurled against the negative plates just as iron filings are attracted to a magnet. The charged particles stay on the collector plates until washed off. Normally, the collected particles include tobacco tar, cooking fats, and other viscous contaminants which act as an adhesive, making it necessary to use hot water and effective detergent to wash the dirt from the ionizer and collector surfaces. A most important feature of the electronic air cleaner is its ability to achieve a high degree of air cleaning efficiency without introducing excessive resistance to airflow. The actual inlet-to-outlet pressure drop across the EAC is minimal. The efficiency of an electronic air cleaner increases with lower air velocity and decreases with higher velocity.

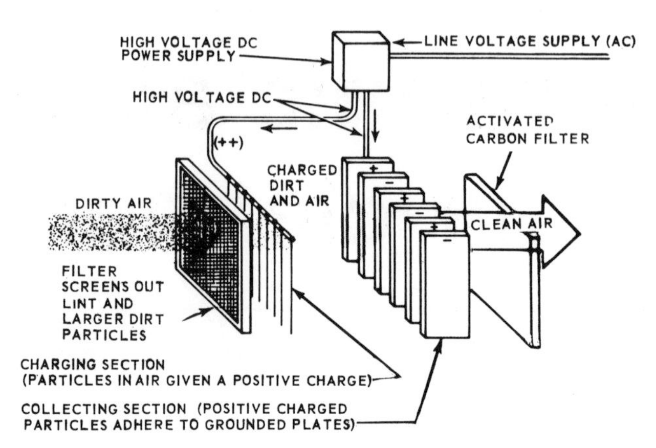

Figure 7-1. Operation of a two-stage electrostatic precipitator.

Usually, a two-stage electrostatic precipitator (Fig. 7-1) is constructed in two sections: a charging (or ionizing) and a collecting section. The charging section consists of a series of fine wires suspended between metal plates; the collecting section is a series of parallel flat metal plates spaced about 1/4 in apart. In practice, a small amount of collecting actually occurs in the charging section of the two-stage EAC, but most particles are carried through the charging section by the airstream to the collecting section where they are attracted to the charged collecting plates. Keep in mind that the airborne particles in either case are first given an electric charge, then they are collected on plates which have an opposite charge in relation to the particles. (Like charges repel, unlike charges attract.)

How an Electronic Air Cleaner Works

As shown in Fig. 7-2, room air is drawn by the fan into the air intake louvers. (This may happen in the front or back of the unit, depending on the design.) As air passes through the three stages of filtering, it is cleaned as follows.

Prefilter. This first stage removes much of the large visible-to-the-eye matter, mostly lint. This

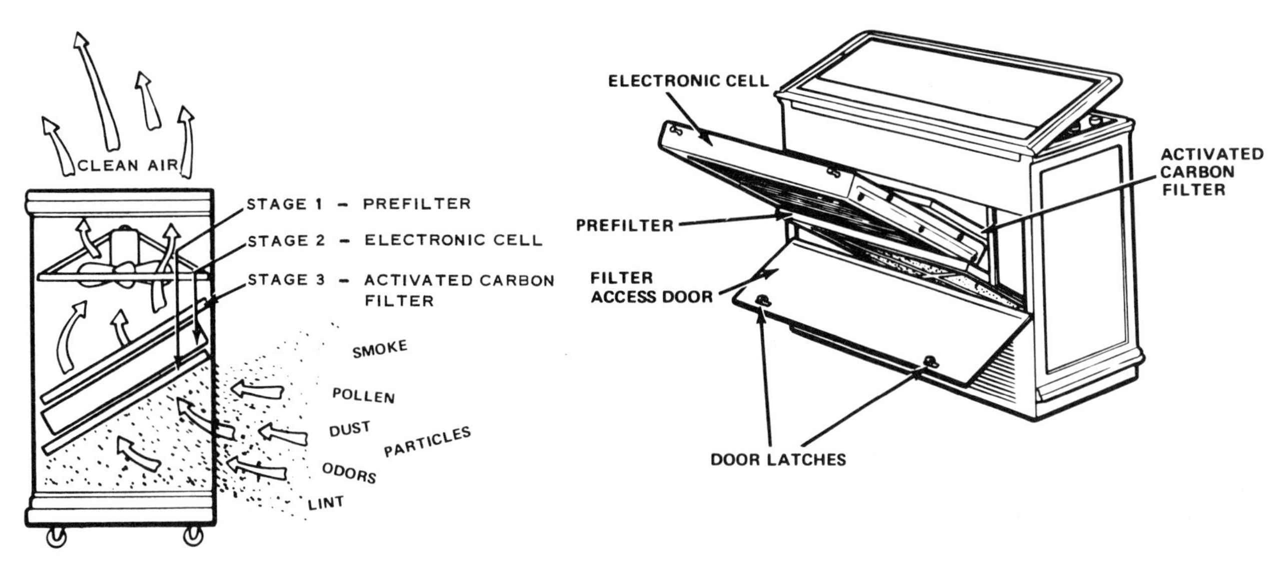

Figure 7-2. How the console electronic air cleaner works: (1) the prefilter; (2) the electronic cell; and (3) the activated carbon filter. The illustration at right shows parts placement.

reduces the frequency of cell cleaning and also minimizes arcing, which produces a sharp, snapping sound. Arcing is a normal occurrence which can result when the cleaner collects unusually large particles. Under abnormal operation, continuous or frequent arcing will occur under the following conditions.

1. Unusually large particles lodged in the electronic cell.
2. Heavy dirt deposits in the cell.
3. Dampness after washing the cell.

Any continuous arcing not caused by these conditions or dirt accumulation usually indicates a malfunction.

Electronic cell. As airborne particles enter the second stage, they are given an intense electric charge. As these charged particles continue through the cell, a series of metal plates, having an opposite electric charge, attract and collect them in a way similar to the way a magnet attracts iron filings. The metal plates hold the particles until washed off.

Activated-carbon filter. In the final cleaning stage, air passes through the activated-carbon filter; this reduces undesirable household odors and ozone. The latter is a fresh, pleasant odor similar to that experienced out of doors after an electrical storm. The strong electric field in the air cleaner similarly creates a trace of ozone usually noticeable only when entering the space. If the air cleaner is operated in a small, confined space, the ozone odor may be objectionable. Opening the doors to adjoining rooms will reduce the ozone odor. If the strong or irritating ozone odor persists, the air cleaner can be regulated.

Control Operation

While each manufacturer has a slightly different arrangement of controls, they are all similar. For instance, the functions of the *power control* are the following.

OFF: Stops all fan and filtering action of the unit.

ON: Energizes both the fan and electronic cell for normal air cleaning operation.

DRY: Energizes only the fan to dry the de-energized cell after washing. This setting may also be used for air circulation only without electronic air cleaning.

The *speed control* regulates the fan speed to control the air volume passing through the air cleaner. Factors to be considered in selecting the best fan speed are noted below.

1. *Area to be cleaned.* To effectively clean a space of approximately 8,000 ft^3 (about 1,000 ft^2 of floor space), the air cleaner should be operated on HIGH. For smaller areas or quieter operation, the speed control may be set proportionally lower.
2. *Desired operating efficiency.* The air cleaner's efficiency in removing airborne particles (other than pollen*) is influenced by the air volume passing through the electronic cell. Maximum cleaning efficiency is obtained by setting the speed control at LOW. It must be considered, however, that operating the speed control at LOW will circulate the air through the air cleaner less frequently, and consequently the unit will not effectively handle as large an area.
3. *Individual preference* (regarding air movement and sound level). The higher-speed control settings will produce a faster movement of air through the unit and consequently a more noticeable sound level; therefore, a lower-speed setting may be used for nighttime operation or whenever reduced air movement is desired.

An indicator light is usually located on the control panel to signal that sufficient power is being delivered by the power pack for air cleaning. The indicator light is not energized when the power control switch is set at DRY or OFF or when the electronic cell is receiving inadequate power.

FUNCTION OF THE PREFILTER

As previously described, the function of the prefilter is to remove the large particles, such as lint, before they enter the electronic cell. This reduces the frequency with which the cell must be cleaned and minimizes arcing. The accumulation of lint or other large particles will hinder the airflow through the unit and reduce the unit's efficiency. It is, therefore, important to clean the prefilter about once each month.

To clean the prefilter, the following steps should be taken.

1. Vacuum the filter thoroughly to remove most of the accumulated lint and dust. Then immerse the filter in a solution of water and mild detergent and agitate it to loosen any remaining lint and dust.
2. Rinse the filter thoroughly with clean water, shake off excess moisture, and let dry.
3. Replace filter in unit.
4. Vacuum the air inlet louvers in the lower back panel.

Cleaning the Electronic Cell

To keep the home electronic air cleaner operating at peak efficiency, it is recommended that the electronic cell be cleaned about every 2 months. Since there are naturally more particles, dust, smoke, etc., in some locations than others, it is advisable that an inspection be made every few weeks until some idea of the normal cleaning interval is established. When the filters and electronic cell look dirty, they require cleaning. If an excessive amount of particles are allowed to build up on the electronic cell, very rapid and frequent arcing may occur which indicates that cleaning is required.

Before the cell is washed, remove the internal particles by vacuuming or brushing the protective screen on each side. Hold it in a vertical position to prevent dirt and dust from falling into the interior of the unit. Do not poke through the screen at internal parts of the electronic cell.

To clean the cell, the following steps usually are taken.

1. Use a laundry tub or any container large enough to completely immerse the electronic cell. Mix hot water (130 to 140°F) with a

* Pollen particles are larger than most airborne matter, and consequently the unit's efficiency for pollen removal is not appreciably influenced by the volume of airflow through the electronic cell.

dishwater detergent in a ratio of 4 tbs of detergent to 1 gal of water. *Caution*: Avoid prolonged contact on the skin or splashing the eyes with detergent solution. Keep detergent out of reach of children.
2. After making sure the detergent is dissolved, immerse the cell completely and soak for 15 to 20 min.
3. After soaking, agitate the cell vigorously from side to side and up and down in the solution. Turn the cell over and repeat sloshing until it looks clean.
4. Rinse the cell thoroughly in clean, hot water, moving it from side to side and up and down. Rinse it again in fresh water.
5. Shake as much water out of the cell as possible, then stand it on a long edge, or one of the corners so that the red terminal board is up. Allow it to drain for approximately 15 min, then shake it again and wipe away any remaining water.
6. Replace the electronic cell in the unit and close the access door.
7. Plug the power cord into the electric receptacle and turn the power control to DRY position. Allow the unit to operate on DRY for a minimum of 2 h at high fan speed. The electronic cell must be dry to prevent arcing.
8. Place the power control in ON position. Electrical arcing may occur for a few minutes. If it continues, the electronic cell is not completely dry. Place the control at DRY and allow the unit to run for another hour, then try cell operation again.

Cleaning the Activated-Carbon Filter

The activated-carbon filter is usually located just above the electronic cell. This filter should be vacuumed or brushed to remove surface dust if the unit has been operated for air circulation only. But, as a rule, it is not wise to wash the activated-carbon filter.

The activated-carbon filter is expendable and should be replaced at least once every year. Under heavy smoke conditions, the life of the carbon will be shortened. Replacement is recommended when the filter ceases to reduce odors.

Trouble Diagnosis

Listed below are some common causes of malfunction and their remedies. If these corrective measures do not restore the air cleaner to proper operation, follow a complete procedure in the typical troubleshooting chart below.

Problem: Air cleaner will not operate.

Possible cause	*Solution*
1. Control switch in OFF position.	1. Turn control to ON.
2. Power cord not plugged in.	2. Plug in power cord.
3. Blown fuse or tripped circuit breaker.	3. Replace fuse or reset breaker.
4. Filter access door open.	4. Close filter access door.
5. Charcoal filter removed.	5. Replace charcoal filter.

Problem: Small volume of air discharged from unit.

Possible cause	*Solution*
1. Dirty prefilter.	1. Clean according to instructions.
2. Activated-carbon filter not removed from polyethylene bag that it was shipped in.	2. Remove filter from polyethylene bag.
3. Speed control set at low number.	3. Set speed control at higher number.

Problem: Motor slows down or stops.

Possible cause	*Solution*
1. Motor bearing oil supply exhausted.	1. Oil motor bearings according to service manual instructions.

Problem: Unit does not effectively clean the air.

Possible cause	*Solution*
1. Control switch	1. Change control

1. (*cont'd.*)

set at DRY.	setting to ON.
2. Electronic cell dirty.	2. Wash according to instructions.
3. Electronic cell not completely dry after washing.	3. Operate unit on DRY cycle.

Problem: Continuous or frequent sound of arcing.

Possible cause	*Solution*
1. Excessive accumulation of dirt on prefilter.	1. Clean according to instructions.
2. Excessive accumulation of dirt in electronic cell.	2. Wash according to instructions.

Problem: Unit does not remove odors as effectively as it once did.

Possible cause	*Solution*
1. Service life of activated-carbon filter used up.	1. Discard and replace with new activated-carbon filter.

Problem: Ozone odor more pronounced.

Possible cause	*Solution*
1. Service life of activated-carbon filter used up.	1. Discard and replace with new activated-carbon filter.
2. Unit being operated in confined space.	2. Suggest opening doors to adjoining rooms.

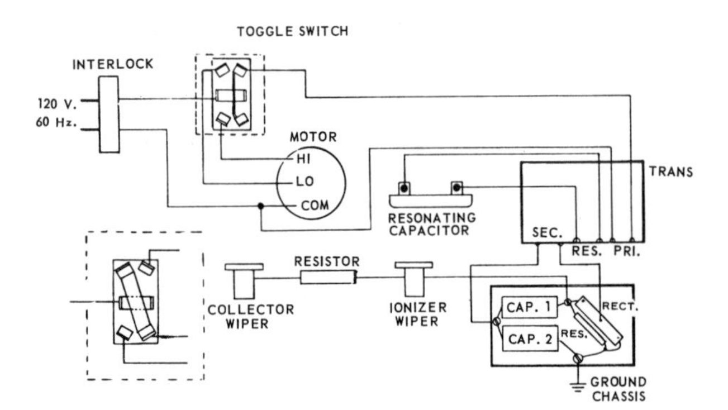

Figure 7-3. Typical high-voltage power supply.

A troubleshooting chart for the high-voltage side of a typical home electronic air cleaner is given on p. 205. This is only an example of how to proceed. The air cleaner's service manual will give exact details. (Most electronic air cleaners have interlock safety switches that automatically cut off power at the switches when the filter access panels or doors are opened.) To avoid the hazard of electric shock, disconnect power to the air cleaner before removing or replacing any internal components.

The chart tests were made on a 120 V ac

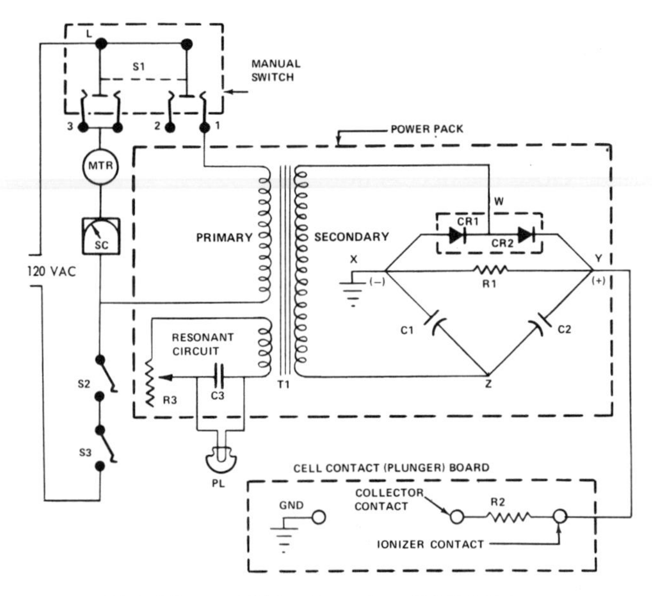

C1 & C2 = .05 Mfd. Capacitor, 5,000 VDC
C3 = .5 Mfd. Capacitor, 660 VAC
CR1 & CR2 = Dual Silicon Rectifier Unit
MTR = Motor
PL = Pilot Light
R1 = 29.4 Megohm Bleeder Resistor ± 10%
R2 = 22.6 Megohm Dropping Resistor ± 10%
R3 = 1500 Ohm Resonant Resistor
SC = Speed Control
S1 = Manual Switch (On-Off-Dry)
S2 = Filter Interlock
S3 = Door Interlock
T1 = Power Transformer

Figure 7-4. Schematic diagram of a typical electronic air cleaner. References in troubleshooting chart are to this schematic.

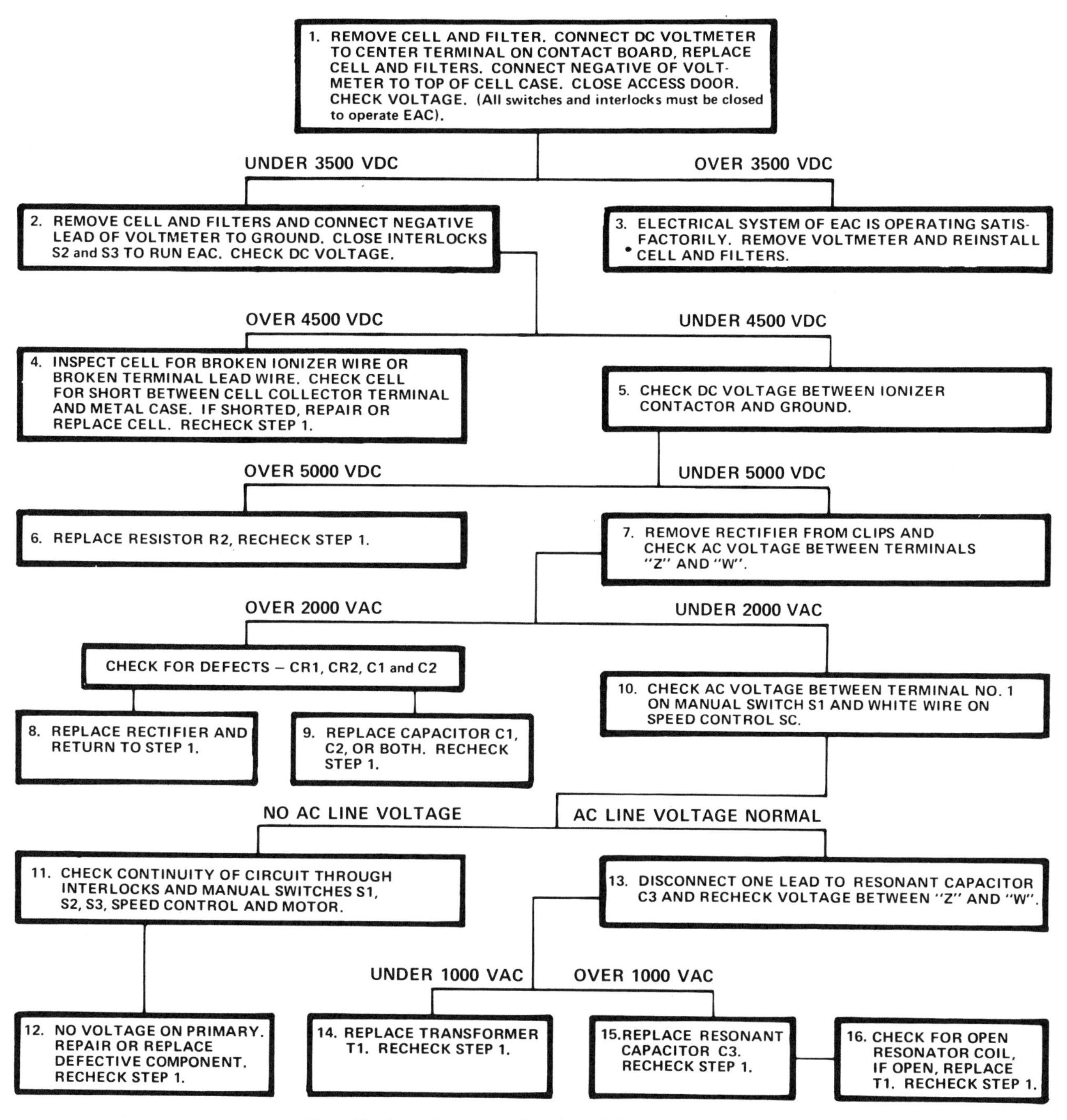

Troubleshooting chart for the high-voltage side of a typical home electronic air cleaner.

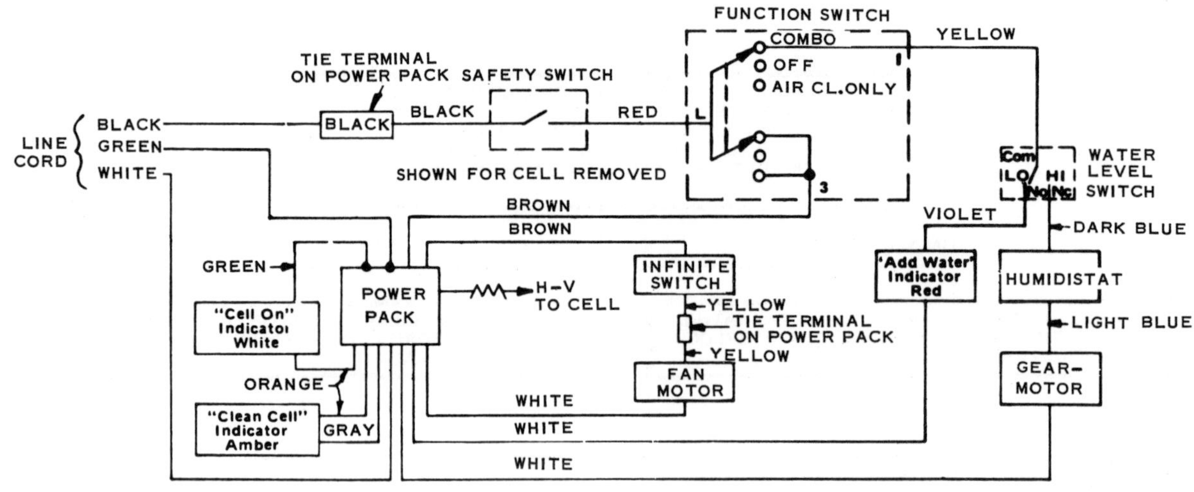

Figure 7-5. Wiring diagram of a typical electronic air cleaner–humidifier unit.

source; allow a slight increase or decrease for variance of voltage. A voltmeter with high-voltage probes in the range of 10,000 V for both alternating and direct current is essential for using this chart and for diagnosing trouble in the home electronic air cleaner. When using the typical troubleshooting chart shown here, refer closely to the schematic diagram given in Fig. 7-4.

ELECTRONIC AIR CLEANER AND HUMIDIFIER

Some manufacturers are producing a combination electronic air cleaner–humidifier. In such a unit, the fan draws room air through the electrostatic air cleaning filter cell and then through the evaporator pad which humidifies the air. Dust particles in the air entering the electrostatic filter cell are charged by high voltage on sawteeth located along edges of high voltage electrodes. As air is drawn through cell, the highly charged dirt particles are attracted to the grounded collector plates. The dirt accumulates on the collector plates. High voltage for the cell is produced by the power pack which contains circuitry for the CELL ON and CLEAN CELL sentinel lights (indicator lamps). Rotation of the drum keeps the exposed pad wet. Humidification is regulated by the humidistat which starts and stops the drive motor in response to changes in the ambient humidity. Rate of humidification varies with fan speed, which is usually controlled by the infinite-speed switch.

As shown in the typical wiring diagram (Fig. 7-5), the power from the line cord flows via the power pack tie terminals to a safety switch, which is operated by insertion of the filter cell. If the filter cell is out, or not properly engaged, unit will be "dead." From the safety switch, power goes to the three-position switch. In AIR CLEANER ONLY, the brown wire is connected to the line, powering the power pack and fan. In the HUMIDIFIER AND AIR CLEANER (COMBO), the brown and yellow wires are energized, powering the power pack, fan, and humidifier parts. The yellow wire on the three-position switch goes to the water level switch. When the water is high, yellow is connected with blue (the humidistat and drive motor). When the water is low, yellow is connected with violet (ADD WATER lamp). The brown wire from the power pack to the infinite-speed switch powers it and the fan. The yellow wire from the infinite-speed switch is tied to the fan motor at the power pack. All neutral connections are tied together at the power pack.

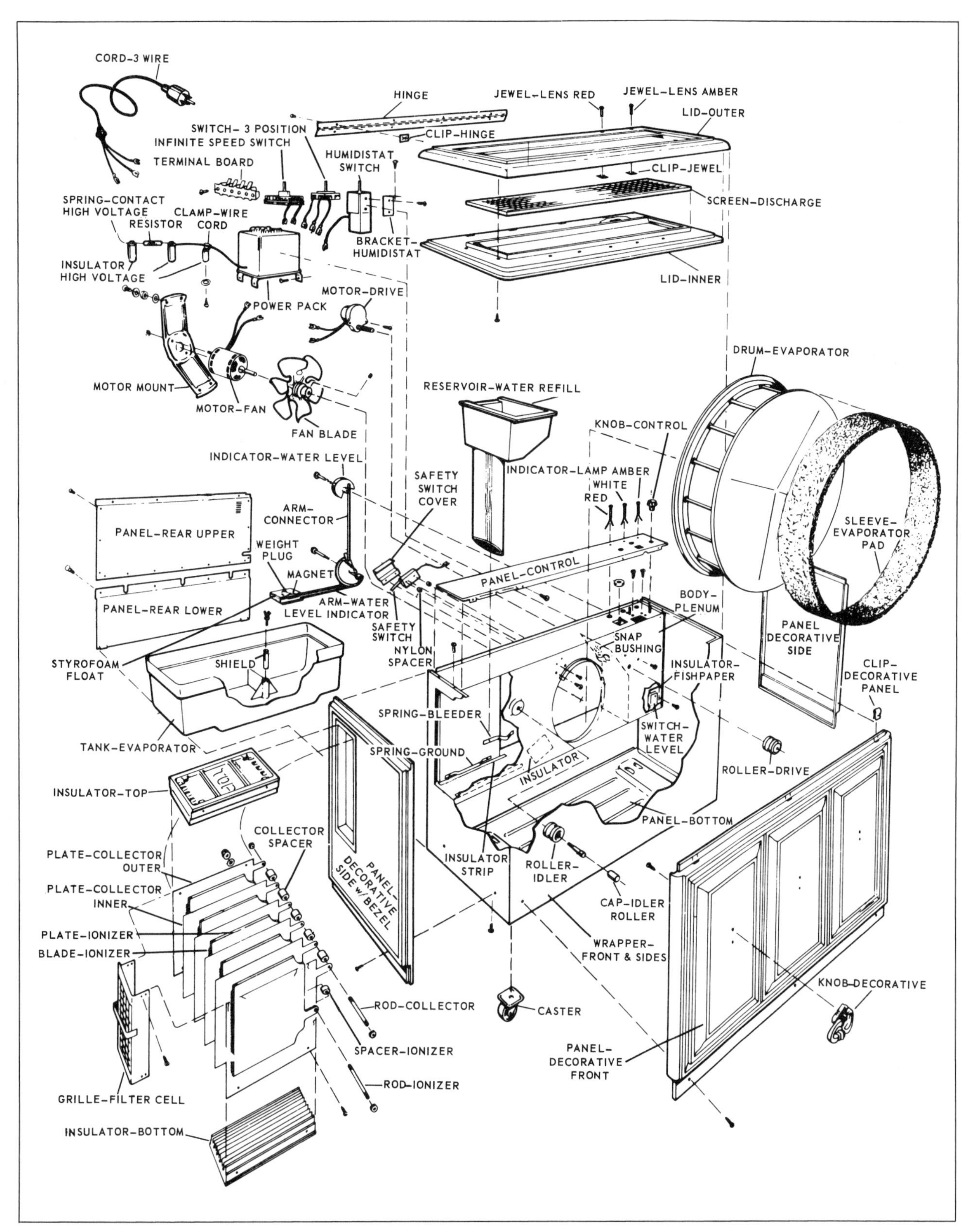

Figure 7-6. Assembly detail of a typical electronic air cleaner–humidifier unit.

Trouble Diagnosis

Table 7-1 gives some of the possible defects that may be found in a typical electronic air cleaner–humidifier unit. The numbers in the chart indicate the order in which the defects should be checked.

Table 7-1. **Trouble diagnosis chart for electronic air cleaner–humidifier units.**

Settings			Operation Indicators				Motors		Symptom	Possible defect																											
Three-position switch	Humidistat	Var. speed (Inf. Spd. Sw.)	Wat. Lev. Ind.	ADD WAT. lamp	CELL ON lamp	CL. CELL lamp	Fan	Drive (Drum)		Line-cord set	Safety switch	Three-position switch	Humidistat	Infinite-speed switch	Water level switch	Wiring error	Float Ga. Assy.	Fan motor	Drive motor	ADD WAT. lamp	CELL ON lamp	CLEAN CELL lamp	Dirty cell	Bent plates	Spring contact H-V	H-V resistor	Power pack	Sludge in tank	Plastic idler wheel	Rubber drive wheel	Low-water level	High-water level	Low-line volts	High-line volts	Bent fan or Mtr. Mt.	Nonconductive dirt	Odorous water
AC*	..	High	..	Off	On	Off	High	Off	Dead unit	4	1	2				5		3																			
AC	..	High	..	Off	On	Off	High	Off	No fan					1		3		2																			
AC	..	High	..	Off	On	Off	High	Off	CELL ON lamp is off							5					2		6		1	4	3										
AC	..	Low	..	Off	On	Off	Low	Off	No fan					3		4		2															1				
AC	..	Low	..	Off	On	Off	High	Off	Speed not variable					2		4		3																1			
AC	..	..	..	Off	On	Off	..	Off	CL. CELL lamp is on														1	2			3									1	
AC	..	..	..	Off	On	Off	..	Off	CL. CELL lamp inop.													2					3									1	
AC	..	..	..	Off	On	Off	..	Off	Arcing in cell														1	2													
AC	..	..	..	Off	On	Off	..	Off	Odor from cell														1	2	3												
AC	..	..	..	Off	On	Off	..	Off	Odor from water																			2									1
HAC†	..	..	0	On	On	Off	..	Off	ADD WAT. lamp off			5			3	6	2			4												1					
HAC	High	High	Full	Off	On	Off	High	On	No drum rotation			7	6		8	10	9		5									4	3	2	1	1					
HAC	High	..	..	..	On	Off	High	On	Fan scrapes orifice																											1	
HAC	High	High	0	On	On	Off	High	On	Wat. lv. Ga. stuck at full								1											2									

* AC—Air cleaner only.
† HAC—Humidifier and air cleaner.

Central vacuum cleaning systems

CHAPTER

8

In *Small Appliance Servicing Guide*, we cover the servicing of portable vacuum cleaners. In recent years more and more homeowners are having their vacuum cleaning systems built right into their houses. Built-in central systems are also being installed in increasing numbers in commercial establishments such as offices, hotels, motels, and schools. While the installation of central vacuum cleaning systems is usually done by electricians, the actual servicing of these units is usually left to appliance service technicians.

COMPONENTS

Most built-in systems consist of a power unit centrally located in a remote area of the home or building which contains the vacuum-producing motor along with the dirt collector, bag or filter. From the unit, tubing is simply routed through walls by way of the attic or under floors to conveniently located wall valves. (Only three to five valves are required in the average home.) To vacuum, the operator inserts a lightweight, flexible cleaning hose into one of the valves. All surfaces in the home or building can be cleaned efficiently using one half dozen or so versatile cleaning tools which are standard with the system. In the average home, the collected dirt needs emptying only a few times a year.

As you can see in Fig. 8-1 a central vacuum cleaning system consists of only three major components: (1) power unit; (2) tubular air-conveying system (suction line); (3) and cleaning tools. The heart of the system is the power unit. While the capacities of the power units vary, they all utilize one of two methods of dirt separation: (1) dry pickup, and (2) wet pickup. In the latter case, the wet pickup is usually combined with the dry to give both wet and dry pickup.

Dry pickup. The dry pickup power unit operates very much like canister vacuum units described in Chap. 3, *Small Appliance Servicing Guide*. Actually, the major difference between the two is that the central system's power unit has a much larger motor and a greater capacity. Otherwise, the servicing techniques are about the same, and the basic information given in *Small Appliance Servicing Guide* holds good for the power unit of the pickup type.

The accumulated dirt must be emptied manually periodically. This is usually done by opening the dirt canister, which is usually located in the power unit, and removing a disposable filter bag. Most dry pickup units also have another filter in addition to the bag, and this must be replaced when it becomes dirty.

Wet pickup. While the wet pickup units also have a filter that must be kept clean, the dirt picked up is carried away by water down the sewer line after each use. Motor and suction

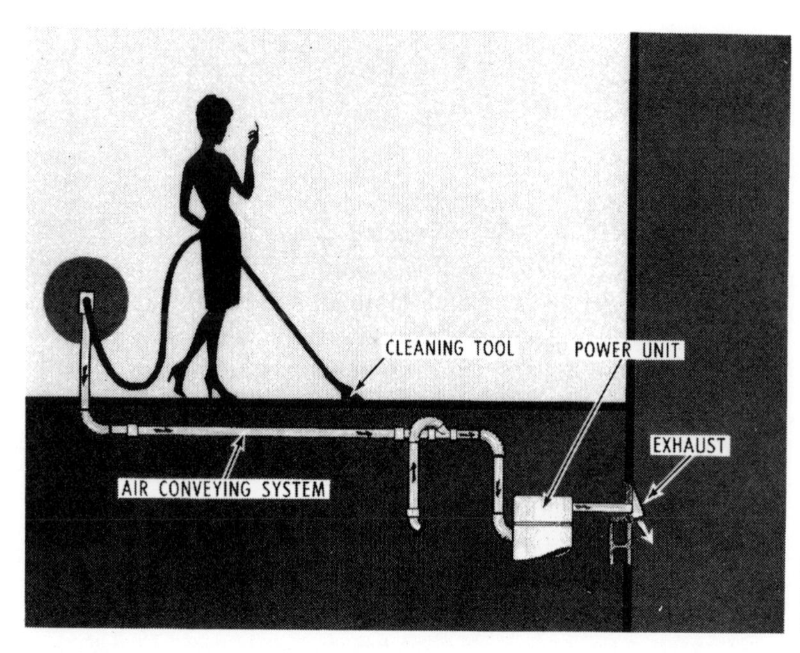

Figure 8-1. Major parts of a typical central vacuum cleaning system.

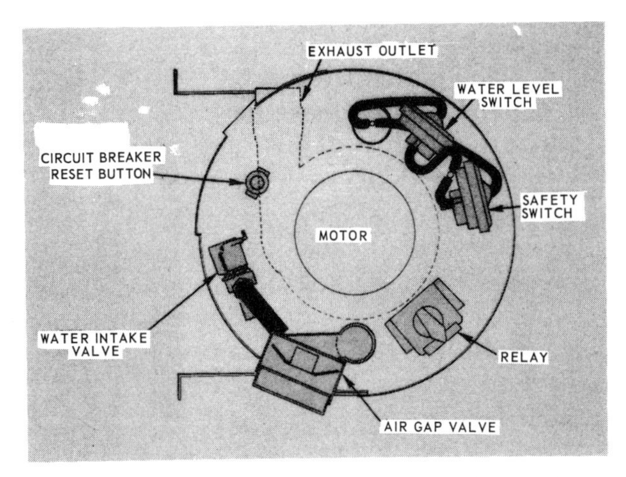

Figure 8-2. Various controls for a typical wet pickup power unit.

arrangement of both types of power units are the same. The big difference in servicing is the number of additional controls necessary in the operation of the wet pickup unit.

The controls for a typical wet pickup power unit are shown in Fig. 8-2. In this arrangement, the water, which is used to flush away the dirt, is held in a container by an extremely light foam type of drain valve. When the wet pickup unit is not operating, the drain valve is open. When the cleaner is started, the rush of air drawn past this valve by the vacuum closes it. At the same time, the solenoid-operated water inlet valve opens, allowing water from the building's plumb-

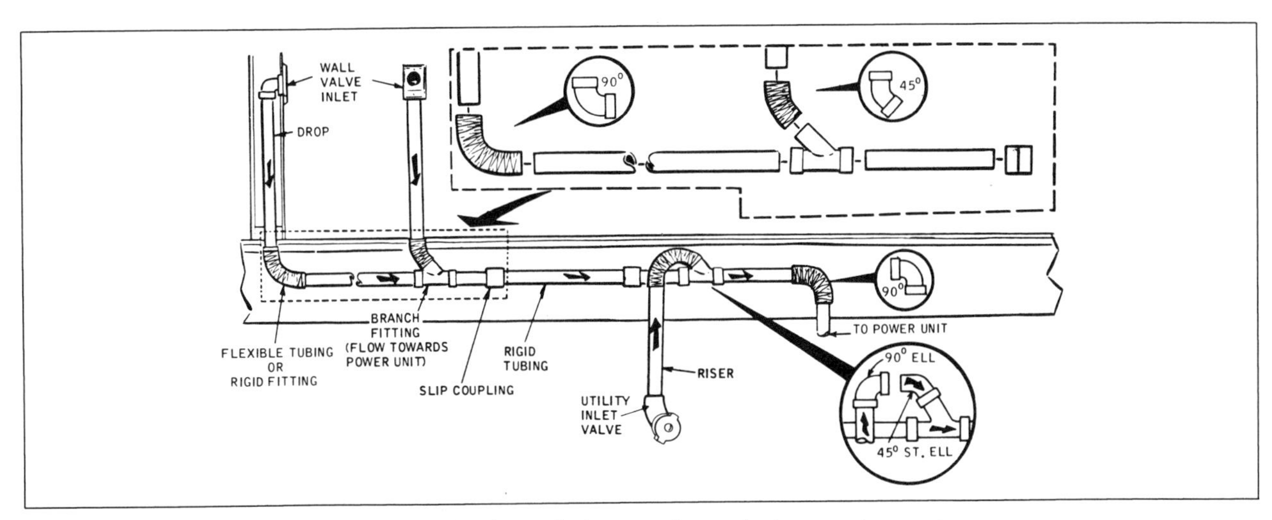

Figure 8-3. Fabrication procedures and schematic layout of a typical conveying system.

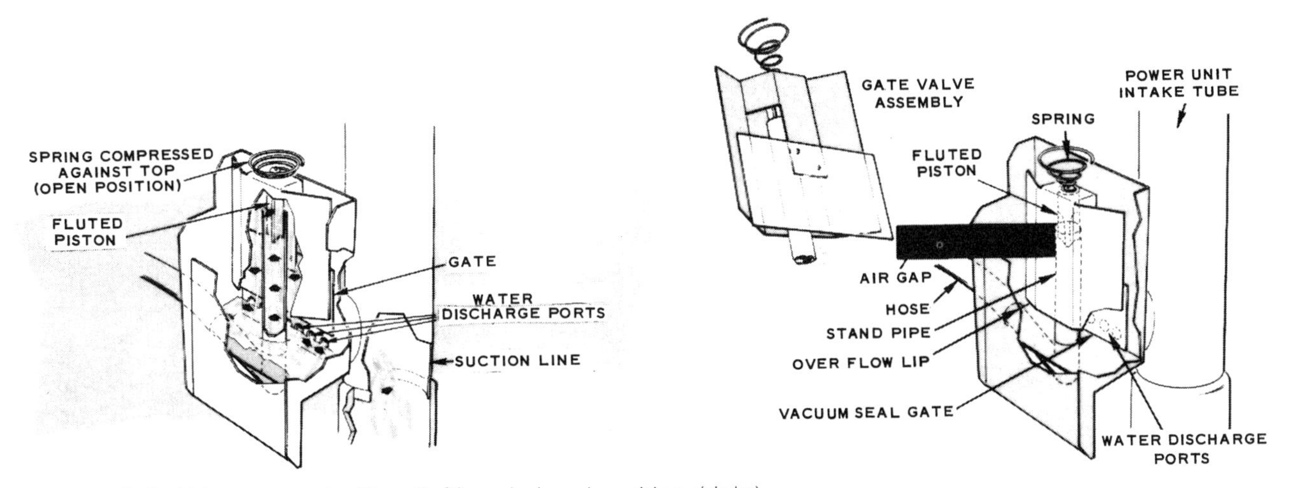

Figure 8-4. Valve—open position (left) and closed position (right).

ing system to enter the air gap valve. The water pressure lifts the fluted piston and the attached gate. Water escaping around the flutes spills into the square container surrounding the valve and is sucked by the vacuum inside the cleaner, through the water discharge ports at the bottom rear, into the suction line. The water level pressure switch, in the unit illustrated in Fig. 8-4, stops the fill at about 3 gal by closing the water valve. With no water pressure the fluted piston with the attached gate drops. The flexible vacuum-seal gate is sucked tightly against the water discharge ports in the rear. This seals the system.

Should the water level pressure switch fail, or if there is an excessive pickup, the safety pressure switch will shut off the entire system at about 7 gal. With the machine shut off and no vacuum, the weight of the water in the water container opens the foam drain valve. The water drains out until the 3-gal level is reached, and then the cleaner again starts. Incidentally, it is the air gap valve (shaded in the drawing) that permits this machine to be used for both wet and dry pickup. That is, the air gap is from the outlet of the water supply to the edge of the surrounding container. Should a water fill escape the tank at the discharge ports, it would spill over the edge. Since the water outlet is higher than this escape, it is impossible for contaminated water to enter the drinking water lines.

While most wet pickup-type units are basically the same, each manufacturer may call its controls by slightly different names and may employ a slightly different principle. For this reason, it is wise to carefully read the maker's service manual before undertaking any service on a wet pickup power unit.

Suction line. In troubleshooting the unit for poor suction, first check the vacuum at the suction line of the unit with a suction vacuum gauge. Low suction here indicates leaking through the canister gasket or a clogged safety filter. A check of suction at each wall valve should read the same as at the unit if the transmission line is 100 percent sealed. Any difference of more than 10 percent should be corrected. Low suction indicates a leaking wall valve or a leaking transmission line. Particularly suspect taped joints. A clogged transmission line can be opened by carefully pushing a smooth-ended plumber's snake or a small garden hose with fittings removed into the transmission lines.

Frequent clogging indicates a Y branch fitting with the arm not running with the flow of air, connecting line not cut square and allowing dirt accumulation, or a plastic-cement bead inside the tube.

For a complaint of exhaust noise, a muffler vent hood is available. The internal fiberglass cover deadens the sounds. For vacuum hose repair, a kit is available. A repair of this type is easy, professional, and permanent and you have a smooth flow through the repaired hose.

ELECTRIC SYSTEM

Most central cleaning systems are made to work on a low-voltage (usually 24 V) inlet relay system, which is activated by inserting the end of the hose into the wall valve; that is, the remote switch at each wall valve is built into the valve for control, ON or OFF, at each outlet. In addition, a few systems have such items as a timer control and touch switches, either 120 V or low voltage. For this reason, it is wise to check the electric system carefully before proceeding with any servicing.

TROUBLE DIAGNOSIS

Here is a general diagnosis list of the trouble you may face when servicing a central vacuum cleaning system.

Problem: Unit will not operate.

Possible cause	*Solution*
1. No power to unit.	1. Check fuse, circuit breaker on house

Possible cause	Solution
	service power line. Check circuit breaker on unit. Check wall switch. Check for loose leads and broken wires. Check low-voltage circuit. Check operation of relay (wet pickup). Check operation of safety switch (wet pickup).

Problem: Motor not operative.

Possible cause	*Solution*
1. Worn motor brushes.	1. Replace brushes.
2. Open or shorted field.	2. Test continuity, replace motor if faulty.

Problem: Repeatedly goes out because of overload — circuit breaker trips — during normal cleaning operation.

Possible cause	*Solution*
1. Electric short.	1. Visually check all electric leads and connections. Check individual components for shorts and continuity.
2. Blocked exhaust or conveying system.	2. Check exhaust vent tube and conveying system.
3. Overheated unit.	3. Provide ventilation in area where unit is installed.

Problem: Machine cycles intermittently during operation.

Possible cause	*Solution*
1. Overfill of water on fill cycle.	1. Check sensing tube in separator bowl sump for restriction caused by accumulation of dirt, then clean out. Check condition of plastic sensing tubes leading to pressure switches. Check function of water level switch. Check function of safety switch. See if water valve is stuck open.
2. Customer picking up a large volume of dirt.	2. This is normal function of machine; explain overfill protection to customer.

Problem: Overfill of water (droplets blowing out of exhaust—wet pickup model only).

Possible cause	*Solution*
1. Safety switch inoperative or out of calibration.	1. Check sensing tube in separator bowl sump for restriction caused by accumulation of dirt, then clean out. Check condition of plastic sensing tubes leading to pressure switches. Check function of safety switch.

Problem: No water enters unit (wet pickup model only).

Possible cause	*Solution*
1. Restricted water lines.	1. Check for restriction in lines and remove.
2. Restricted filter screen in valve.	2. Clean screen in water valve.
3. Inoperative water level switch.	3. Replace switch.
4. Inoperative solenoid coil on water valve.	4. Replace coil.
5. Valve in water line not open or line clogged.	5. Open valve, check flow rate and pressure. [Generally 32/min to 15 lb/in^2 (min).]
6. Low water pressure.	6. Suggest correction to system.

Problem: Low water fill level (wet pickup model only).

Possible cause	*Solution*
1. Water level	1. Check calibration;

Possible cause	Solution
1. (*cont'd.*) switch out of calibration.	water fill can be measured by disconnecting drain hose and draining into a pail. Replace switch if out of calibration. Do not attempt adjustment.

Problem: Water overflowing at air gap valve or similar valve (wet pickup models only).

Possible cause	*Solution*
1. Low water pressure fails to actuate gate valve assembly on air gap.	1. Adjust spring tension on gate valve. (Reduce tension by compressing spring.) Suggest corrective measures to water system.
2. Air gap gate valve assembly stuck closed.	2. Check valve for freeness of movement, clean off foreign matter which may be restricting movement of piston and gate.
3. Low suction.	3. See next problem.
4. Clogged filter screen.	4. Clean off screen.
5. Top hood not in proper position.	5. Check top hood for position and tightness. Only one screw holds it in place. The hood controls spring pressure of gate valve assembly.

Problem: No suction (all types).

Possible cause	*Solution*
1. Complete restriction in line.	1. Check for restriction and remove.
2. Separation chamber not latched securely.	2. Check latches.
3. Separation chamber gasket missing.	3. Replace gasket.
4. Safety filter completely plugged.	4. Clean filter and check for adequate water fill on wet pickup models.
5. Fan impeller blades clogged up.	5. Clean out fan area; more frequent cleaning of collector required.
6. Blocked exhaust.	6. Clean exhaust; advise more frequent cleaning of collector.
7. Hose clogged.	7. If there is suction at inlet valve but not at hose end, hose is clogged. It can usually be unclogged by reversing hose and putting wand end into inlet valve, forcing the obstruction backwards and out.
8. Power unit lid not sealed tightly.	8. Be sure lid is tightly closed.
9. Check filter bag and be sure it fits tightly (dry pickup models only).	9. Clean or replace.

Problem: Low suction (all types).

Possible cause	*Solution*
1. Valve covers not seating.	1. Check for bind or obstruction.
2. Gasket missing on valve cover.	2. Replace gasket.
3. Warped valve cover and/or valve base.	3. Replace valve assembly; do not torque base mounting screws too tightly as this will cause a bow.
4. Leak in line or valve connection.	4. Check joints in line and O ring in valve to elbow connection.
5. Partial restriction in line or	5. Clean out restriction.

Possible cause	Solution
safety screen.	
6. Improperly seated separation chamber gasket or motor mount gasket.	6. Check gasket locations, also clamps and condition of the canister rims.
7. Dirt collector full.	7. Empty collector.
8. Branch fitting installed backwards.	8. Check installation instructions for proper direction and correct.
9. Low line voltage of unit.	9. Advise user to have corrective measures taken.
10. Blocked exhaust.	10. Clean exhaust, check vent flapper.
11. Air gap valve or similar valve stuck open (wet and dry pickup models).	11. Free up movement of valve.
12. Drain valve not closing (wet and dry pickup models).	12. Clean drain and separator bowl and free up valve, adjust wire counterbalance.

Problem: Noisy valves (vibrations of valve covers on base).

Possible cause	*Solution*
1. Valve covers not seating.	1. Check for obstruction or bind.
2. Warped valve cover and/or base.	2. Replace valve assembly; do not torque base mounting screws too tightly as this will cause a bow.

Problem: Vibration in unit (all types).

Possible cause	*Solution*
1. Unbalanced fan due to dirt buildup or damaged fan blades.	1. Clean fan; more frequent cleaning of dirt collector needed. Replace motor if damaged.
2. Unit not mounted properly. Some vibration is normally expected during wet pickup.	2. Consult installation instructions; tighten lag screws.

Problem: Water dripping from inlet valves (wall and utility; wet pickup units only).

Possible cause	*Solution*
1. Branch fittings incorrectly positioned.	1. Reposition for proper directional flow and upward position of branch lag.
2. Transmission lines not clear.	2. Customer should allow the machine to run briefly after completing pickup.

Electric ranges and ovens

CHAPTER

No other category of appliances displays the variety that electric ranges and ovens do in sizes, configurations, and features. But all electric ranges are alike in having two separate sets of heaters: the surface units and the oven unit or units. The more popular configurations of these units that a service technician will be called upon to service are as follows.

Free-standing. This type of range stands by itself, independent of the wall or cabinets on either side. It can be a single-oven model, with broiler, or a double-oven with eye-level oven or broiler, even with a microwave oven. The controls may be in the front, on top, along the side, or on a backsplash.

Built-in. The oven, with either a single or double cavity, must be fitted into a wall cabinet built for this purpose, with or without a microwave component. Companion to the wall oven is the built-in cooktop or surface unit.

Slide-in. This is basically a free-standing range, but with the side panel left off and engineered to fit snugly against the countertop, or even overlap it, for a built-in look. The slide-in ranges are available in the same configurations as the free-standing styles.

Drop-in. A variation of the slide-in, the drop-in, does not rest on the floor. Flanges rest on the countertop on either side, and it is supported from there. Special cabinets are available to fit under it for a more built-in look, although it might extend all the way to the kick space.

Regardless of the physical construction of the various ranges and ovens, their operating principles are essentially the same. All conventional electric ranges use units (resistances) to produce heat for cooking. Depending upon unit design, this heat is applied through conduction, convection, or radiation. For instance, all top or surface units are designed with a large amount of surface area for greater conduction.

Oven bake units use the principle of convection for heat transfer. Units are designed to fit the oven cavity and occupy a minimum of space. Air circulation around the unit is most important. On the other hand, the basic function of a broiling unit is to emit large amounts of radiant energy (radiation) uniformly over the surface of the broiler grille. This usually results in longer coils. Most ranges locate the broil unit in the top of the oven cavity.

The principle of operation of ranges and ovens is the same as those described in Chap. 5, *Small Appliance Servicing Guide.* That is, heating is achieved by connecting a resistance of the proper amount across an electric potential, thus creating current flow. Variations in heat are generally obtained by connecting two or more elements in parallel or series and/or by varying the amount of voltage supplied to the heating units. Control of these units (resistances) to meet the needs of the various cooking requirements is by means of switches. The units and controls vary according to the requirements of the design and the control to be used.

INSTALLATION OF RANGES

In order that customers receive full value for their investment a few general steps should be followed during installation of all ranges. If the range is installed and checked out properly, the customer will receive greater satisfaction from the new appliance and the seller will encounter fewer service problems.

Receiving Inspection

The range should be inspected for transportation damage at the time it is accepted from the carrier (as should all electrical appliances). In fact, there are two types of transportation damage: visual and concealed. When a visual damage is evident, a damage claim should be filed at once with the carrier's claim agent. It is important that this be done without delay and that the claim agent sees both the range and the crate.

When concealed damage (in which the product is damaged but there is no damage to the shipping crate) is noted, the consignee should request an inspection by the last carrier, and this claim would be covered on a consignee concealed damage claim form. After the claim agent has made an inspection and accepts the responsibility, the damage, visual or concealed, can be repaired so the range is in salable condition and the bill submitted with the claim against the carrier. Transportation damage is the responsibility of the carrier.

Uncrating and Handling

The exterior finish of most ranges is either porcelain enamel or steel, or baked enamel on steel. (A few models are available in stainless steel.) With proper usage, these finishes will last indefinitely; however, porcelain is inherently fragile to impact and to bending strains, and so a porcelain-finished range must be uncrated and moved with extreme care.

Careful handling of the range in the user's home is also very necessary since most kitchen floor coverings scratch easily. If it is necessary to slide the range against the wall, raise the front edge to avoid marking the floor. Do not slide the range sideways.

The range is usually packaged in a carton-type crate and fastened to a wood base by four shipping bolts. These bolts are the range leveling feet. First, cut the bottom metal band, then lift the carton off the range. With the range on its back, remove the wood base, and install the leveling feet.

Now, check the following.

1. Inspect the wiring and all electric connections for tightness, as it is possible for some of the connections or screws to loosen in shipment. Where applicable, make certain the cap of the sensor unit at the automatic surface unit moves upward and downward freely.
2. Make sure that all shipping material has been removed from the range, and that booklets, warranty, and other instructions have been removed from the oven or platform surface and handed to the user.
3. All ranges have a serial plate giving model and serial numbers. Location of the serial plate is usually identified on the specification chart in the user's manual.

Current Supply

Most ranges and built-in cooking equipment on the market today are manufactured to one of three standard electrical specifications.

1. For 120/240-V, three-wire, single-phase ac supply. This was formerly called 115/230-V or 118/236-V supply.
2. For 120/208-V, four-wire, three-phase ac supply. When this source is used, it must be adapted for use to the range by using only three wires of the four-wire system (two "hot" leads and the neutral).
3. Dual-rated, 120/208 V and 120/240 V. These ranges or built-in equipment will operate on either electric source. Units must be connected to a supply circuit of the proper voltage and frequency as specified on the rating plate. The electric range should be installed in accordance with the National Electric Code (NEC), state and local ordinances, and the rulings of the local power company.

The local power company should always be consulted before a range installation is made. In the average residence, unless an electric range has previously been used, the existing service and meter may be found to be of insufficient capacity. Under such circumstances, a new service and meter with adequate capacity for both range and lighting circuits must be installed.

Sizes of power line wires and fuses should

Ranges

NEC fuse rating	Maximum kilowatt rating		
	208 V	236 V	240 V
35 A		12.4	12.4
40 A	12.4	15.4	16.0
50 A	17.4	21.4	22.0

Wall-mounted ovens or counter-mounted cooktops.

NEC fuse rating	Maximum kilowatt rating		
	208 V	236 V	240 V
20 A	4.2	4.7	4.8
30 A	6.2	7.1	7.2
35 A	7.3	8.3	8.4
40 A	8.3	9.4	9.6
50 A	10.4	11.8	12.0

be determined by the total connected load wattage (refer to the model's specifications in the service manual) and by the National Electric Code, state and local ordinances, and the rulings of the local power company. The tables on p. 218 show the minimum fuse requirements specified by the National Electrical Code.

The branch circuit load for one wall-mounted oven or one counter-mounted cooktop is the rating on the nameplate of the appliance. The branch circuit load for a counter-mounted cooktop and not more than two wall-mounted ovens —all supplied from a single branch circuit and located in the same room—shall be computed by adding the nameplate ratings of the individual appliances and treating this total as equivalent to one range.

In all cases the range body must be grounded. Most utilities permit grounding of a range body to the neutral of the range service. Otherwise, ground the range body to a water pipe. A ground strap is provided at the main terminal block of the range, just below the neutral terminal. When the ground wire has been connected to the neutral at the factory, it is so noted on the wiring diagram on the back of the range.

Making the Installation

When installing wall-mounted ovens or counter-mounted cooktops, be sure to follow the mounting instructions given in the installation booklet. In the case of free-standing models, place the unit in the position requested by the customer and level it properly. All ranges should be shimmed or adjusted so that the oven racks are level. Place one oven rack in the center glides of the oven and place a level sideways on this rack. Ranges are equipped with two slotted leveling feet, one each in the front bottom corners of the range body. On these models, remove the bottom drawer and adjust the feet with a screwdriver until the oven rack is level from side to side. Then place the level lengthwise on the rack, and, if necessary, put shims under the two rear corners of the body to level the rack from front to rear.

With most ranges, a terminal block is provided for the connecting of the unit to the service line. In many cases these terminal blocks have been approved only for direct connection of copper house wiring. All built-in models with a flexible conduit are generally equipped with copper power leads. Improper connection of aluminum house wiring to either can result in a serious problem. That is, if aluminum house wiring is used, a copper-to-aluminum splice must be employed, using only connectors designed for joining copper to aluminum and approved by Underwriters Laboratories Inc. Be sure to follow the connector manufacturer's recommended procedure closely. This means that cabinet ranges will require a length of copper building wire to each of the three terminals on the range terminal block to accommodate the copper-to-aluminum connector. Built-in power leads are approved by Underwriters Laboratories for connection to larger-gauge household wiring. The insulation of these leads is rated at temperatures much higher than the temperature rating of household wiring. The current-carrying capacity of a conductor is governed by the temperature rating of the insulation around the wire rather than the wire gauge alone.

A few built-in wall ovens and counter-mounted cooktops are equipped with aluminum flexible conduit. The following warning is usually attached to the end of the conduit:

CAUTION
WHEN APPLYING SETSCREW TYPE CONNECTORS, CARE SHOULD BE TAKEN NOT TO DAMAGE CONDUIT

Instructions to User

After the range is installed and ready, explain to the user how the range operates. Lack of knowledge of the principles of electric cooking is responsible for the majority of complaints. Complete this part of the installation with care and tact. True, most of this necessary operational information is given in the care, use, and recipe book furnished with the range; but because many service calls are instructional, it will pay you to explain and show the following points.

1. How to use the timer for automatic control of the oven and appliance outlet.
2. How to reset the timer for manual operation of the oven and receptacle.
3. How to replace platform lights.
4. How to remove oven heaters for cleaning and how to put them back again.
5. How to remove plug-out units, lift-out units, pan, and rings for cleaning.
6. The advantages of the infinite-type switches.
7. The self-clean operation, if range is so equipped.
8. The operation of any accessory or special feature that may be supplied with the range.

SURFACE UNITS

The surface units of either range or built-in unit consist of two basic components: surface heating elements and surface unit temperature controls. Each heating element has its surface unit control or switch.

Surface Heating Elements

Early electric ranges used an open resistance coil placed in a ceramic block for surface units. This type of unit was subject to relatively easy shorting and breaking, but component servicing was easy. Later improvement in surface units put the ceramic block and coil inside a metal sandwich. This eliminated most shorting and breaking, but made component repair impossible. These designs (ceramic block and sandwich) have a comparatively large mass, causing some cooking problems because they are slow to heat and cool.

Modern surface units eliminate much of the mass by placing a Nichrome resistance wire, insulated with magnesium oxide (MgO), inside a small Inconel tube which forms the sheath. Most modern electric ranges use variations of this for surface, bake, and broil units.

The wattage, or heat-producing capacity, of the surface unit is directly related to the number and size of the wire coils, which set up an ohmic resistance to the electric current. In most ranges there are two surface units (6-in size) of approximately 1,250 to 1,500 W each, and two larger units (8-in size) of about 2,000 to 2,800 W. A few range models feature a 4-in surface unit.

As a rule, the heating or resistance wire extends only through the part of the unit which is flattened for contact with utensils. Long lead terminals carry current but do not heat that section of the unit which is bent down. In fact, No. 16-gauge stranded wire with special heat resistance is often used for surface unit lead wires. The factory process of crimping terminals to these wires is rigidly controlled to prevent cutting, nicking, or overcompression of the strands of wire. If crimping is not properly done, it may result in high electric resistance, thus creating heat and eventual fracturing of the wire. Because of this, crimping of the surface unit lead wires in the field is not recommended. The preferred procedure is to strip back the insulation about $\frac{1}{2}$ in, solder the wire strands

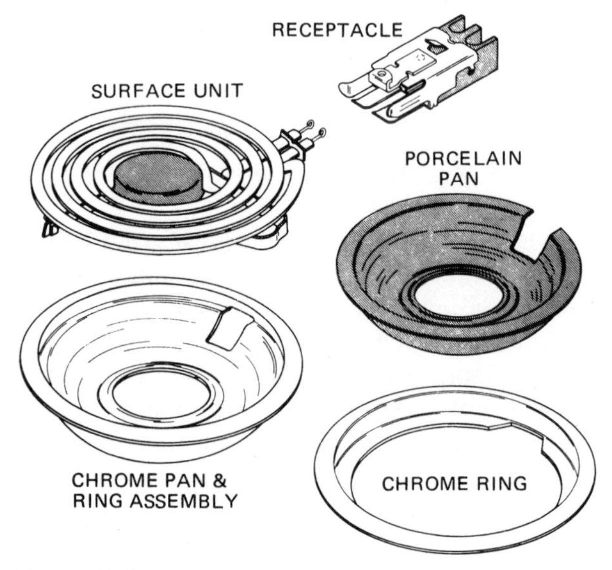

Figure 9-1. Surface unit assembly.

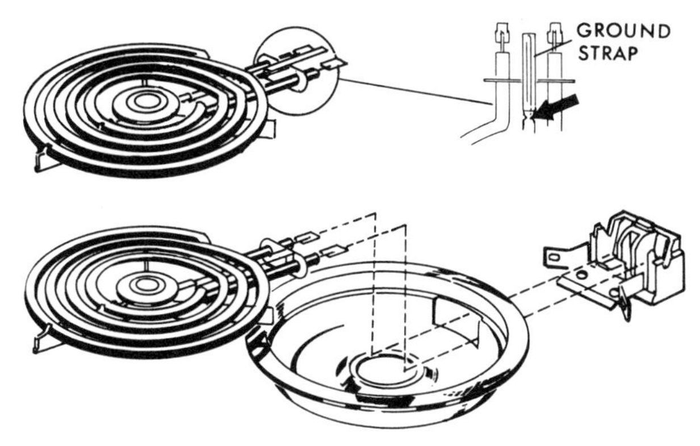

Figure 9-2. Plug-in (top) and liftout (bottom) arrangements for single-tube surface units.

together, form a U-shaped loop around the terminal screw, and tighten securely. The wire loop under the screwhead should not overlap. Always loop the wire around the screw in the direction it tightens.

The heating elements are usually supported in the range top by a trivet that rests on a trim ring partially recessed into the range top surface. Drip bowls of porcelain on steel, aluminum, or stainless steel catch boilovers or spilled foods. The element may be of the hinged style or the plug-in style. The hinge mounting bracket on surface units may include provisions for "stand-up" operation of the unit for cleaning purposes.

Plug-in units can be completely removed from the range to facilitate cleaning under the surface unit. This style of unit is removed from the range by lifting slightly and pulling straight from the receptacle. The receptacle hinge action will allow the unit to be raised approximately 2 in on the side opposite the receptacle. Grounding for the plug-in unit is accomplished by means of a ground clip which is crimped to the unit sheath. When the unit is plugged into the receptacle, the ground clip contacts the receptacle bracket which is grounded to the rest of the range by its mounting screw. This ground path to the unit is established approximately $\frac{1}{4}$ in before the unit terminals touch the spring contacts in the receptacle during plug-in. The ground clip also maintains proper spacing and overall alignment of the electric terminals plus being the mechanical stop when the unit is being plugged into the receptacle.

Heating elements may be of three types: single coil, two coil, and three coil. Single-coil units are controlled by thermostatically operated switches usually of the infinite-heat type (see Infinite-Heat Switch below). When this control is used, the entire surface unit cycles on and off at the full 240 V. Total wattage input is determined by the setting of the control and may vary from approximately 6 percent of the total "on time" at the WARM setting to 100 percent at the HIGH setting.

Two-coil surface units are controlled by fixed-position rotary or push-button type switches. In the double-coil type the heat is controlled by connecting the coils in series or in parallel, or by energizing only one of the coils, depending on the amount of heat desired. At the same time, the voltage applied to the heating element is switched between 120 and 240 V, depending on the heat necessary. The units are usually manufactured in two configurations: "spot heat" and interwound. Typical switching circuits are shown in Fig. 9-3. *Note*: Leads numbered 1 and 2 may be reversed at the switch or surface unit to give a different heat pattern at MEDIUM-HIGH and LOW switch selections.

In some applications, two-coil units are controlled by infinite switches. When this is done, the coils are paralleled with an external shunt or jumper across the terminals of the two single elements.

The three-coil unit allows the user to select heat patterns of 4-, 6- or 8-in diameter. The coils are cycled FULL ON or FULL OFF by an infinite-heat switch or automatic surface control.

Checking surface heating elements. If any element does not heat up, the first step in checking it is to be sure that power is being supplied to the range. This can be determined quickly by checking to see if the other surface units or the oven heats. After you are sure that the range is receiving power, the element can be unplugged from the unit. Set the surface control for that unit on the HIGH setting.

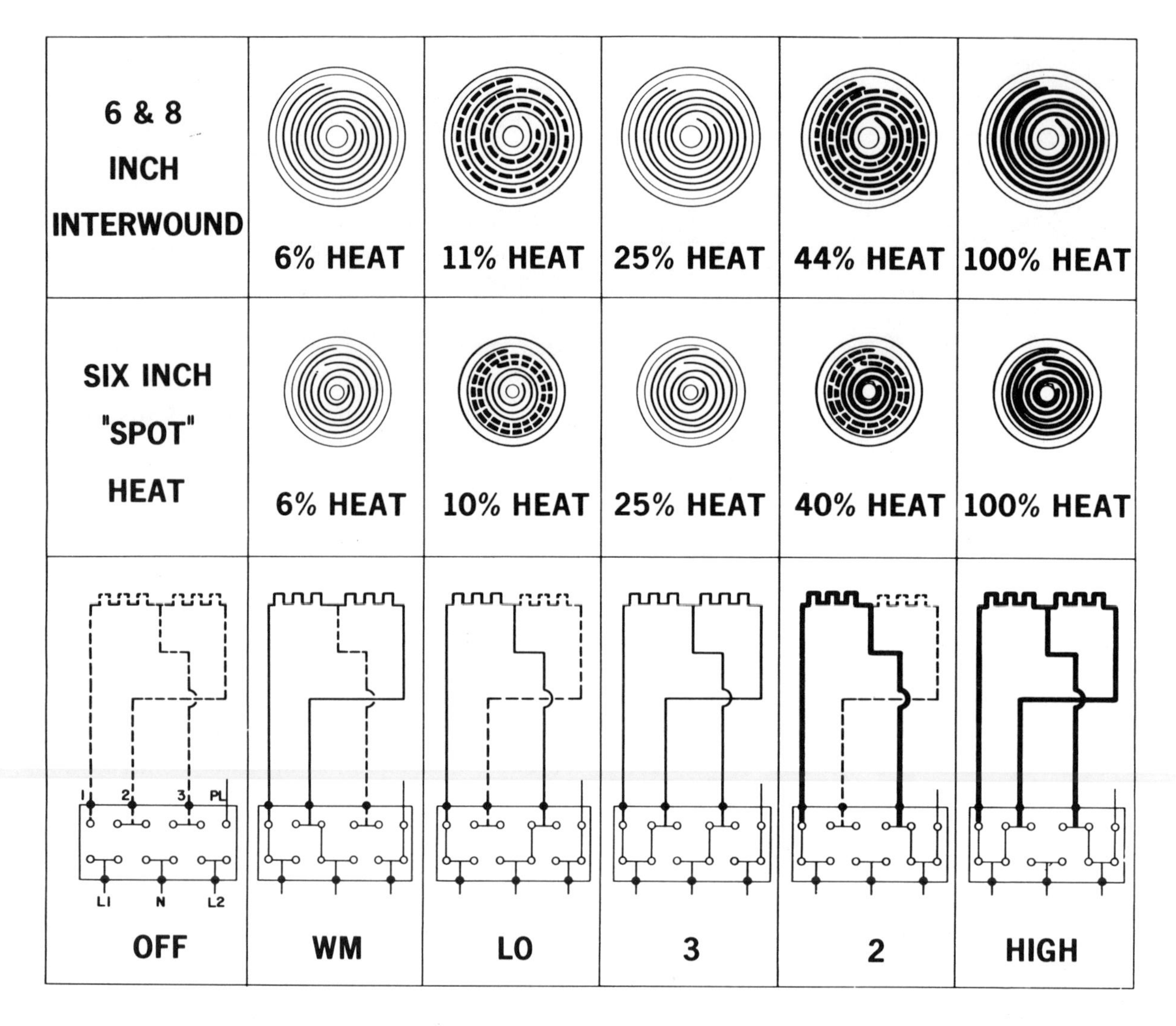

Figure 9-3. Two-coil surface unit circuits at various switch positions.

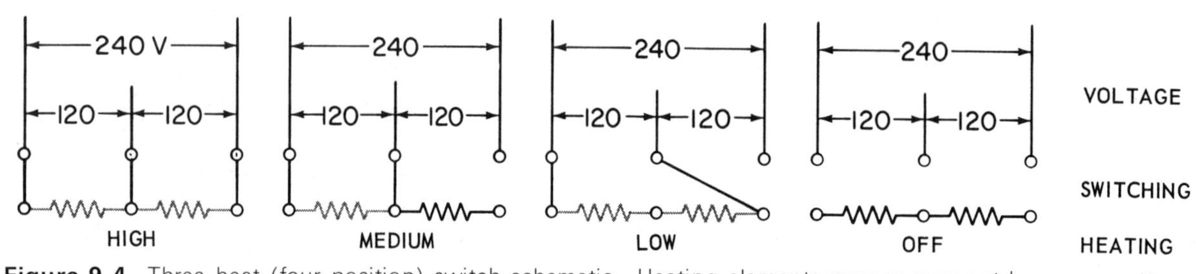

Figure 9-4. Three-heat (four-position) switch schematic. Heating elements may or may not be same wattage.

Using a voltmeter set on the 250- or 300-V scale, check for voltage at the surface unit terminal block. If there is voltage at the terminal block, then check the surface element.

To do this, check the element for continuity. If you are using a multimeter, set the selector for ohm or resistance reading and zero the dial. The resistance should be very low, generally 50 Ω.

Some elements, as just described, have two or three heat patterns. Be sure to check the resistances of each pattern for continuity by applying the ohmmeter probes to every other terminal. If the element shows no resistance, it should be replaced. Also if for any reason the element shows extremely high resistance, it should be replaced because it may have a short.

Surface Unit Temperature Controls

Like surface heating units, controlling switches have undergone considerable change through the years. Today, as already mentioned, there are three principal types of switches in use: (1) the fixed-position switch, (2) the infinite-heat switch, and (3) the automatic control.

Fixed-position switches. The fixed-position switch is the oldest method of controlling surface units. Early top units were either open-coil or sandwich type. They generally used two differing resistances. Control was with a three-heat, four-position, manually operated switch (Fig. 9-4). The fixed positions were OFF, HIGH, MEDIUM, and LOW, and the jumps between were rather large. While they were better than no control, the exact heats required for certain types of cooking were difficult to maintain.

An improvement, still used on some modern ranges, was the development of the five-heat fixed-position switch (Fig. 9-5). Heat control was accomplished by taking advantage of the

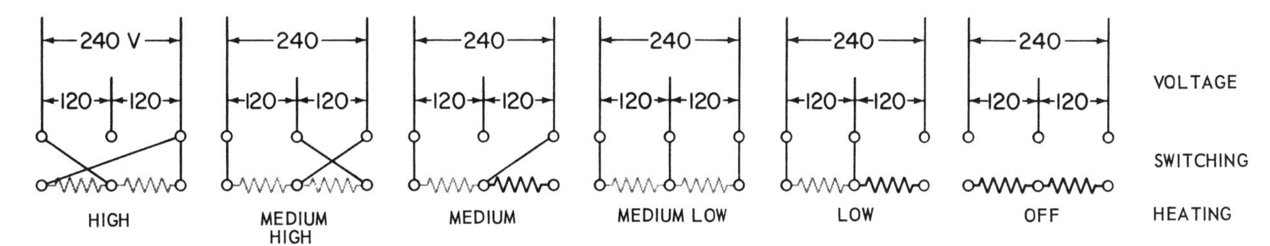

Figure 9-5. Five-heat (six-position) switch schematic. Heating elements may or may not be same wattage.

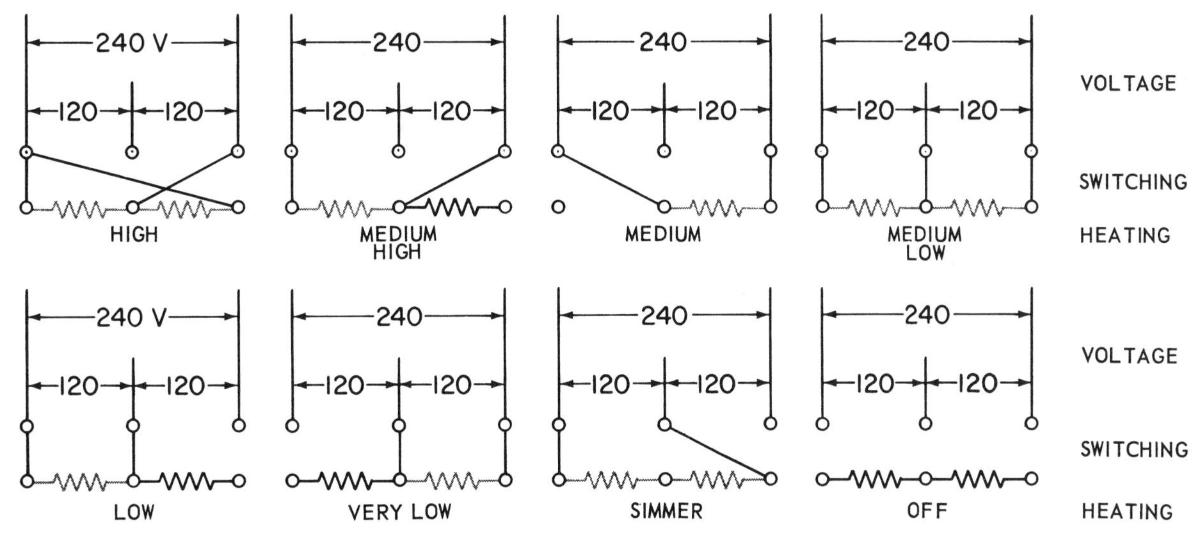

Figure 9-6. Seven-heat (eight-position) switch schematic. Heating elements must be of unequal wattage.

two voltages (240 and 120 V) available in a three-wire, 240-V system. A parallel and series circuit of the two resistances at 120 V was added. The six fixed positions are OFF, HIGH, MEDIUM HIGH, MEDIUM, MEDIUM LOW, and LOW.

Today, most of the ranges that employ a fixed-position switch use the seven-heat type. This manually operated switch adds two more heats by putting each resistance in a 120-V circuit. Illustrated in Fig. 9-6 is the circuitry for a typical seven-heat (eight-position) fixed switch using dual-resistance units of unequal wattage. At HIGH heat, both coils are connected in parallel to 240 V. The MEDIUM HIGH position connects the outer coil with 240 V. The MEDIUM position connects the inner coil with 240 V. In the MEDIUM LOW position, both coils are connected in parallel to 120 V. The LOW position connects the inner coil to 120 V. On the VERY LOW position, 120 V is connected to the outer coil. In the SIMMER position, 120 V is applied across both coils in series. The voltages at surface units and switches would be as follows:

Switch position	Surface unit terminal check Contacts	Volts	Switch position	Switch load terminal check Contacts	Volts
OFF	1 & C	0	OFF	(1 & C) (R & W)	0
	2 & C	0		(2 & C) (B & W)	0
	1 & 2	0		(1 & 2) (R & B)	0
HIGH	1 & C	240	HIGH	(1 & C) (R & W)	240
	2 & C	240		(2 & C) (B & W)	240
MEDIUM HIGH	1 & C	240	MEDIUM HIGH	(1 & C) (R & W)	240
	2 & C	0		(2 & C) (B & W)	0
MEDIUM	1 & C	0	MEDIUM	(1 & C) (R & W)	0
	2 & C	240		(2 & C) (B & W)	240
MEDIUM LOW	1 & C	120	MEDIUM LOW	(1 & C) (R & W)	120
	2 & C	120		(2 & C) (B & W)	120
LOW	1 & C	120	LOW	(1 & C) (R & W)	120
	2 & C	0		(2 & C) (B & W)	0
VERY LOW	1 & C	0	VERY LOW	(1 & C) (R & W)	0
	2 & C	120		(2 & C) (B & W)	120
SIMMER	1 & 2	120	SIMMER	(1 & 2) (R & B)	120

The fixed-position switch may be of the rotary type, which makes its contact selections by a round cam, or the push-button type, which slides a bar with protrusions that press against the moving contacts.

Infinite-heat switch. The surface units used with an infinite-heat switch are heated with their full-rated wattage whenever the switch is making contact. On the HIGH setting the switch keeps the surface unit on all the time. At a lower-heat setting the unit is cycled on and off by the switch. That is, at the HIGH detent setting the surface unit is energized continuously; at WARM detent setting the surface unit is energized only approximately 6 percent of the total ON time. The chart in Fig. 9-7 compares the percentage of total surface unit wattage obtained with typical infinite-heat controls and fixed-position rotary switches.

As you can see, the infinite-heat switch can be set at any of an infinite number of settings between HIGH and WARM—hence its name. The ON-OFF cycling of the surface unit is accomplished by a bimetal strip in the switch. This strip is

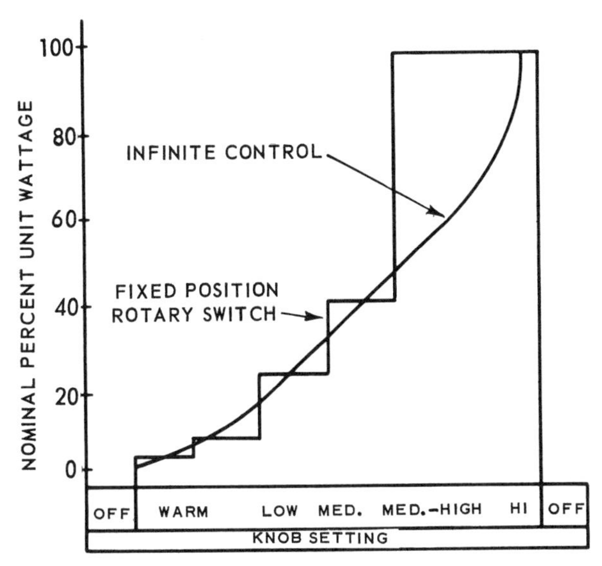

Figure 9-7. Resultant surface unit wattage infinite-control versus fixed-position switch.

made of layers of two metals, one of which expands more than the other when heated, thus causing the strip to bend. The resistance-heater strip is connected in series with the surface unit (Fig. 9-8) so that the same current which flows through the surface unit will flow through the resistance-heater strip. The bimetal strip therefore will be heated and bent in direct proportion to the heating of the surface unit.

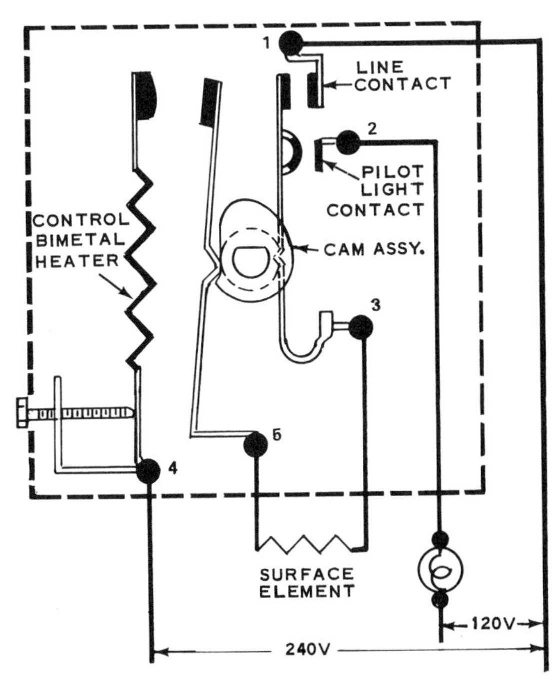

Figure 9-8. Typical infinite switch.

One of a pair of contacts is attached to the end of the bimetal strip and is moved away from the other contact by the bending of the bimetal. The relative distance between the contacts is changed when the switch setting is changed by a cam attached to the shaft of the control. This cam moves the second "stationary" contact closer to or farther from the cycling contact on

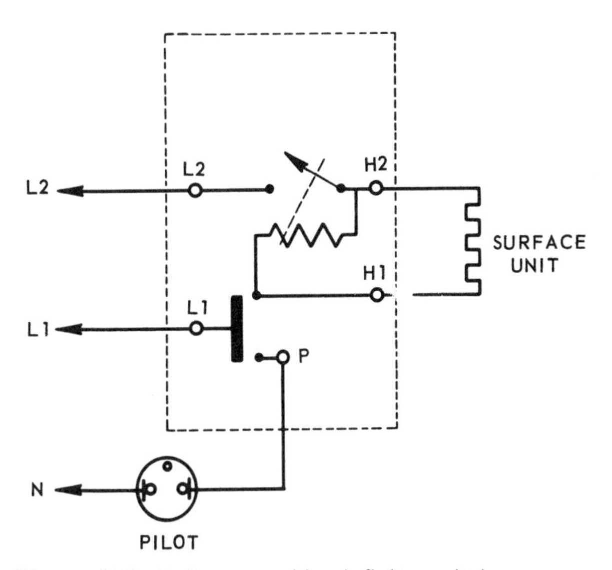

Figure 9-9. Voltage-sensitive infinite switch.

the heated bimetal arm and in this way determines the amount of bending of the arm which is required to open the contacts. Since more heat is required for more bending, the ON time is increased or decreased by the setting of the switch. The pilot light is turned on by a separate set of contacts in the switch. It is on all the time when the switch is turned to any position but OFF.

There are two basic designs of infinite-heat switches: voltage-sensitive and current-sensitive. The heater in the voltage-sensitive control (Fig. 9-9) is shunt (parallel) wired with the load. The controls are rated at either 208 or 240 V.

It is not necessary for the control to be associated with a specific wattage unit, but the control must be matched to the application voltage.

The heater in the current-sensitive control (Fig. 9-10) is connected in series with the load. All the load current flows through the bimetal heater. Heater design characteristics are determined by load current. These switches must be matched to a specific wattage unit. The electrical load on the control affects its performance, since the bimetal heater operates in series with the load. It is therefore necessary to have a load connected to the control when checking its performance.

Automatic-control switch. The automatic control constantly provides the temperature selected on the control dial. The sensing element, mounted in the center of the surface unit, measures the temperature of the utensil and regulates the amount of heat accordingly. The control is essentially a hydraulic thermostat. A bimetal heat anticipator inside the control opens and closes the cycling switch, by means of the heat produced by the surface unit current that it carries. This portion of the control functions only when the temperature of the hydraulic system has reduced the contact pressure, which occurs near the preset temperature of the control dial.

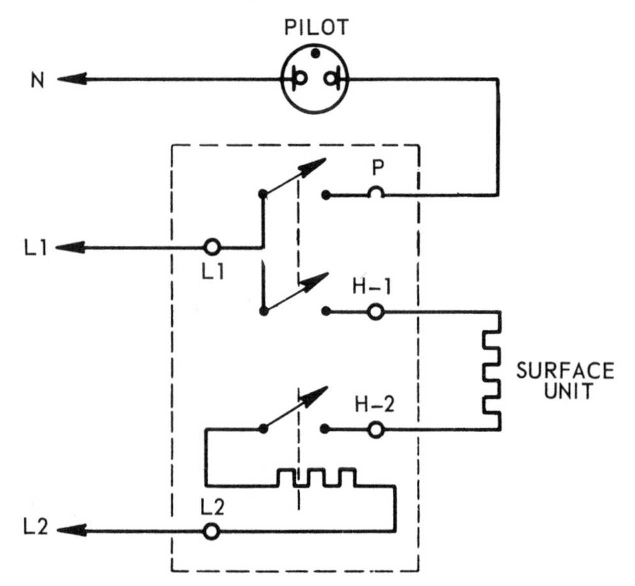

Figure 9-10. Current-sensitive infinite switch.

There are several methods of obtaining automatic control, but in the most popular way, the temperature sensor is mounted beneath the surface unit and includes the bulb, which is held up in the center of the unit so that it touches the bottom of any utensil that sits on the unit and thus "senses" the temperature of the utensil. The bulb is filled with a liquid which expands when heated. Some of the liquid is forced through a capillary tube to the switch, where it forces a diaphragm outward and, at a temperature determined by the setting of the switch shaft, opens a contact in the circuit to the unit. One of the contacts is surrounded by a magnet which gives a snap action to the opening and closing of the contacts (Fig. 9-11). Close control of the operating temperature of the automatic control is facilitated by a bimetal strip which transmits the movement of the center of the diaphragm to one of the cycling contacts. The bimetal is heated and warped (as in the infinite-heat switch) by conducting the current which flows through the surface unit and thus causes quicker opening and closing of the switch.

This combination of hydraulic thermostat and the hydraulic wafer heat anticipator allows the surface unit to impose the maximum heating rate while preheating or during heavy loads; however, once the utensil temperature comes within the operating range of the temperature selected, the bimetal cycling switch takes over and cycles the surface unit current ON and OFF at relatively short intervals, to produce nearly constant temperature on the utensil, as evidenced by the pilot or signal light which cycles with the control.

As with the infinite-heat switch, there are two designs of automatic controls: voltage-sensitive and current-sensitive. The latter is always wired in series with the unit, while the voltage-sensitive control is connected in parallel with the load. Many ranges feature one automatic control (the 8-in element) along with three units controlled by infinite-heat switches.

The calibration of the hydraulic automatic controlled unit may be checked by the following procedure.

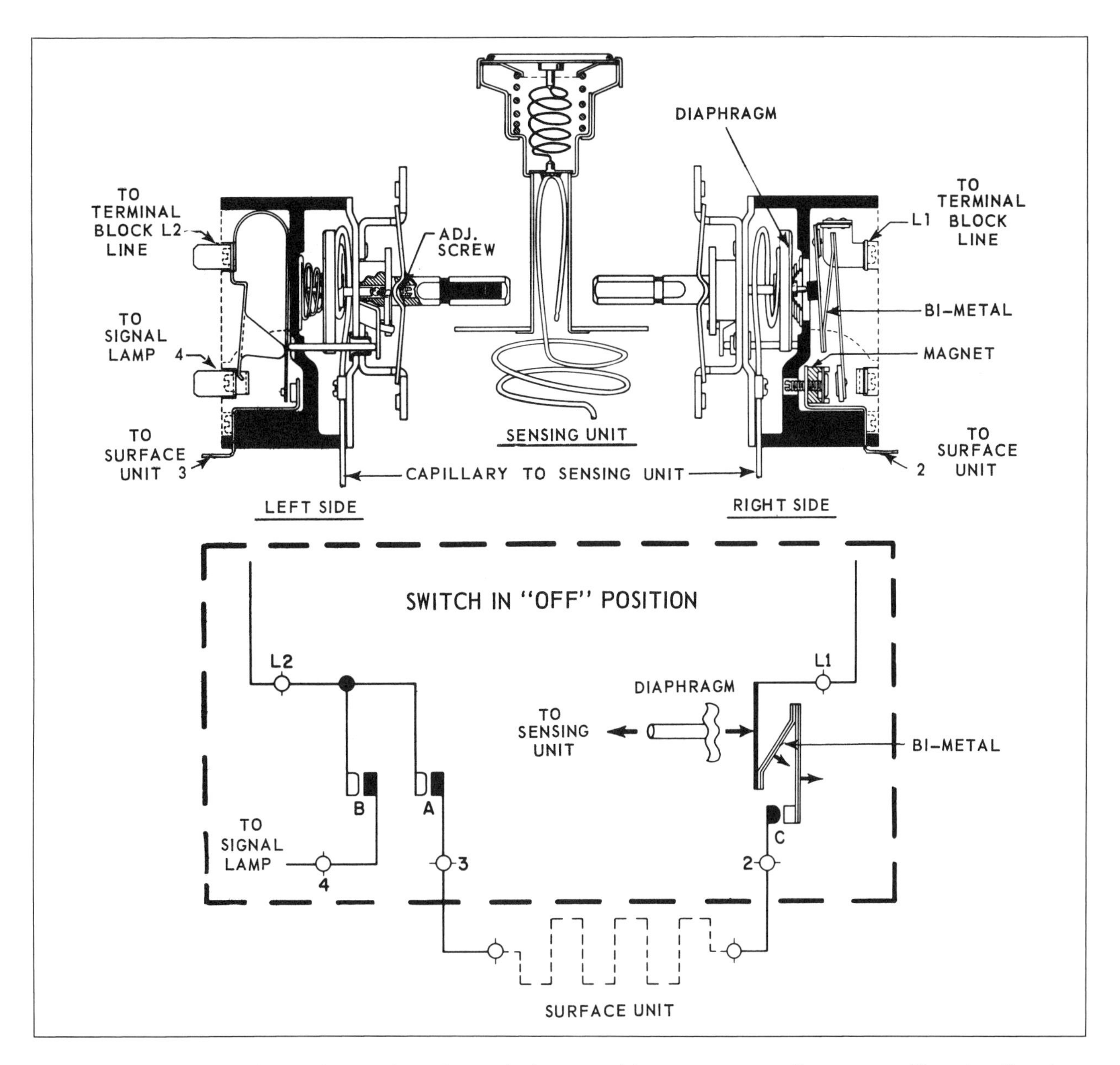

Figure 9-11. Construction and operation of a typical automatic control. A and B contacts are closed in all positions except OFF. C contacts will cycle off and on according to switch setting and utensil temperature.

1. Place an empty 8-in saucepan on the automatic unit.
2. Place the tip of an oven tester thermocouple lead in the utensil over the sensor. Hold firmly in place with a wooden rod.
3. Set the control dial to the LOW BOIL or 225°F position.
4. The dry-pan temperature should be between 215 and 245°F. (Allow about 5 to 6 min, after the initial overshoot, for the temperature to stabilize.) *Note*: A voltage lower than 236 V would increase the heating time required. With control knob settings above 225°F, while boiling liquids, the signal light cannot cycle. This is because after the water reaches a temperature of 212°F, the temper-

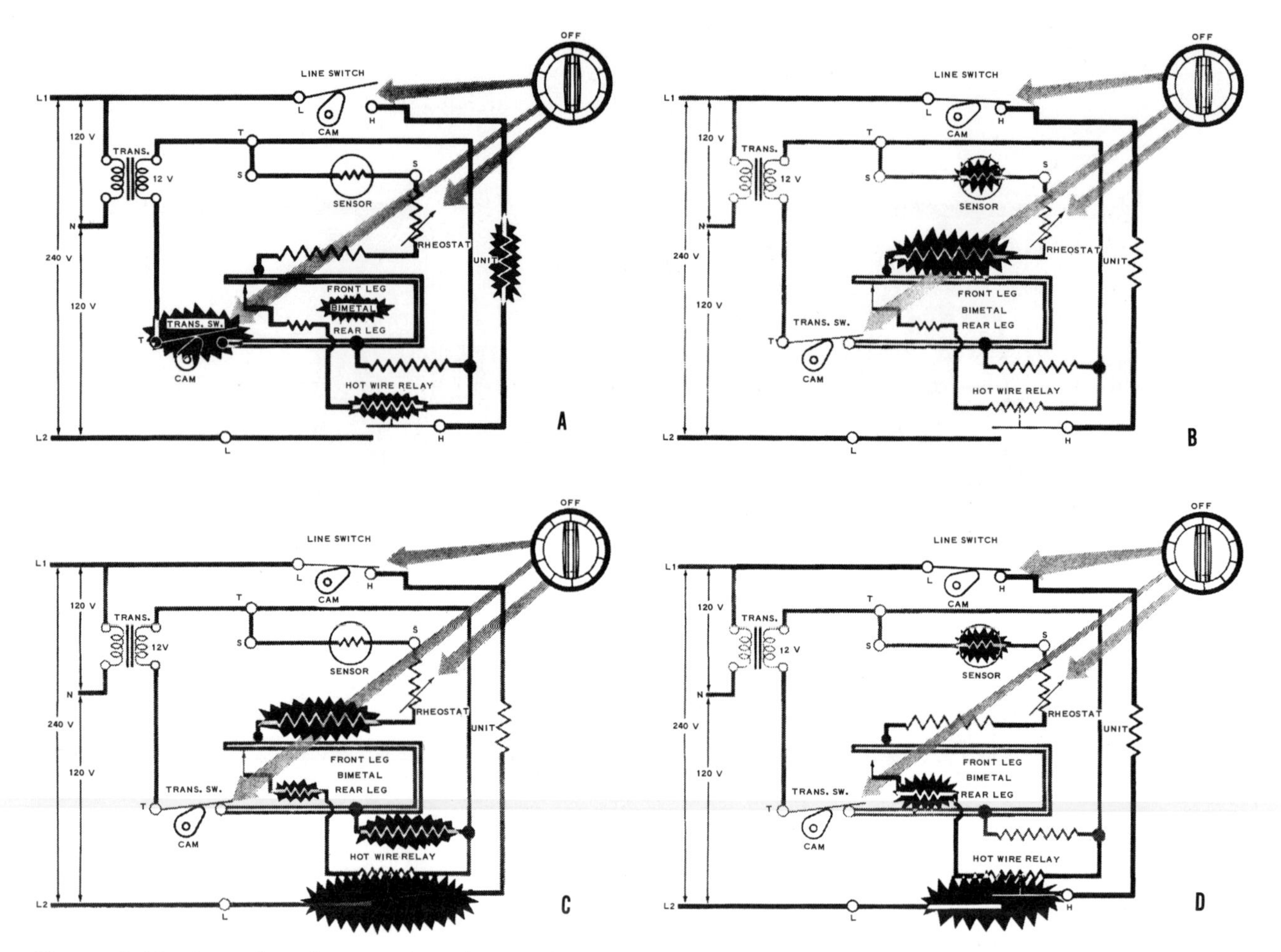

Figure 9-12. Operation of automatic resistance thermostat.

ature can go no higher. A higher control knob setting will result only in more vigorous boiling.

The adjustment for calibration of more automatic controls is a small calibration screw usually located in either the face of the control adjacent to the shaft or in the center of the shaft itself. As a rule, turning the screw counterclockwise increases the temperature; turning it clockwise decreases the temperature.

Another automatic control device that is frequently used features a resistance thermostat. The sensor in this arrangement is mounted much like the hydraulic wafer. It is, however, a resistance embedded in an aluminum sandwich. The connections to the main control body are two electric wires.

The resistance has the property of varying with temperature. At a low temperature, it has a lower resistance to current flow than at a high temperature. As shown in Fig. 9-12A, the main control consists of a rheostat, a U-shaped bimetal switch arm controlled by a heater, low-voltage switch, and hot-wire relay. These components along with the sensor are in a 12-V circuit. The control cam closes line No. 1, and the hot-wire relay closes line No. 2 of a 240-V circuit to the controlled unit. Turning the control dial to any temperature setting (Fig. 9-12B) closes the line switch and the transformer switch through a cam. A control resistance is set up through the rheostat at the same time, for the temperature selected. The 12-V transformer energizes the sensor, rheostat, bimetal heaters, and hot-wire relay. The hot-wire relay quickly closes line No. 2, energizing the unit.

Because of the U shape of the bimetal, the heaters on the "back leg" and "front leg" tend to oppose each other, keeping the normally closed bimetal switch closed (Fig. 9-12C). The unit, at full wattage, heats the pan bottom and the sensor contacting the pan. As the sensor heats, it becomes more resistive, reducing the current to the front leg heater. At a point some 65 to 75°F below dial setting, the back leg heaters overcome the bimetal switch, breaking the contact. This de-energizes the hot-wire relay and the heater in series with it, opening line No. 2 to the unit. With the heater on the rear leg off, the heater on the front leg quickly recloses the contacts, recycling the unit. Since the sensor is continuing to heat and become more resistive, the front-leg heater becomes progressively less effective (Fig. 9-12D). This changes the cycle so that there is more OFF time and less ON time. Finally the unit balances out to produce enough heat to maintain dial temperature at the point where the sensor contacts the pan. The result of this gradual reduction in power helps to control overshoot. It makes possible a relatively wide "boil" range from low boil to high boil, for a selection suited to the contents of the pan.

With either type of automatic control thermostatic system, the sensor height is usually adjustable and should be only about 1/8 in above the surface unit. Gradually the sensor can be adjusted up or down by bending its mounting bracket.

The shape of the cooking utensil that the consumer uses can affect the life of the surface element. For example, if a pan that has either a concave or convex bottom is used, all the rings of the element will not make contact with the utensil, causing parts of the element to overheat and shortening element life. Extended use of utensils with convex bottoms will actually cause the element to warp, and using either concave utensils or oversize utensils on an element will cause the element pans, or bowls, to discolor.

Checking surface unit temperature controls. Most control switches are not serviceable in the field and must be replaced if not functioning correctly. Using a voltmeter set on the 300-V scale, set the control on the HIGH setting. Check for voltage at both the input and output terminals of the switch. Generally the input terminals are marked L1 and L2, and the output terminals are marked H1 and H2 (Fig. 9-13). When in doubt, check the wiring diagrams. If there is voltage at the input terminals and no voltage at the output terminals, replace the switch. If there is voltage at the output terminals, one of the wires leading to the element

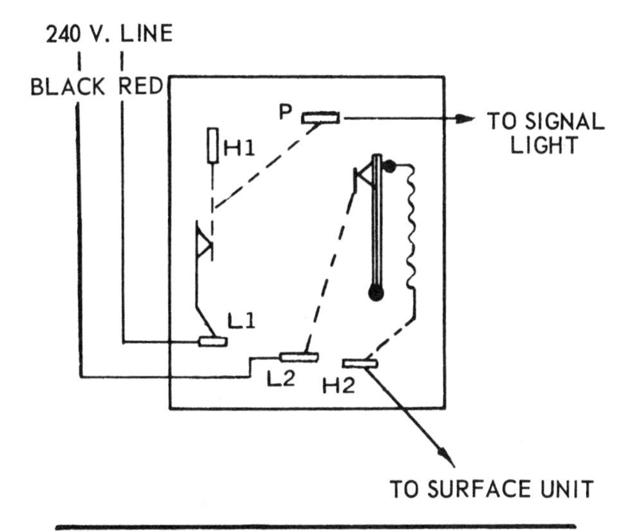

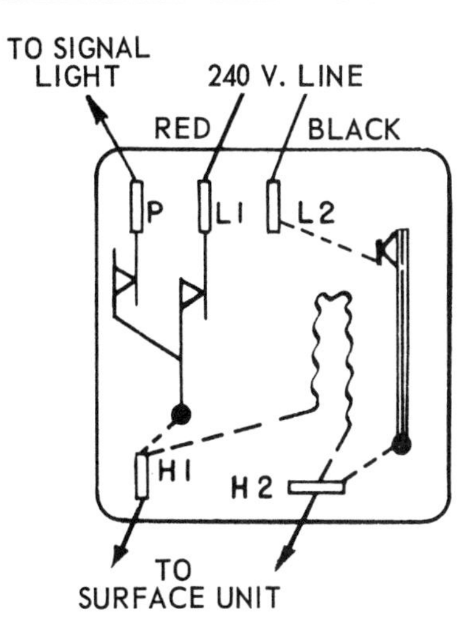

Figure 9-13. Switch electrical continuity of two most popular makes of controls.

does not have continuity and will have to be repaired.

A heat control switch can also be checked for continuity with a test light or an ohmmeter. When doing this, be sure to disconnect one of the leads to the switch. If the switch is open at room temperature it should be replaced. But, because the design and testing methods vary from one make of range to another, it is never wise to attempt to service any type of control switch mechanism until you have studied the service manual for the model on which you are working.

Since the wiring in all ranges is polarized, work on a surface unit and its connecting wire can be made without fear of suffering a shock if the surface unit is turned off. If there is any doubt about a certain switch, that is, whether it is off when it is supposed to be off, test each of the surface unit terminals to ground with a test lamp before handling the wires or terminals. When replacing a switch, however, the entire range must be shut off before handling the wiring which serves the switches and controls.

Other Features

Range manufacturers are constantly devising new ideas for special uses for their surface units. Automatic stirrers, deep-well cookers, automatic grills, charcoal heater units, and pressure cookers are some of the more popular ones. However, in a book of this size, it would be impossible to describe them in detail. Many of these "special" features appear in only a few models and then are dropped from the manufacturer's line. Thus, when you are required to service one of these special features, you must refer to the service manual.

Ceramic-Glass Cooktops

Glass cooktops of a ceramic-type material are now being used as surface units for both ranges and built-in counter units. With most cooktops, heat is applied to the ceramic top by any of or all four heating elements mounted under the surface and properly positioned by element retainer bars. The two types of heating element assemblies most commonly used with cooktops are:

Type	*Size, in*	*Wattage*
Large	9	2,200 W at 240 V alternating current
Small	$6\frac{1}{2}$	1,100 W at 120 V alternating current

The heating element assembly consists of a helix resistance wire heater semiencased in an alumina silicate casting. Electric connections to the helix are extended through the side of the casting by stainless steel strips which are welded to the external leads. All leads and their terminations within the heater box area (under the cooktop) are of high-temperature type materials and are considered nonrepairable. If a heater lead or its terminal fails, the heating element assembly must be replaced.

The elements of the cooktop (Fig. 9-14) are usually controlled by infinite switches on the console. (These switches operate in the same manner as those mentioned on p. 221.) A high-temperature limit switch mounted beside each surface unit interrupts the circuit to the unit if the surface temperature is excessively high. To warn the user when the surface is hot, many units have a HOT SURFACE light which is turned on by either of two thermostatically operated switches mounted under the cooking top between the elements. When the top is hot, one or both of the switches will close the circuit to the light and the words HOT SURFACE will appear in red at the rear of the cooking top. To turn the light on quickly, a 200-W biasing heater in contact with the thermostat is energized whenever an infinite-heat switch is turned on. This heater is turned off when the thermostat turns on the light. The thermostat keeps the light on as long as the surface is hot (above 120 to 150°F).

Some cooktops employ a slightly different method of warning the user that the unit is on. In these, when heated, the decorative design (frequently a sunburst) on the cooking top turns a color. When the switch is turned to OFF, the corresponding cooking area cools and returns to the original white color.

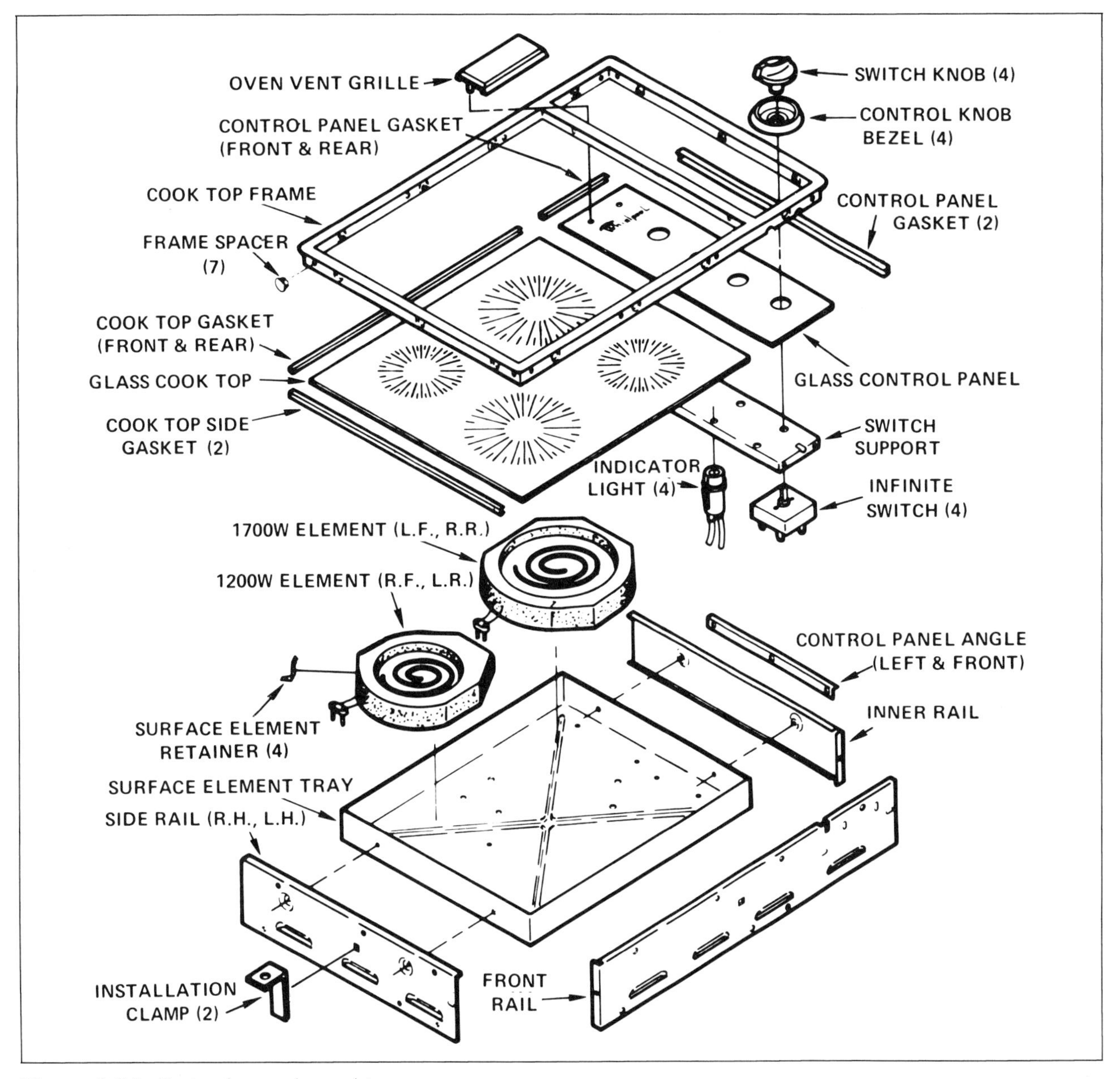

Figure 9-14. Parts of ceramic cooktop.

High-temperature control. All cooktop units, as already mentioned, have some type of high-temperature control which is used in conjunction with each heating element assembly and serves to control the maximum temperature of the cooking panel. This control may cycle during any of the following conditions:

1. Liquids in the utensil boil dry.
2. Warped utensils are being used.
3. The heating element is energized with no utensil on it.

The high-temperature control (Fig. 9-15) is usually mounted to the heating-element assembly with the temperature-sensing portion of the control positioned between the heater and the cooking panel. One of the most popular of the various sensing arrangements in use consists of a glass tube and a metal rod stretched through it.

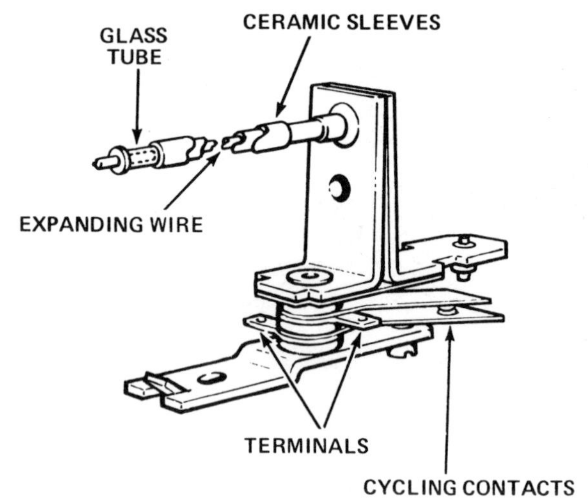

Figure 9-15. Typical high-temperature control.

The rod and tube are mechanically linked to a set of normally closed contacts which are connected electrically in series with the element heater. As the temperature increases, the metal rod expands more than the glass tube and opens the control contacts. With power to the heater interrupted, the rod cools enough to close the control contacts and the cycle repeats. The glass tube is encased in ceramic sleeves to reduce control cycling due to radiant heat from the heater.

The high-temperature control must be checked for proper calibration if any of the following circumstances occur.

1. The cooking panel is broken by high temperature.

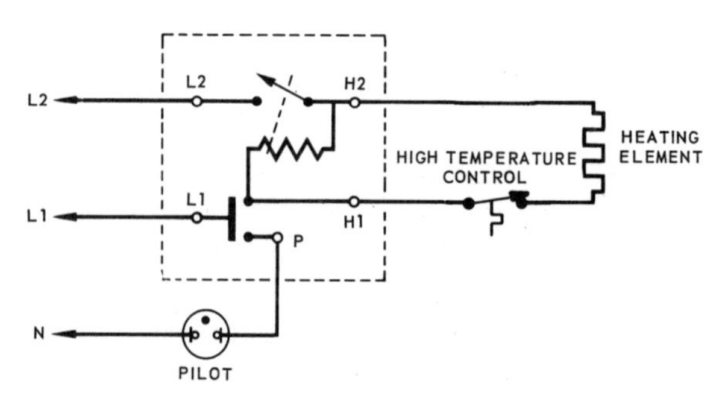

Figure 9-16. Typical high-temperature control circuit.

2. The heating element is replaced.
3. The high-temperature control is replaced.

Since designs of the heating element and control vary to some degree, the method of checking the calibration and making any adjustments must be done as directed in the service manual. As a rule, replacement and/or calibration are the only service functions to be performed on the high-temperature control. Any failures of the sensing device, the contact assembly on the spade terminals, usually require replacement of the entire control.

Ceramic cooking panel. The cooking panel is approximately $\frac{3}{16}$ in thick and is usually secured to the stainless steel-trim frame by screws and clips extending through the heater box. A self-adhesive silicone gasket generally seals the edges of the cooking panel to the trim frame to prevent spills from entering the heater box area. Most replacement panels are furnished with the gasket attached.

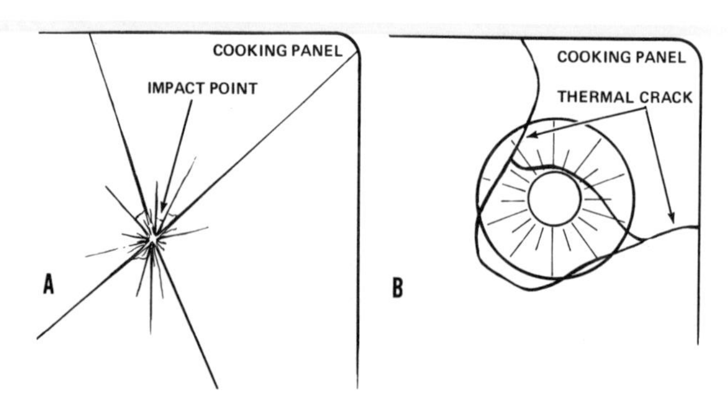

Figure 9-17. Cooking panel problems: (A) impact failure; (B) thermal failure.

The cooking panel may be broken by either impact or excessively high temperature. Impact failures generally may be identified by a star pattern with radial cracks extending from the impact point. Thermal failures may be identified as cracks that start at the edge of the panel and extend to another edge. In addition, thermally produced cracks generally pass through or around a heating-element location. In cases where a thermal failure has occurred, *all* high-temperature controls must be checked

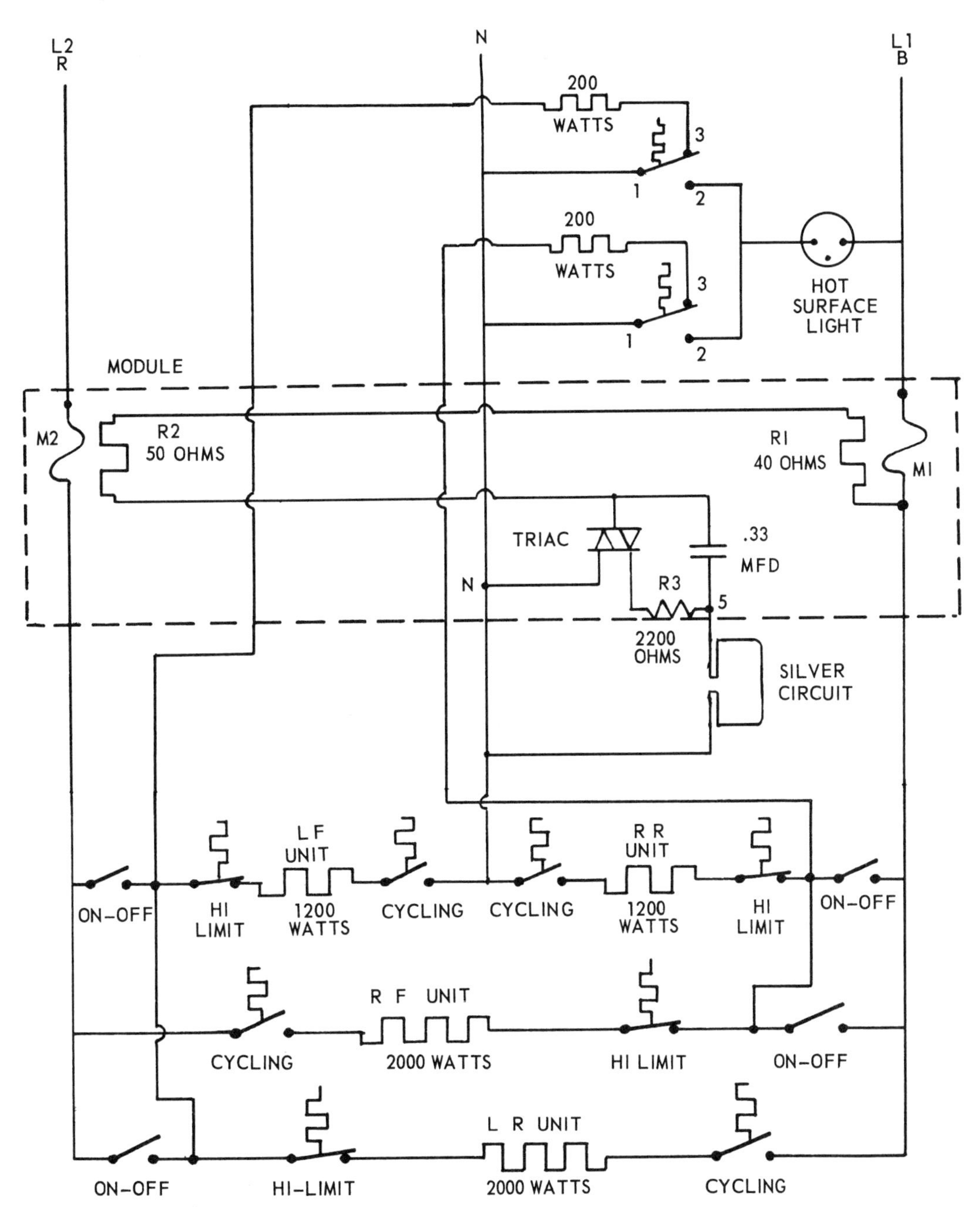

Figure 9-18. Simplified schematic of typical ceramic cooktop. *Note*: The cycling and the ON-OFF contacts of the infinite-heat switches are shown separately.

for proper calibration when the replacement panel is installed.

Many ceramic cooktops have a special safety circuit to protect against a damaged panel. A typical sensor for a broken safety circuit is a ribbon of silver film deposited on the underside of the ceramic. The silver film completes a circuit from the neutral line to the gate terminal of the triac. If the ceramic top and the silver film are broken, an ac voltage will be applied to the gate of the triac through a capacitor and a resistor (Fig. 9-18). (The triac is a solid-state device similar to a transistor.) As long as its gate is connected to the neutral line through the silver film, it will not conduct electric current, but when the silver is broken and the ac voltage is applied to the gate, the triac will carry a specific current. This current flows through both resistors on the module, heating the resistors and melting the thermal fuses which are enclosed in the resistors. These fuses are in the circuits to the infinite-heat switches and the heating elements. With this safety circuit, therefore, no heat can be supplied to the cooking top if the silver ribbon has been broken. A new top must be installed and new fuses installed on the module to restore operation.

Proper care and cleaning procedures are essential for a long and satisfactory life of the glass-ceramic cooktop. A cleaner-conditioner has been especially formulated to remove soil from the surface and build a protective silicone coating. This coating helps to prevent scratches and abrasions in which food particles can collect. Regular use of the cleaner-conditioner makes future cleaning easier.

Heating element assembly. To check a heating element assembly, follow the same procedure as for a conventional surface element (see Checking Surface Heating Elements, p. 221). But, when handling heating element assemblies, remember to take care to avoid damage to the alumina silicate casting. The casting is fragile and will break when subjected to undue pressures. Do not install an assembly if there is a break or void in the heater casting. Also keep in mind that if a heating element assembly requires replacement, the high-temperature control must be checked for proper calibration with the replacement assembly installed.

CONVENTIONAL OVENS

There are four basic types of ovens in use: (1) the conventional oven, (2) the continuous-cleaning oven, (3) the self-cleaning oven, and (4) the microwave oven. The modern oven, unlike surface units, incorporates features such as automatically controlling the time the food is cooked and measuring cook temperature. In other words, a conventional oven contains automatic control devices. To see how they work, let us take a look at various electric components of a conventional or non-self-cleaning oven.

Heating Elements

Most ovens employ tubular heating elements of the same design and materials as the surface units. Generally two tubular heating elements are used: a broil element located at the top and a bake element located at the bottom. Both elements operate at 240 V and require anywhere from 2,000 to 4,000 W, depending on the model. The determination of bake is made by a switch either on the range or the oven.

When broiling with most models, the user should keep the door open to the broil stop position. The broil heat is controlled by raising or lowering the oven shelf. The broil element should stay on continuously while the control is on BROIL, although in some instances it may be cycled off and on by the thermostat. This is apt to happen under one or more of the following conditions.

1. Prolonged broiling
2. Broiling with the door closed
3. Shiny surface near the thermostat sensing device (aluminum grid or aluminum foil)
4. Light broiling load

In some areas, electric ovens are used to heat the kitchen on a chilly day. This will cause the oven element to fail prematurely. Elements that have been used in this way will have a rough appearance as compared with those that have been properly used. The customer should be advised that the life of the element will be shortened if the oven is used for heating the kitchen.

Checking heating elements. To check the heating elements, first observe the elements when they are hot for even redness throughout their heated lengths. This check must be made with the oven control set on BAKE when checking

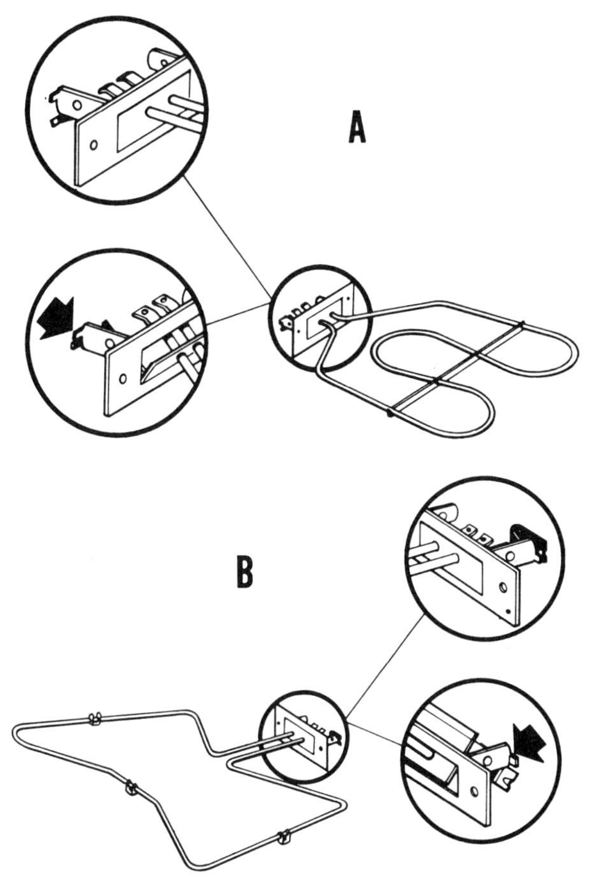

Figure 9-19. (A) Typical tilt-down broil element and (B) typical tilt-up bake element.

the bake unit and on BROIL when checking the broil unit.

Most oven heating units are completely front-servicing; that is, both the broil and bake elements may be removed from the inside of the oven by removing a screw at each side of the element and pulling the element out of the oven cavity. Be careful not to pull too hard when removing the oven element because it is possible to accidentally disconnect the spade terminals if care is not taken. Bake and broil elements have very low resistance and may be checked at the terminals in the same manner as surface units (see Checking Surface Heating Elements, above).

Temperature Control

The temperature control circuit consists of an oven control thermostat in series with a selector switch. Once the user sets the selector switch to the desired cooking operation and turns the unit on, the thermostat senses the temperature in the oven, and then by opening and closing the heating element circuit, maintains a relatively constant oven temperature for the cooking operation selected.

Thermostats. Most oven thermostats used on non-self-cleaning ovens are the hydraulic-actuated, bellows (diaphragm) type. The hydraulic control system is operated by the expansion and contraction of a liquid in a long tubular bulb suspended in the oven. This bulb is connected by a capillary tube to a bellows or diaphragm in the thermostat. Changes in liquid

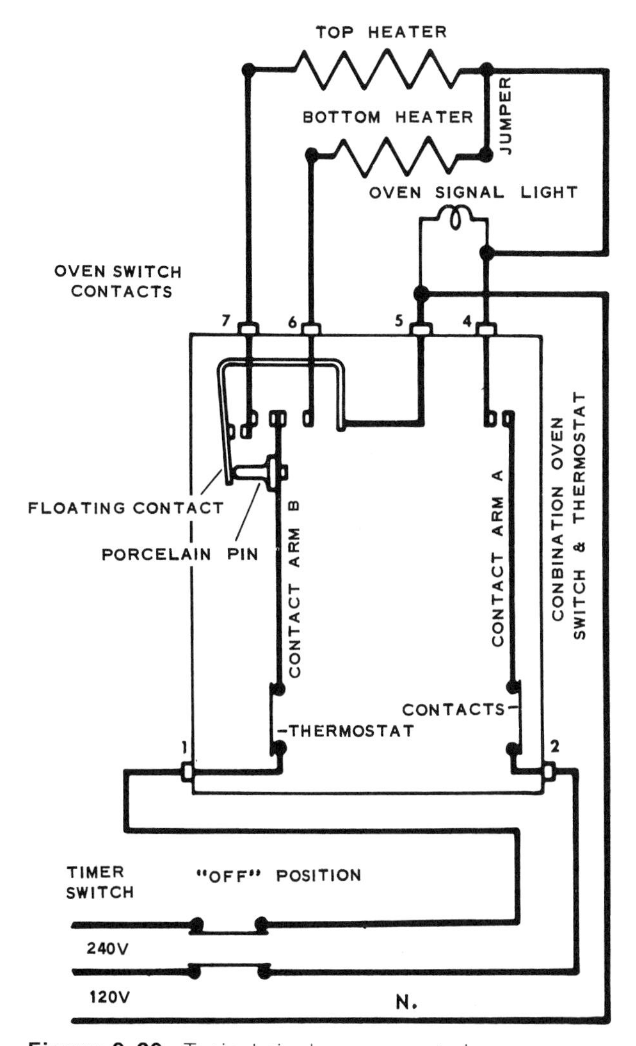

Figure 9-20. Typical single-oven controls.

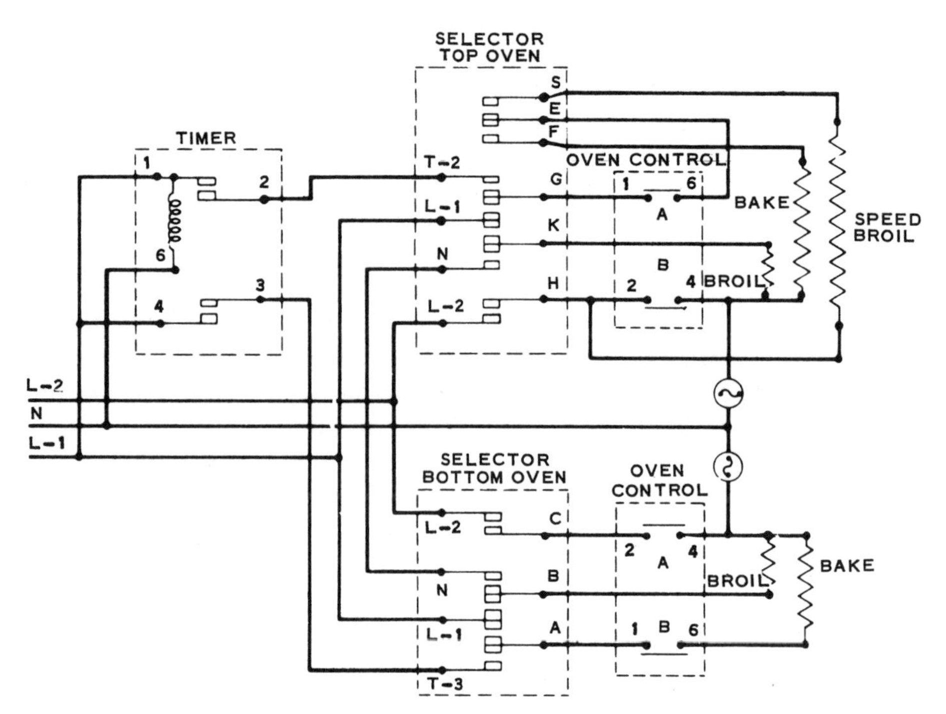

Figure 9-21. Typical double-oven controls.

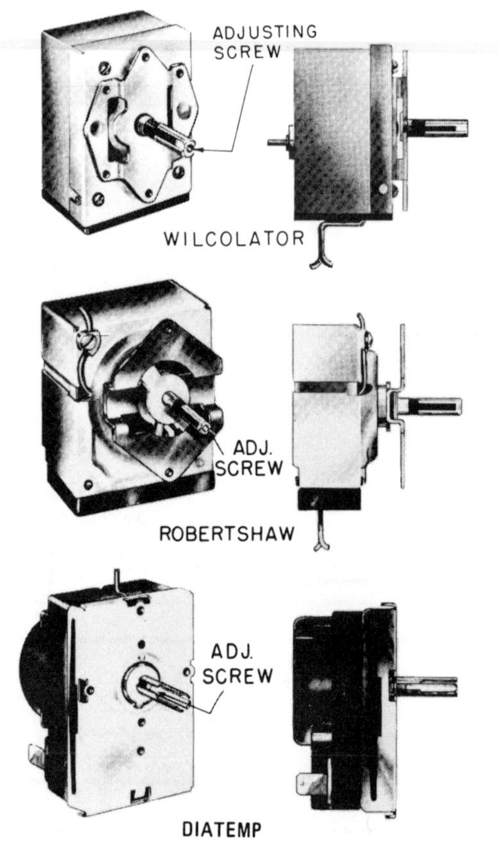

Figure 9-22. Popular types of oven thermostats.

pressure in the bellows caused by changes in oven temperature expand or contract the bellows to open or close a double-pole switch in the circuit to the elements. The special fluid used in the hydraulic system has full stability (linearity) in its expansion-contraction characteristics and maintains this stability in the entire operating range.

A few models employ a thermostat which is practically identical to the surface unit resistance thermostat described under Surface Unit Temperature Controls, p. 223. The main difference is in the sensor. Instead of the resistance being placed in a conductive sandwich, it is placed in a relatively long tube or rod. This provides more surface area for the heat transfer, which is by convection. Remember that rod-type sensors must be placed so that air can circulate around them.

Checking an oven thermostat. One of the major oven complaints that a service technician hears is that the unit is not heating properly. To check out this complaint, make certain that the thermostat sensor is not contacting the oven liner. The sensor must be away from the oven

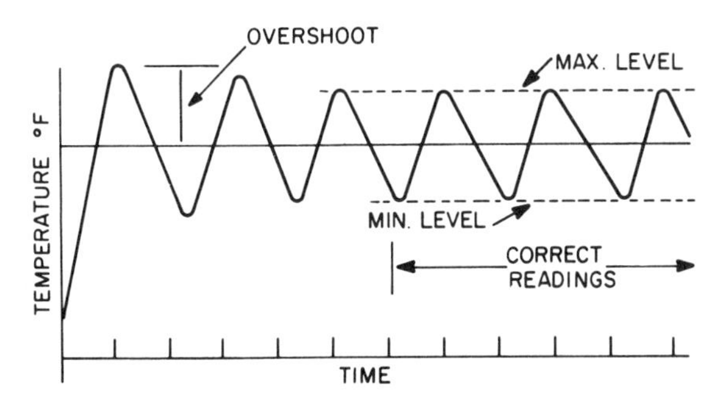

Figure 9-23. In the initial heating of a cool oven, the air temperature rises rapidly and will go beyond the selected dial setting. This "overshoot" is normal and is typical of actual oven operation. Overshoot continues until absorption of heat by the oven contents and oven walls is approximately equal to that given off by the units to maintain the selected temperature.

liner so that air can circulate around it to allow for proper readings. The oven liner generally becomes much hotter than the dial setting on the thermostat. The bending of capillary tube clips will usually remedy the problem.

In models using a rod-type sensor, it should be checked to be sure that the rod is perfectly straight. This type of sensor has an interior rod that moves back and forth. If the rod is bent, the inner rod cannot move freely, causing a complaint of improper heat.

A good temperature tester is required to determine if the oven thermostat is functioning correctly. Inexpensive thermometers, often used by customers, can be very inaccurate and should never be used to determine if the thermostat is performing properly. Using a good temperature tester, place the thermocouple lead in the center of the oven (Fig. 9-24). It is best to use a weighted thermocouple lead because it will give you more stable temperature readings. If you do not have a weighted thermocouple lead, take a piece of aluminum foil, fold it at least five times, and place it around a regular thermocouple lead. Turn on the oven, and set the thermostat at 400°F. Allow the oven to cycle four times or wait 20 min before taking a reading to allow the temperature in the oven to stabilize. The average temperature of the high-low readings must fall between 390 and 430°F. Do not place the temperature tester on the range because the body of the tester may sense higher surface temperatures created by the oven heating. Incidentally, with most ovens, the maximum temperature will occur about 30 s after the cycling light goes out. The minimum temperature occurs as the light comes on.

If the temperature readings do not fall between 390 and 430°F, the thermostat must be recalibrated. Remove the control knob with the thermostat still set at 400°F. There is usually a calibration screw in the center of the thermostat stem or near it. By turning the calibration screw either clockwise or counterclockwise, the thermostat can be calibrated (Fig. 9-25). Check the service manual to determine which direction the

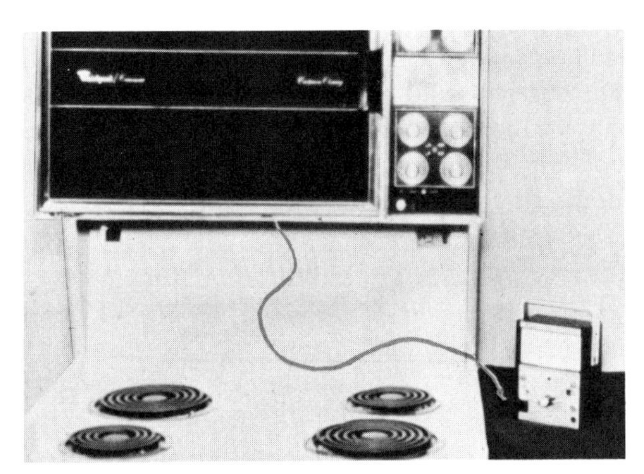

Figure 9-24. Checking the temperature of an oven.

Figure 9-25. Making a temperature adjustment.

calibration screw should be turned; it varies from model to model.

Whenever it is found necessary to replace either a heater wire or oven thermostat, it will be necessary to remove the thermostat capillary tube from the oven liner. The heater wire is usually held to the capillary bulb by two metal clips, one clip at the front where the winding begins, and another clip at the rear where the winding ends. The special resistance heater wire terminates outside the oven liner, and is connected to range wiring by electric connectors. To replace a heater wire, cut the range wires as closely as possible where they connect to the heater wire. Connect a new heater, using electric connectors. It is important to follow carefully the heater windings specifications in the service manual when replacing an oven thermostat or heater wire. This also holds true when replacing a rod-type sensor.

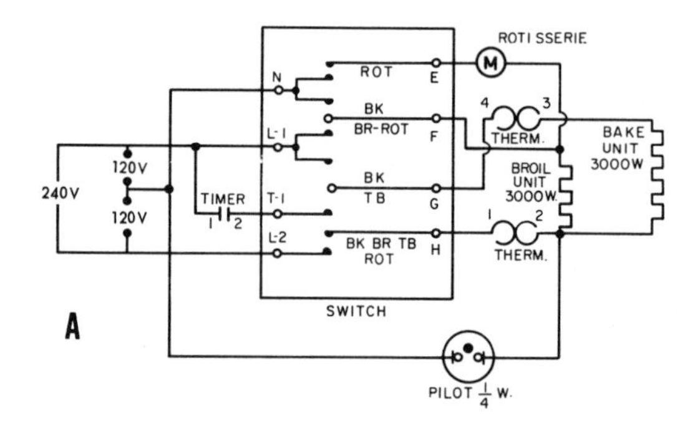

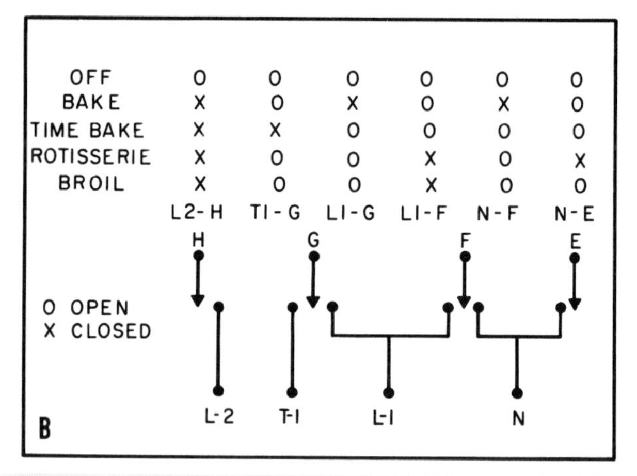

Figure 9-26. A typical five-position oven-switch circuit.

Oven selector switch. While a few models employ push-button switches, the vast majority use the rotary type. Four basic types of oven switches are usually used on non-self-cleaning ranges:

1. Three-position: OFF, BAKE, BROIL
2. Four-position: OFF, BAKE, TIMED BAKE, BROIL
3. Five-position: OFF, PREHEAT, BAKE, TIMED BAKE, BROIL *or* OFF, BAKE, TIMED BAKE, ROTISSERIE, BROIL
4. Six-position: OFF, PREHEAT, BAKE, TIMED BAKE, ROTISSERIE, BROIL

The oven selector switch is usually located on the oven control panel or the range backsplash or console.

Checking temperature control circuitry. Most "no-heat" complaints that can be traced to the temperature control circuit are generally due to problems with either the selector switch or the thermostat. If the proper contacts are not closed in either of these components, current cannot flow to the elements and relays. Because of the variety of selector switches and thermostats, you will have to check the wiring diagram to determine which terminals to check. But before replacing an oven unit control switch, closely inspect the lead and quick-connect terminals for corrosion or thermal breakdown. Each terminal should be clean and fit tightly on the switch spade. Any terminal that does not meet these standards should be replaced. Should the wire be oxidized or show signs of over-temperature, it should be cut back before replacing the terminal. The entire lead assembly should be replaced if severe oxidation is evident. Because of the high electric currents encountered in range operation, it is important that all wire splices and quick connects be soldered. Following these instructions should eliminate unnecessary call-backs due to poor wiring connections.

On some models both the bake and broil elements are controlled by relays (Fig. 9-27).

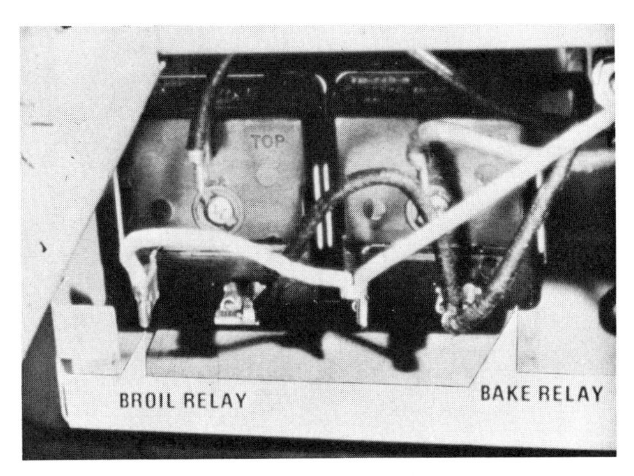

Figure 9-27. Bake and broil relays.

On a "no-heat" or "no-broil" complaint, check the relays to be sure that they are functioning properly. To do this, turn the oven on and place the probes of a voltmeter set on the 250- or 300-V scale on terminals H2 and L2. At first the voltmeter should read 240 V. After a period of 15 to 30 s, the meter should read no voltage because contacts inside the relay have closed.

Sometimes the "no-heat" complaint can be caused by incorrect use of the controls. Be sure that the selector switch is not set on the timed position for manual use of the oven. Another source of such a complaint could be the improper use of the range clock or timer. For example, to use the range timer with most models, the user should turn the selector to TIMED and set the thermostat dial to the desired temperature. More on the operation of range timers can be found in Automatic Timers, p. 240.

Meat Thermometers

The meat thermometer is designed for perfect roasting of meats and poultry. It is particularly helpful in cooking these foods since doneness is difficult to determine from external appearance. To be accurate, the probe must be inserted a minimum of 3 in.

Several meat thermometer arrangements are in use today. In the typical circuit shown in Fig. 9-28, the operation is controlled by a probe (thermistor) pushed into the meat and plugged into a receptacle on the side of the oven. A thermistor has more resistance at low temperatures and less at high temperatures; that is, the thermistor probe has less resistance as it heats. At room temperature the resistance should be more than 100 Ω; in boiling water, less than 25 Ω. As the probe heats, more current flows through the heater on the variable switch arm. This eventually closes the switch.

With the probe below temperature, the variable switch arm in the meat probe is not closed. The heater for the nonvariable switch arm is across the 12-V transformer as indicated by the dashed line. This switch pulsates more or less rapidly, depending on voltage. In this way it acts as a voltage regulator across the probe thermistor and heater. With the probe at temperature, the variable switch heater closes the circuit to the 12-V relay coil. The coil closes a holding circuit through contacts 6 and 4 of this relay. This holding circuit causes the relay coil to remain energized regardless of probe temperature. The only function of the roast probe is to energize the 12-V relay. Unplugging the probe breaks the circuit to the relay coil. As you can see, this circuit is designed to turn the oven off. The contacts at the bottom of the relay would be in the line to the thermostat.

As was already stated, this is a basic meat

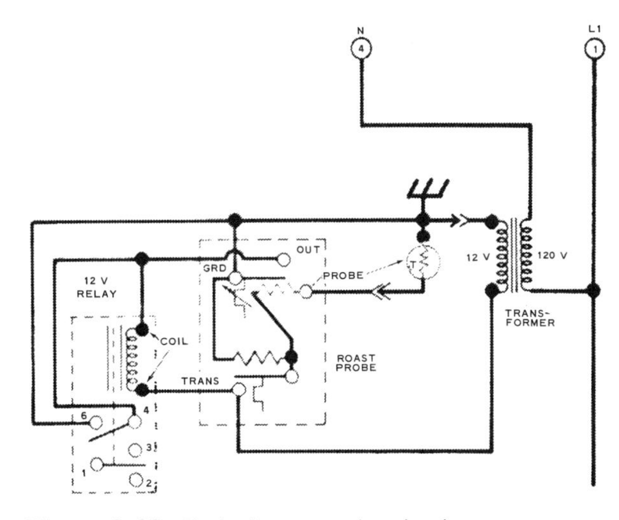

Figure 9-28. Typical roast probe circuit.

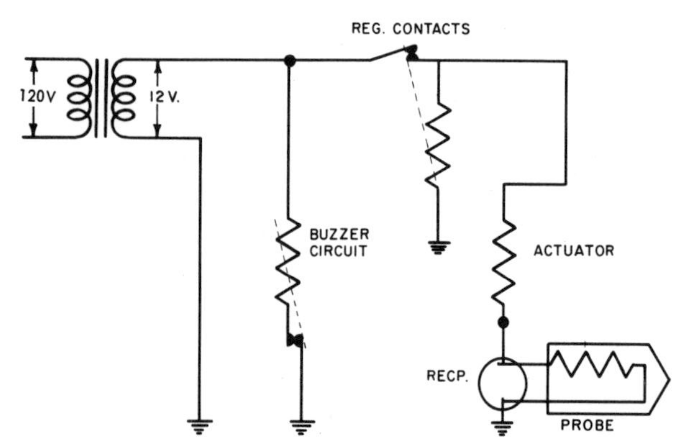

Figure 9-29. Roast probe circuit with buzzer.

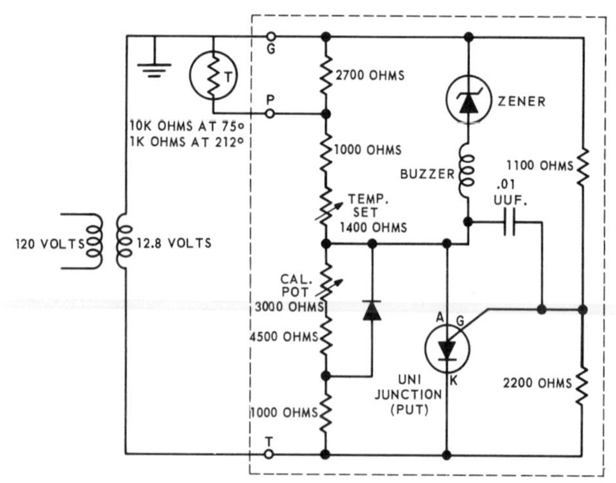

Figure 9-30. Typical solid state meat thermometer circuit.

thermometer circuit, and changes are made to meet various design features. For instance, many manufacturers do not shut off the oven with the meat probe—instead, the probe sounds a buzzer unit until it is unplugged or turned off (Fig. 9-29). A few manufacturers are even featuring solid-state control meat thermometers (Fig. 9-30). These, as well as the various other models, function in much the same manner as the typical one we described.

Rotisseries

Many ovens offer the versatility of cooking with a rotisserie. A typical rotisserie assembly consists of a motor, connected through the oven selector switch or a separate switch to operate the motor, and a drive sprocket attached to the shaft of the motor. The other parts usually consist of one square shaft spit, two U-shaped skewers with clamp screws, one removable handle, and the rotisserie frame. The user's instruction manual advises on the operation of the rotisserie. Be sure it is read carefully.

If the spit, when it is inserted in the drive sprocket or chuck of the motor, does not rotate, proceed as follows.

1. Gain access to motor leads.
2. Be sure the gears are not broken or jammed. If they are, replace the motor assembly.
3. With a test lamp or voltmeter across the motor connections, turn on the rotisserie.
4. A light or voltage indicates the motor is open and requires replacement.
5. Absence of light or voltage indicates that no voltage is getting to the motor, and the selector switch should be replaced.

Automatic Timers

Most modern ranges have some type of automatic timer. Most timer assemblies consist of a set of cams which operate in much the same way as those in an automatic washer (see Chap. 2, *Major Appliance Servicing*). A series of terminals open and close the appropriate circuits to turn ovens on and off automatically. (On some models, timers are also used on surface units and appliance outlets; a few ranges have a separate timing device for surface units.) Automatic control can extend for as long as 11 h.

Most timers are simply electrical clocks that are fitted with a built-in switch setup. In fact, most timers are the same, with changes only in the face position. As a rule, they have PUSH FOR MANUAL printed near the stop control knob. This is required for all ranges that use a combination ON-OFF and temperature oven control, or when there is only an AUTO appliance outlet. The stop control knob must be pushed in, in these cases, so that the oven or appliance outlet may be operated or controlled manually.

With a great many oven models all operations

can be timed at one setting. Two ovens, or one oven and an appliance outlet, cannot be operated at different time settings. With a selector switch oven control system, one oven can usually be operated manually while the other is timed. In the same type system, the appliance outlet can generally be timed, and the oven or ovens operated manually. In a nonselector switch system, all ovens or outlets are operated automatically or all are operated manually—one operation cannot be timed while the other is manual.

In a selector switch system the oven signal light will come on when the switch is set to any bake or broil position. With timed bake, it indicates only that the oven is ready to operate when the timer contacts close at the preselected time setting. The signal light will operate as long as the oven is cool enough for the thermostat contacts to remain closed. The signal light will cycle when heat is sufficient to open the thermostat contacts. With most combination ON-OFF temperature controls, the oven signal light will come on only when the oven heaters are energized, regardless of whether automatic or manual operation is being used.

For proper operations the time of day must be set correctly and the user must set the START TIME dial to the time that the oven is to begin operation. The STOP TIME dial should be set to the time that the oven is to stop operation. This is done by pushing in on the knobs on the dials and advancing them to the desired times. For example, if dinner is to be served at 6:00 o'clock and the food should be cooked for 2 h in a 350°F oven, cooking should start at four o'clock. The user should place food in oven, close the door, and proceed as follows.

1. Set START TIME by pushing in and turning it to 4. Oven will turn on at four o'clock.
2. Set STOP CONTROL by pushing in and turning it to 6. Oven will turn off at six o'clock.
3. Set OVEN CONTROL by turning the knob to the 350°F setting. Set the selector switch to the TIME BAKE position. The oven signal light will come on.
4. At four o'clock the oven will turn on, cook for 2 h at 350°F, and shut off at six o'clock. The oven signal light may come on when the oven cools, unless the selector switch is turned to OFF.

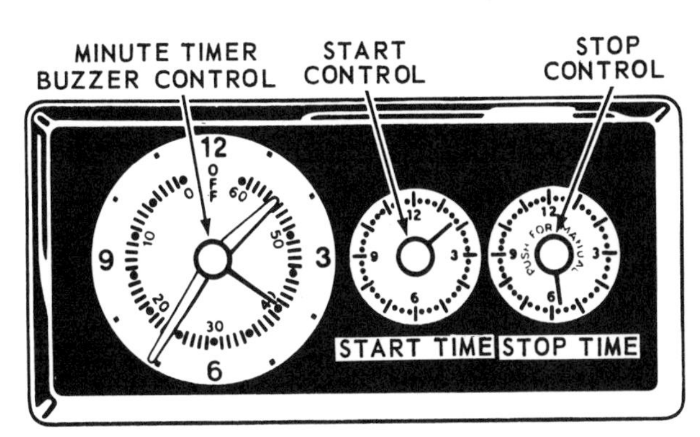

Figure 9-31. Typical timer setting.

Some ranges do not have oven selector switches. The clock assembly will look very similar to the one just described. However, a manual position will be indicated on the STOP TIME or cook hours dial on the clock. The indicator on that dial must be in the manual position for manual operation of the oven. Most ranges have a buzzer to signify the end of the timed cooking period.

While automatic timers seldom are faulty, replacement of the entire timer unit is usually necessary if one is found defective. Be sure to continuity-check the timer most carefully, however, before condemning it.

Other Range Components

Such components as lamps, appliance outlets, vent fans, and the mechanical parts at some time or another may require attention by the service technician.

Appliance outlets. Convenience outlets are frequently incorporated into ranges to supply 120-V current for the operation of small appliances on or near the range. For normal use of the outlet on a typical range equipped with a control switch, the switch must be set at the manual position, as shown in Fig. 9-32. Power flow is from one side of the line L1 through the control switch, through the fuse or

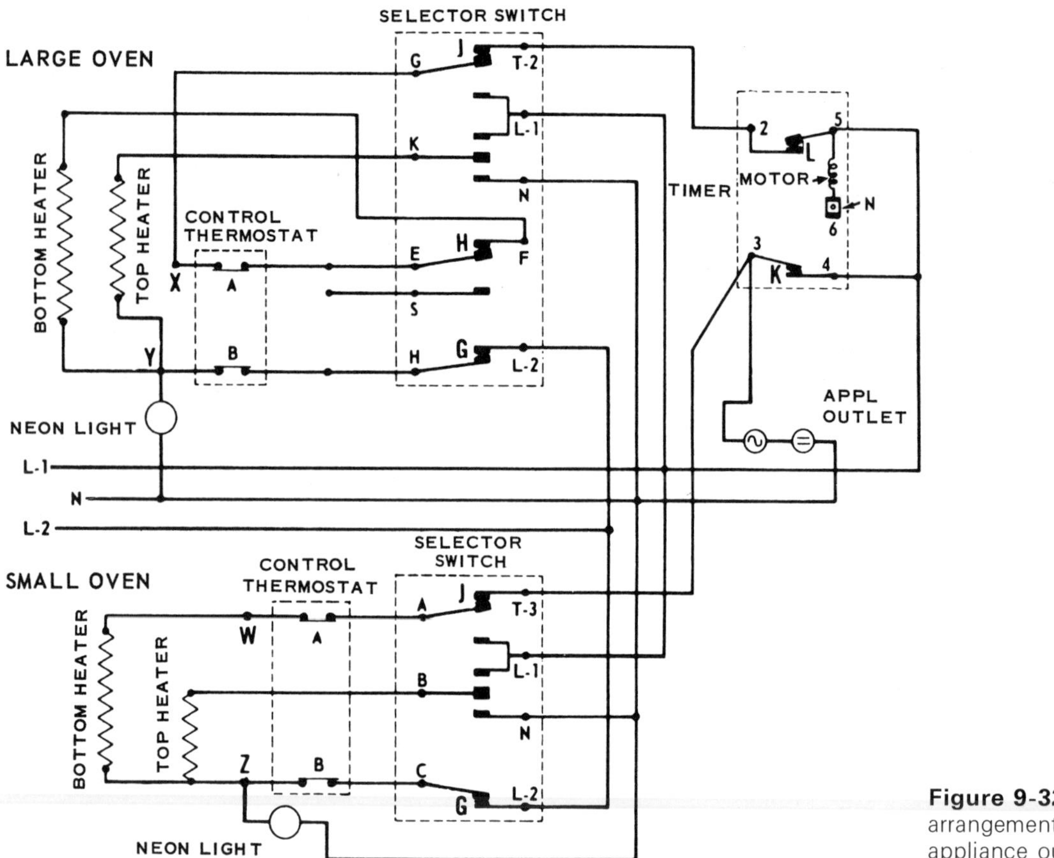

Figure 9-32. Double-oven arrangement with an appliance outlet.

circuit breaker within the range, outlet, and plugged-in appliance to the neutral side of the line.

For automatic use of the appliance outlet, set the START time and STOP time on the timer. Setting the START time opens the contacts of the timer switches. Then set the control switch at automatic and plug the appliance into the outlet. At the START time the contacts of the timer switch close at L and K, completing the circuit from L1 through K, through the control switch, fuse, and the plugged-in appliance to the neutral side of the line. This circuit will be maintained until the STOP time is reached when contact K is broken or the user removes the appliance from the outlet. In either case, the control switch should be set to manual. If during a timed bake setting the customer wants use of the appliance outlet, the control switch must be set at manual.

Signal lights. Signal lights are most important to the operation of an automatic electrical range. They are used to show the cycling of an oven, to indicate which surface units are used, and for various other functions, depending on the model. If a signal light should become nonfunctional, it should be replaced.

Platform light. Many models are equipped with a fluorescent platform light located at the rear of the platform beneath the top oven or in the rear console. The platform light is usually operated by a toggle switch or a momentary start switch.

If the fluorescent lamp does not attempt to light, proceed as follows.

1. Check the lamp to ensure that it is properly mounted in its sockets.
2. Check to ensure that the lamp pins are not bent and that they are clean and free from corrosion.

3. Replace the lamp with one that is known to be good.
4. Replace the starter with one that is known to be good.
5. Check for continuity of the ballast switch and all wiring.
6. When you find the inoperative part, replace the test parts with the original good parts.

When the ends of the lamp light, but the lamp will not light fully, proceed as follows.

1. With the lamp turned on, remove the starter from its base. If the lamp lights, the starter is inoperative.
2. Check all wiring for improper connection or ground.

When blinking occurs in the lamp, check for the following.

1. Possible faulty lamp. Replace, using a good one.

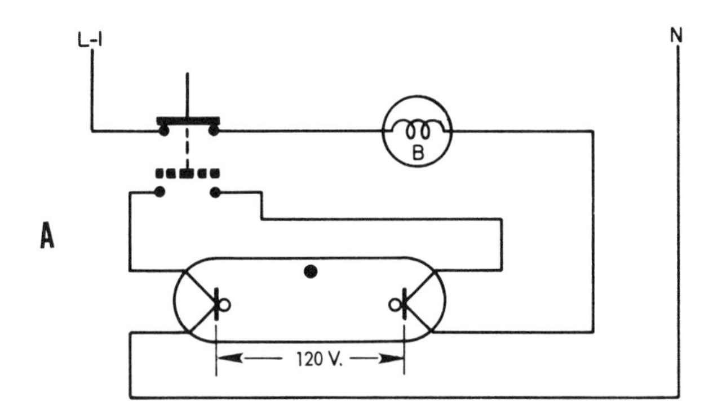

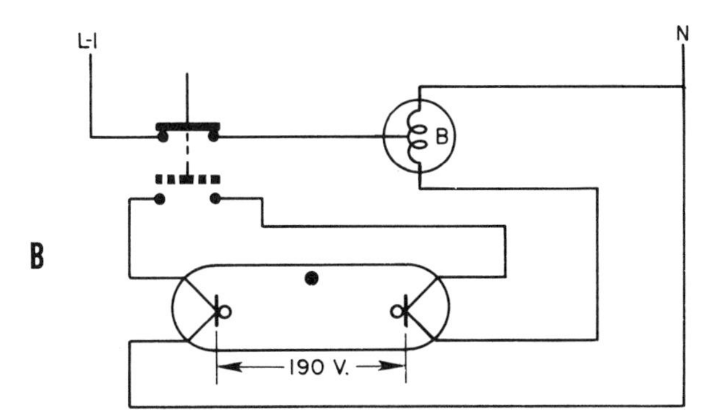

Figure 9-33. (A) Typical two-wire ballast circuit and (B) three-wire ballast circuit.

2. Possible inoperative starter. With the lamp turned on, remove the starter. If the lamp then operates properly, replace the starter with a good one.
3. Low temperature, below 50°F.
4. Low voltage (usually below 110 V).
5. Loose connection at the lamp sockets.

When a lamp gives one big flash, then no action, check for faulty ballast or wiring which has destroyed the lamp. Do not try another lamp until you find the defect and correct it.

When a lamp emits a humming sound, check for the following.

1. Ballast mounting screws loose.
2. Ballast overheated. Overheating can result from:
   a. Prolonged blinking.
   b. A shorted capacitor in the starter.
   c. A short in the leads to the ballast.

Interior lights. Most ovens are equipped with an oven interior light (usually an incandescent bulb). Most frequently a plunger-type switch for the oven light is operated by the door, and is so mounted that the light comes on as soon as the door is opened. Some ovens are also equipped with a "peak" switch, located on or near the control panel, so that the user can operate the oven light with the door in a closed position.

If the interior light system does not work, conduct a continuity check of the components and replace any faulty items. When replacing the bulb, be sure to use one of the same wattage.

Vent system. Certain models come with a built-in vent system; that is, the vent system is part of the range, since it uses channels behind the top oven to the platform. Since such models are available both with ducts and ductless, it is most important to follow the installation instructions that come with the range.

The servicing of the range vent fans is the same as that described for the exhaust-type fan described in Chap. 3, *Small Appliance Servicing Guide.* To keep vent fan service calls to a minimum, inform the customer that the filter given beneath the upper oven should be removed

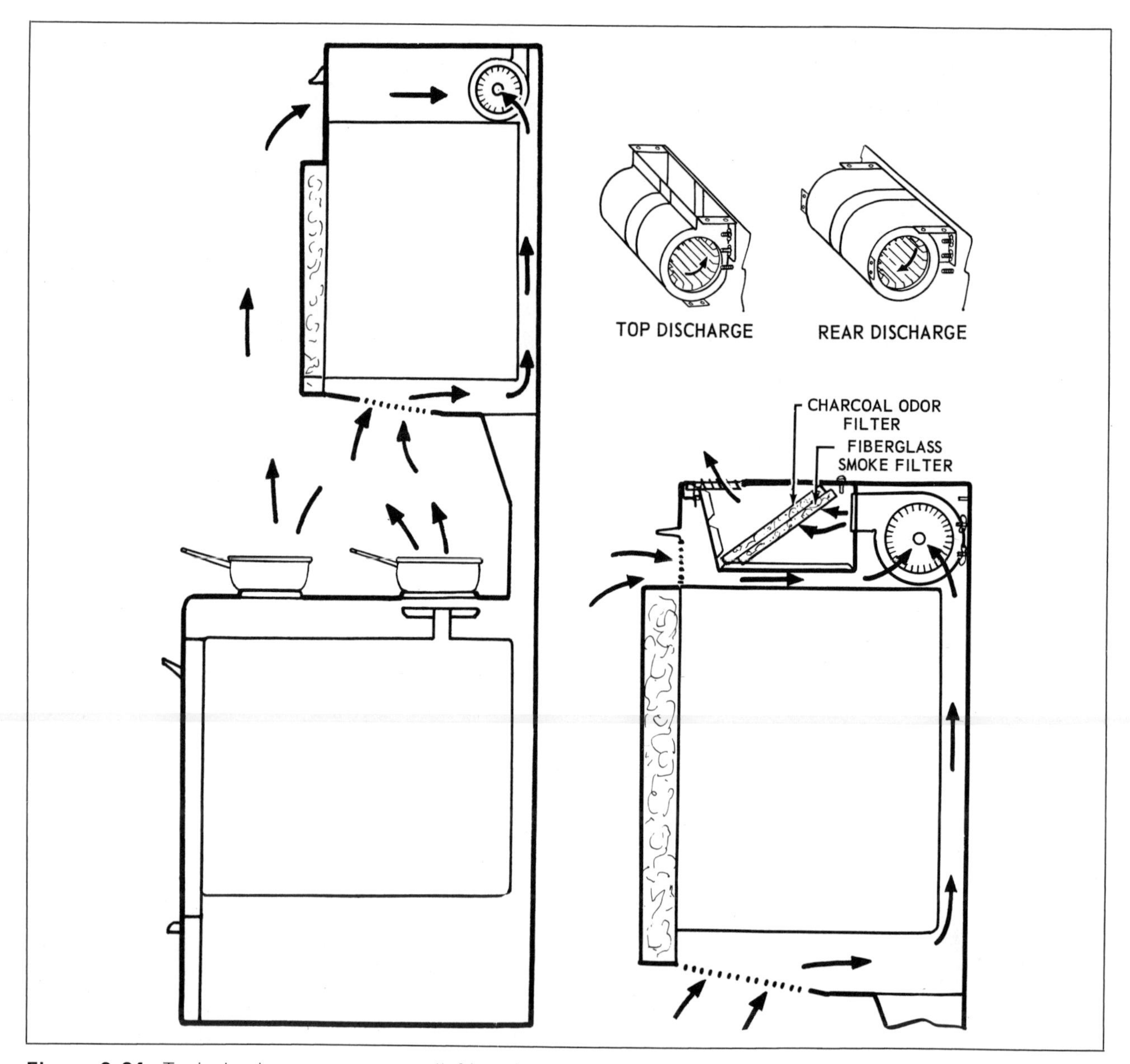

Figure 9-34. Typical exhaust vent system (left) and no vent recirculation or ductless vent system (right).

and washed in soap and water weekly. If the oven is used often, the filter may have to be cleaned more often, because it is really only effective when clean. The filter at the vent door requires less care. If a ductless fan is required, greater care of the vent filters will prolong the effectiveness of the charcoal filters. The more grease that reaches the charcoal, the sooner it will have to be replaced.

Mechanical parts. About the only mechanical components that give any trouble are the oven doors and oven liners. Oven door springs do not require attention in modern ranges, but keep in mind that both the springs and hinges should work smoothly and with no suggestion of crunching. Occasionally it may be necessary to adjust the spring tension or perhaps to apply a spot of lubricant to points of friction to eliminate squeaking. Door adjustment must be carefully checked, too, since many cooking problems (see Trouble Diagnosis, p. 257) can result from poor oven door fit. Because of the various designs of doors and different hinging methods, it is important to check the range's installation booklet or the service manual for complete details.

Many oven doors have glass panels. If the panel should break, it can be replaced with an exact replacement part by following the directions given in the service manual. This is also true of the oven liner if it needs replacement.

CONTINUOUS-CLEANING OVENS

When continuous-cleaning and self-cleaning ovens were introduced, some were referred to as *catalytic* self-cleaning ovens. The use of the word *catalytic* was an error. The proper terminology is *continuous cleaning* and *pyrolytic self-cleaning*. What is the difference between pyrolytic self-cleaning ovens and continuous-cleaning ovens? In the pyrolytic oven, high temperatures ranging from 850 to 1100°F are used. The oven is held at these temperatures from 1 to 3 h, and the soil in the oven is broken down by a chemical process called *pyrolysis*. In a continuous-cleaning oven, normal oven temperatures are used. The oven is cleaned during normal usage and the soils are dispersed by the process of organic oxidation.

The continuous-cleaning oven requires no special insulation, vents, controls, or thermostats. (Pyrolytic self-cleaning ovens require them.) Thus, continuous-cleaning ovens are serviced in the same manner as conventional ovens. Then, what makes the continuous-cleaning oven clean continuously? The answer is found in the type of porcelain used. Porcelain used on both the regular ovens and the pyrolytic self-cleaning ovens is very smooth, and the surface might be compared with the texture found on a glass pane. The porcelain finish used in the continuous-cleaning oven has a texture that is best compared with that of a brick. It is dark in color and contains many white stipples.

When a fat spatter strikes regular porcelain, it tends to form a drop just as water or liquid striking a piece of waxed paper would form a drop, which dries, causing the oven to soil. On the porcelain in the continuous-cleaning oven, the spatter is dispersed in much the way that a drop of moisture is in hitting a paper towel. In other words, it tends to wick-out. With the spatter wicked-out (broken down into small particles), it is decomposed at normal oven temperatures. The continuous-cleaning surface works well on fat spatters—the soil that generally occurs during roasting, broiling, and oven frying. Spatters created by spillovers from pies and casseroles are not decomposed on the continuous-cleaning finish. This type of soil generally collects at the bottom of the oven, and many manufacturers recommend the use of a heavy foil liner on the bottom of continuous-cleaning ovens.

What happens if, for some reason, a continuous-cleaning oven does become heavily soiled? What can the customer do to clean the oven? In time, most of the soil will disappear because the more the oven is used, the more decomposition of the soil will occur. However, if the customer would like to speed up the process, the interior of the oven may be cleaned with soap and water, or soap-filled pads. The use of chemical oven cleaners is not recommended because they actually reduce the efficiency of the continuous-cleaning porcelain. If there is very much spatter or spillover onto the surface when it is hot, it can cause a permanent stain. This is best described as permanent varnishing. No method is known for removing this type of stain.

The continuous-cleaning porcelain can be repaired if chipped. A repair kit is available from your authorized parts supplier. The kit consists simply of a bottle of liquid porcelain that dries and cures at room temperature. Clean the area to be repaired thoroughly and dampen it with water. Apply the porcelain repair material with a small brush and allow approximately 10 to 15 min to dry. Do not apply heat to speed up the drying process of this porcelain.

SELF-CLEANING OVENS

The pyrolytic method of oven cleaning is a chemical decomposition of cooking soils (which consist of compounds of hydrogen and carbon present in cooking oils, greases, sugar, starch, etc.) by the application of high heat. The process begins to occur at temperatures above 700°F. The temperature required for a satisfactory cleaning job at reasonable speed is in the area of 850 to 1050°F; the top limit depends on the model. Actually, the main difference in controls from conventional oven controls is a design to withstand these temperatures and control them accurately. That is, range models with self-cleaning ovens can, in most instances, be considered standard ranges from the service viewpoint. They have the additional feature of self-cleaning, but basically use the same components to bake and broil. During self-clean, they change the circuitry and add only a few extra components to control the operation. Therefore, only the special added requirements for the self-cleaning oven and circuitry will be covered here. But first let us examine the principle of operation of the self-cleaning cycle.

As already mentioned, cleaning begins in the temperature range between 700 and 750°F, initially removing moisture and then breaking down the hydrocarbons into smoke and gases. These by-products are further heated, as they pass through the oven vent system, by a smoke eliminator, which breaks down the smoke into a clear, odorless gas which resembles a "heat" odor. These relatively high oven temperatures are above the flash point, or burning point of oven greases, cooking oils, etc., and therefore the cleaning process is controlled by limiting the amount of oxygen entering the oven. Although this oven cleaning is commonly referred to as "burning off the soil," the process is a controlled temperature/oxygen decomposition of oven soil. During cleaning, the time element is also important, and is based on the degree of soil, light, medium, or heavy.

Although the oven is capable of cleaning many times the normal soil load without a noticeable smoke odor, it is not an incinerator, and for best results, the cleaning process should be performed frequently according to soil loads. *Typical* approximate cleaning times are as classified below.

1. Self-clean oven alone, 2 h for light soil, 2 to 3 h or more for a moderately to heavily soiled oven.
2. Self-clean oven plus aluminum reflector pans, 2 to 3 h or more, depending upon amount and type of soil.
3. Self-clean oven plus removable panels from companion oven, $2\frac{1}{2}$ to $3\frac{1}{2}$ h or more, depending on amount and type of soil.

The above cleaning times are based on soil evenly distributed throughout the oven. Whenever heavy spillovers occur, they should be wiped up to remove the bulk of soil from concentrating in one area of the oven. The type of soil present will also affect the cleaning time. Baked-on barbeque sauces are the most difficult to remove. Fruit acids such as pie drippings are the next most difficult to remove, with cooking oils, grease, and fats being the easiest. At the end of the cleaning cycle, a light residue of white ash remains which can be wiped out with a dry cloth, if desired.

While the procedure may change slightly with the various models, here are the typical instructions given the customer to start the self-cleaning operation.

1. Close the door; turn the oven set to CLEAN. Push and hold the latch release while moving the latch to CLEAN. Be certain the latch is moved as far as possible toward CLEAN. Check the following if the latch does not move easily: (a) Be certain the shield is

locked in place over the window; (b) be certain the oven set is at CLEAN; and (c) be certain the lock light is not glowing. The latch is designed not to move easily unless all these conditions are met.

2. On the automatic oven timer, set the pointer ahead, on the stop dial marked CLEAN, for as many hours as are needed to clean the amount of soil in the oven.
3. The oven cleaning light will glow when all the steps have been set up properly. Also, the sound of a fan will be heard sometime during the cleaning cycle. The lock light comes on when the oven heats to temperatures above those usually used for cooking. The light stays on during the cleaning time and until the heat decreases in temperature (to about 500°F), when it goes out.
4. When the lock light is off (at below 350°F), push and hold the latch-release button while sliding the latch to the COOK position. (Lower the window shield by pushing the handles toward the bottom of the window.)
5. Turn the oven set to OFF.

Bimetal-actuated High-Temperature Thermostat

Three basic types of temperature controls are used in conjunction with the self-cleaning operation: (1) bimetal-type thermostat; (2) responder-type control circuit; and (3) solid-state control. Let us first look at a bimetal-actuated high-temperature thermostat control circuit—the first of all self-cleaning control systems.

Actually there are two basic types of bimetal-actuated high-temperature thermostats in common use today. One uses the same controls as in conventional oven thermostats or top unit thermostats. The high-limit control in this design is a rod and tube of two different materials. The differing rates of expansion provide the force to open a set of contacts at the desired high temperature (anywhere from 850 to 1100°F). This control is in series with the low-voltage hot-wire relay. Opening the contacts causes the hot-wire relay to open. The cycle is repeated over and over to maintain the desired high temperature.

In a few models, a conventional hydraulic-

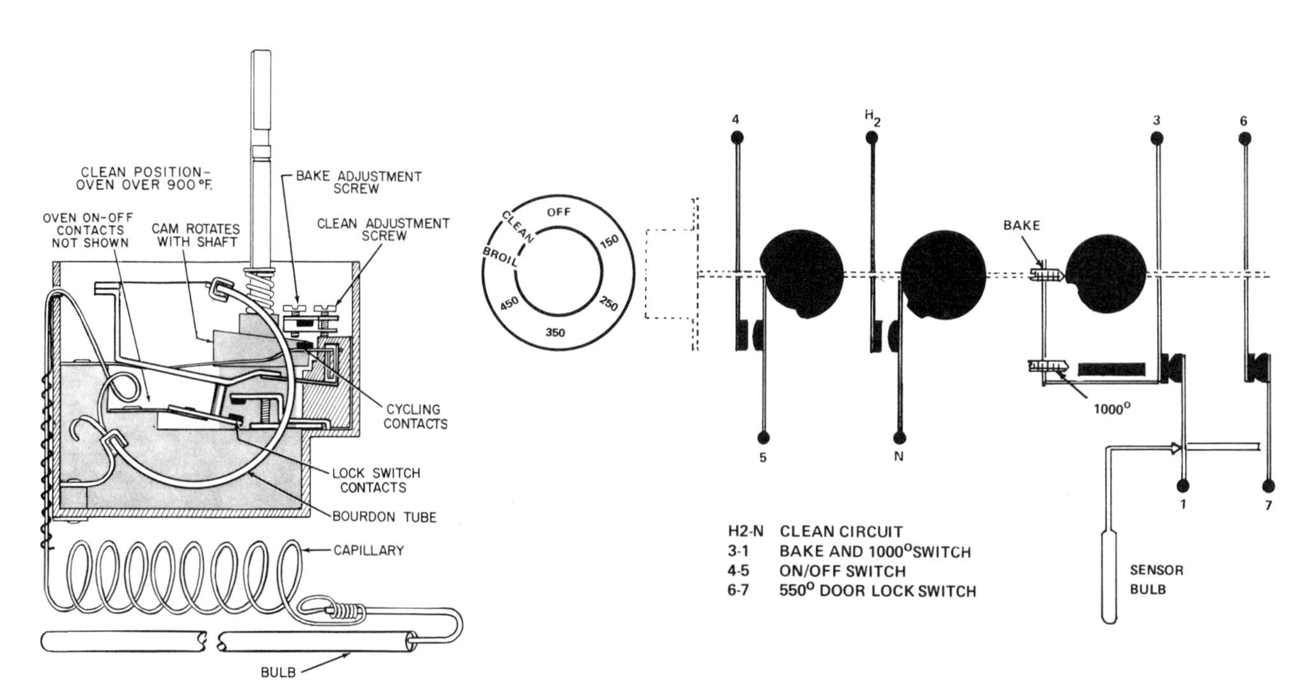

Figure 9-35. Typical oven temperature thermostat that can be used for baking, broiling, and cleaning.

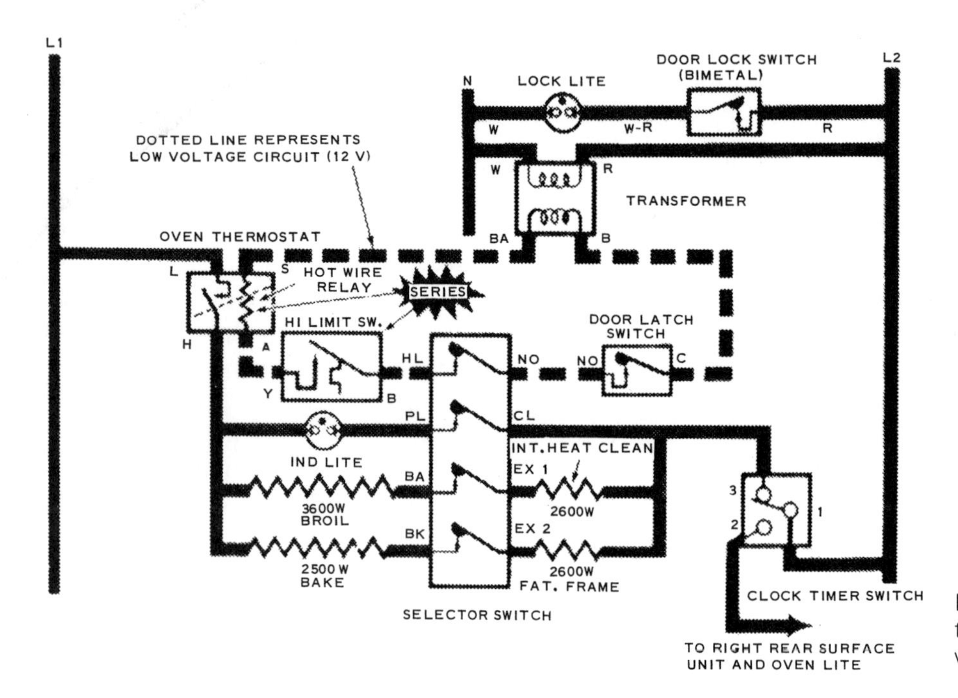

Figure 9-36. A high-limit thermostat in series with the low-voltage hot wire.

type thermostat (Fig. 9-35) is used, but to handle the high temperature of the oven clean cycle, the temperature control uses helium gas instead of a liquid as in the nonelectrical-clean ovens. The helium gas operates a curved bourdon tube which tries to straighten as gas pressure is built up by expansion of the gas in the thermobulb. This straightening action opens and closes the cycling contacts at temperatures determined by the angular position of a cam which is rotated when the user turns the knob. A pair of lock switch contacts is usually opened by the bourdon tube at a temperature of 500 to 600°F to interrupt the circuit to the latch motor or solenoid and to energize the "locked" light, letting the user know the oven door cannot be opened. Because the temperature-control contacts cannot handle the high current experienced during cycling of the oven unit, a thermal expansion hot-wire type relay is used to make and break the circuits to the units.

The typical circuit shown in Fig. 9-36 employs a clock to time the cleaning operation in conjunction with the bimetal-actuated thermostat. A selector switch is used to put the broil unit in series with an additional upper cleaning unit mounted in the broil unit frame and the bake unit in series with an element mounted inside and contacting the oven front frame (often called a *mullion heater*). Being in series, this cleaning cycle uses a total wattage of about one-fourth the total rated wattage. For example, a unit rated at 12,000 W at 240 V would reduce to about 3,000 W in a series 120-V arrangement such as the one shown here. But remember that since the mullion unit is usually used in bake and time-bake cooking cycles (the mullion unit in series with the broil unit), the switch heater and resistor combination is energized by the mullion unit voltage at any time the oven units are in the ON cycle. When the units cycle off, the switch heater also cycles off. The temperature operating limits of the switch are set high enough so as not to be affected by normal baking temperatures or normal cycling of the units. *Note*: Since this is a temperature-time relationship, the oven will cycle on the temperature limit switch if the oven door is held open for unusual lengths of time during a bake cycle.

The step-down transformer shown in our typical circuit reduces the 120-V supply from one side of the 240-V line to the 120 V needed to operate the high-limit control and the door latch switch. During the cleaning operation the oven sensor circuit controls the hot-wire relay. This relay physically operates the high-voltage contacts to cycle the oven units on and off. Incidentally, the timer clock terminates the cleaning cycle by opening the line to the heater units. In most cases, the clocks for self-cleaning ovens are the same as for standard ovens. The one difference is that there is no manual position. The selector switch bypasses the clock in normal operation other than TIMED BAKE and CLEAN. The setting for clean is by advancing the COOK HOURS or STOP TIME dial.

A variation of this design (Fig. 9-37) used by many manufacturers has a bias resistor in parallel with the standard oven sensor (see Thermostats, under Conventional Ovens, p. 235). This in effect acts to fool the sensor so that its operating limit is increased. It is designed to raise the high limit to upward of 1100°F. The external rheostat in series with the resistor is the adjustment for high temperatures in the clean cycle only.

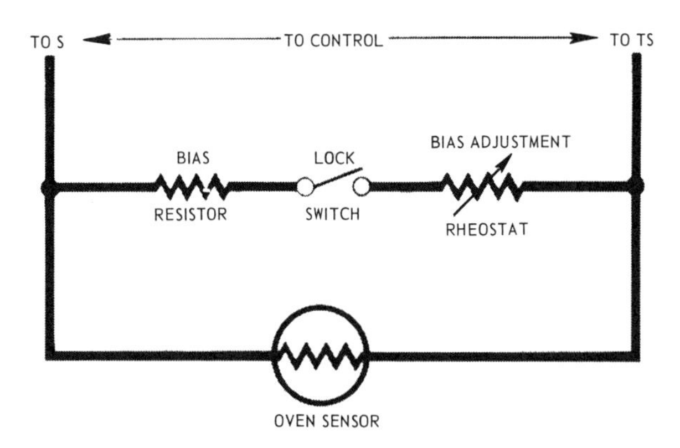

Figure 9-37. The use of a biased oven sensor.

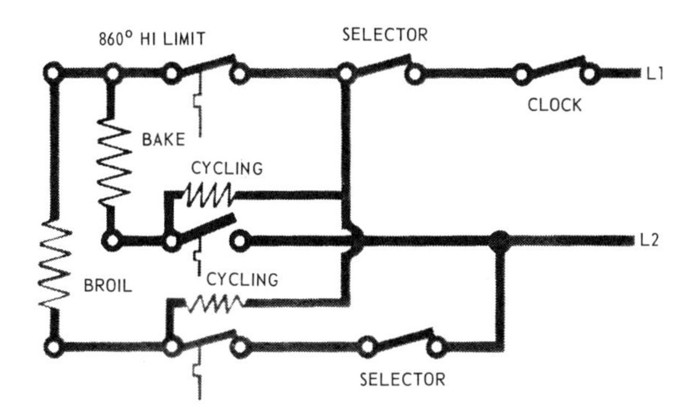

Figure 9-38. Another method of using a bimetal-actuated high-temperature thermostat.

The second method of employing a bimetal-actuated high-temperature thermostat generally uses only the bake and broil elements (Fig. 9-38). The selector places a preset infinite control (similar in construction to the type used in surface units) in the lines to cycle both these 240-V heating units. This preset control cycles at about 50 percent "on" time. This means that the units are producing about 50 percent wattage during the cleaning operation. As a rule, no field calibration is possible with these infinite switches.

With the second method, the high limit control (thermostat) is completely separated from the oven control. That is, with most models of this type, the oven thermostat is a conventional single-pole hydraulic type located on the outside of the oven. Inside the oven an expanding rod and tube or push-rod type of bimetal sensing element is used. In most designs, the rod presses against a liquid bellows found on the end of the hydraulic line; this operates a switch in the control. It is this push-rod switch that opens at the high-temperature limit. Additional switches may be added to activate at preset temperatures to control oven locking and/or a KEEP WARM circuit, if one is employed.

In most models, the push-rod, bimetal-actuated thermostat is fixed at the factory, and no field calibration is possible. On the other hand, the calibration procedure for the oven sensing device is the same as for a standard thermostat (see Checking an Oven Thermostat, p. 236, under Temperature Control above).

Responder-Type Control Circuits

In a responder-type circuit, a transformer is employed to furnish low voltage to the controls which determine oven temperature during the bake, broil, and cleaning cycles. This trans-

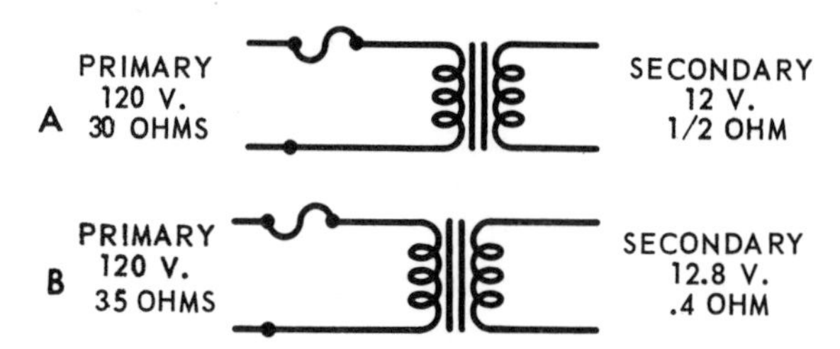

Figure 9-39. Typical characteristics of (A) responder-type transformer and (B) solid state type transformer.

former (Fig. 9-39) is usually a 120/12-V step-down type. In most responder transformers, a fuse is usually connected internally in the primary winding under the wrapper. The fuse protects against overheating of the transformer in case of a short circuit in the low voltage or secondary winding. Since the fuse is internal, it generally cannot be replaced. By the way, noisy transformers can be a source of complaint, especially on wall ovens.

The oven sensor, which senses the oven temperature and applies the proper signal to the control, is a variable resistor which is mounted in a metal tube and located in the oven, usually at the rear.

The internal operation of the oven control and sensor is shown in Fig. 9-40. As you can see, both the potentiometer (pot) and the sensor are in series with the responder, which opens and closes as temperature changes. The potentiometer resistance is set according to the temperature selected on the knob attached to the potentiometer.

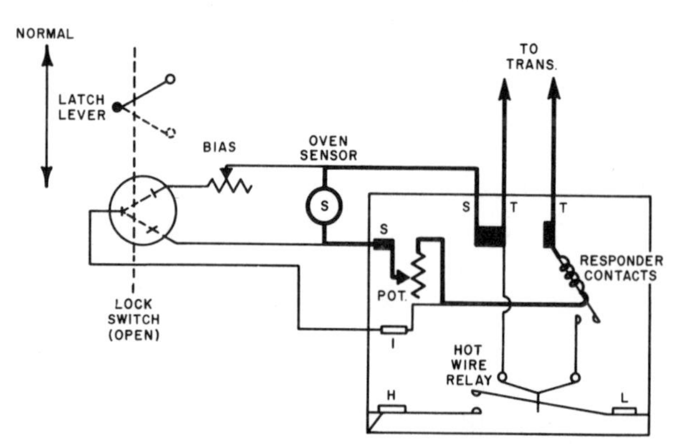

Figure 9-40. The normal bake-broil circuit.

The controlling circuit is a balanced system, which means that, regardless of the temperature selected, the control point is reached when the combined resistance of the sensor and potentiometer is equal to the resistance of the responder winding. The variable involved is the sensor which changes in resistance as the oven temperature changes.

Basically the warm setting of the control is close to the maximum resistance of the potentiometer; the broil setting is close to the minimum resistance. When the control point is reached, the responder winding side of the bimetal opens the responder contact and breaks the circuit to thc hot-wire relay, which in turn opens the hot-wire relay contacts and interrupts power to the oven units. The reverse operation occurs when the oven sensor cools off during the OFF cycle of the control. When the resistance of the sensor decreases enough to call for heat, the bimetal closes the responder contacts and finally the hot-wire relay.

When the self-cleaning cycle is started by the locking of the latch handle or lever, the lock switch is then closed. This puts the bias potentiometer (rheostat) in parallel with the sensor. This action increases current flow and changes the value of the sensing circuit to allow the heater units to remain on for a longer period of time, thus creating the higher temperatures needed for cleaning. In other words, this parallel circuit merely fools the temperature control (responder or solid-state control) to sense a much lower oven temperature than is really present. The bias circuit, therefore, increases the operating temperature range of the temperature control to a nominal clean temperature. By proper calibration of the bias rheostat, the clean temperature limits can be controlled. When the clean temperature point is reached, the internal electrical operation of the control, from a resistance standpoint, is the same as it would be for the control point of any baking temperature. The control still operates around a bal-

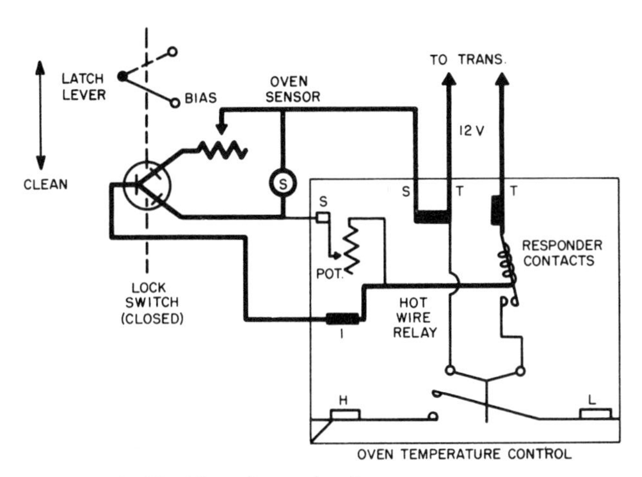

Figure 9-41. The clean circuit.

ance point of about 50-Ω combination of the sensor and bias resistance (Fig. 9-41). The calibration of the bias rheostat should never be changed except when the actual oven temperature is known.

As just described, a bias resistance is switched in parallel with the oven sensor to obtain temperatures high enough for cleaning. If the bias circuit does not get energized during the CLEAN cycle, the oven temperature will reach only about 550 to 600°F. This results in hard baked-on soil with the appearance of black varnish. To test most bias circuits, proceed as follows.

1. Place the oven switch in the OFF position.
2. Latch the oven door.
3. Connect the ohmmeter test leads to the terminals and on the bias board. Use the R × 1 scale.
   a. The ohmmeter should read 10 to 30 Ω (approximately).
   b. Rotate the oven responder knob slowly and observe the meter. The resistance should not vary as the knob is rotated.
   c. Unlatch the door. The ohmmeter should now read about 100 to 200 Ω.

The CLEAN cycle temperature (the exact degree should be checked in the service manual) can be checked during an actual cycle with a temperature meter having the proper temperature scale. The maximum or stabilized temperature will be reached in about 1 h. There are several alternate methods of "calculating" the clean temperature, but these are for specific models. Check the service manual for these methods.

Solid-State Controls

The same general principles which applied to the just-described responder and sensor system also apply to the solid-state oven control. In fact, four of the five basic parts—the transformer, the sensor, the hot-wire relay, and the rheostat (potentiometer)—are the same. The only major addition is the printed circuit board.

The printed circuit board contains all the solid-state components and is the heart of the system. The components are a silicon diode, zener diode, unijunction transistor, and silicon-controlled rectifier (SCR). In addition to these parts, it frequently also contains two factor calibration rheostats, one for the clean temperature adjustment and one for bake temperature adjustment.

Circuit operation. The logical way to troubleshoot most electronic circuits is to start with the end result, where the work is done, and then work backwards to determine how it is done. The following discussion is based on this method.

The work required in this case is to heat an oven. As in any oven, heat must be controlled at various temperature levels. This means that the heating units must be cycled on and off to hold an average temperature. Therefore, an electric contact is required which will cycle the power on and off to the heating units. A typical solid-state control circuit is shown in Fig. 9-42; its basic operation is as follows.

In the responder system previously described, a hot-wire relay was used to cycle the units on and off. A hot wire is also used in this system to operate the high-voltage line contacts. Also, as in the responder system, the hot wire is operated by a low-voltage circuit. To obtain

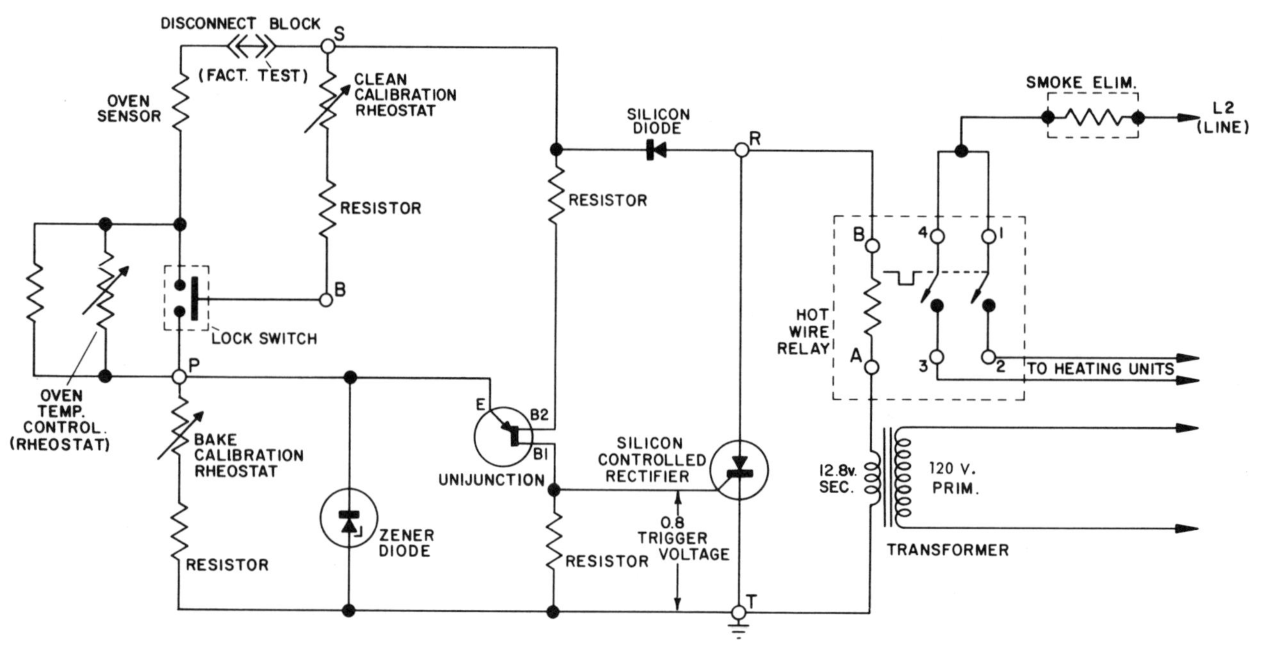

Figure 9-42. Typical complete solid state circuit.

the low voltage, a transformer is used for the solid-state system.

The silicon-controlled rectifier acts as a switch to close and open the circuit to the hot-wire relay. If a positive voltage is present at the anode and 0.8 V or more is present between the gate and cathode, the SCR will be triggered and will conduct from cathode to anode. This will provide the hot wire with a half-wave dc voltage during the positive swing of the voltage across the secondary of the transformer. When the gate voltage is less than 0.8 V, the SCR will not conduct, and the circuit to the hot-wire relay will be open.

A trigger voltage is now required to operate the gate of the SCR. In order to generate a voltage, a resistor is added between the gate and cathode. However, there must be a method to control the current through the resistor in order to provide the voltage drop which will be used as the SCR gate voltage. The device used to control the signal through the resistor is the unijunction transistor. Since the unijunction is a dc device, a diode is added to the circuit, as shown, which converts alternating current to half-wave direct current.

When the unijunction conducts, emitter current flows from base B1 through the emitter and the 100-Ω resistor connected to B1. A voltage is developed across the resistor which triggers the SCR and closes the circuit to the hot wire. However, additional circuitry is required to tell the unijunction what to do and when to do it. This is accomplished by adding a resistive load in parallel with the unijunction which develops the voltage and current necessary to operate the unijunction. The basic resistive load added is the 15-Ω oven sensor, a 0- to 29-Ω rheostat, and a 0- to 35-Ω calibration resistance. The rheostat is set by the temperature control knob which the user sets to select cooking temperatures. The combination of the oven sensor, two rheostats, and resistor provides the input signal to the unijunction. This portion of the circuit constitutes a voltage-divider network. Whatever voltage is present across the network will be divided between the sensor-rheostat combination and the bake calibration rheostat-resistor combination.

The firing point of the unijunction depends upon the ratio of the emitter voltage to the B2 voltage. This is termed the *stand-off ratio.* The

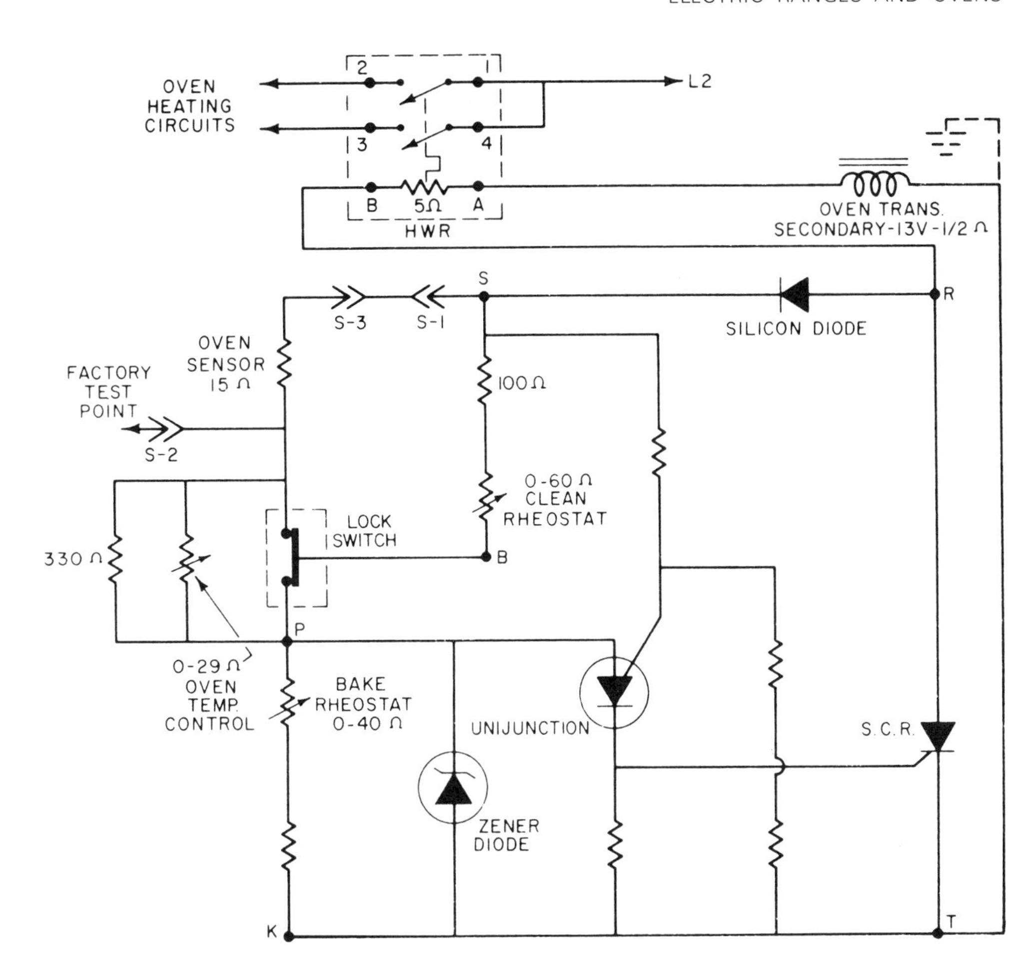

Figure 9-43. Bias circuit clean cycle.

stand-off ratio of the unijunction used in this application is 0.67. Therefore, when the voltage drop across the bake calibration rheostat-resistor combination is 0.67 of the total voltage present across the divider network, the unijunction will conduct. When the unijunction fires, the voltage across the 100-Ω B1 resistor triggers the SCR which conducts and completes the circuit through the hot wire. This closes the relay and turns on the oven heating units.

Another method of understanding the basic circuit operation is to examine the series combination of the oven sensor and 0- to 29-Ω rheostat. When the total resistance of the sensor and rheostat is less than 50 Ω, the stand-off ratio is greater than 0.67 and the unijunction will fire. When the total resistance increases to 50 Ω, the unijunction will not fire. This is the same principle which operated the responder.

The same principle of responder control also is employed in a great many solid-state systems as to self-cleaning operation. That is, when the lock switch is closed to start the cleaning cycle, another resistor(s) (the bias) is placed in parallel with the oven sensor and locks out the temperature control (Fig. 9-43). This changes the value of the sensing circuit and permits the oven to reach the necessary high temperatures for self-cleaning.

Solid-state diagnosis procedure. The following test procedure is intended to isolate faults with the transformer, hot-wire relay, oven sensor, rheostat, printed-circuit board, or lock switch. It assumes that other general circuits and components such as oven switch, heating units, transformer primary circuit, and power circuitry are functioning properly. Remember, the oven will not operate unless the transformer is getting primary voltage.

To make a general test, turn the oven on for 30 s or so to determine whether the oven units heat; then turn the oven off.

1. If units heat, this means the transformer, sensor, and basic control circuit are working; the question is, How well? The nature of the complaint will guide you. If a baking problem is involved, check the oven temperature and calibrate on the knob whether the *units cycle*. If the *units do not cycle* but operate in a constant-heat manner, follow the diagnosis information for Oven Will Not Turn Off below.
2. If units do not heat, follow the diagnosis information for Oven Will Not Heat below.
3. If a cleaning problem is involved, check bias circuit and clean calibration.

Rapid cycling of the solid-state oven control system may occur because of an interference signal being present on house wiring circuits or lead dress within the range. One of or all the following may be required to resolve a specific complaint.

1. Reverse incoming power leads (L1 and L2) at range terminal block.
2. Reverse oven transformer secondary leads at oven transformer.
3. If condition still exists after Steps 1 and 2 have been completed, replace printed circuit board. Replacement boards incorporate a filter capacitor used to eliminate false triggering of the unijunction transistor.

Other Components of Oven Clean System

There are several other important components of the oven cleaner system. They include the following.

Latch assembly. Because the self-cleaning oven reaches temperatures between 850 and 1100°F, the door must be locked for user safety. As a rule, it cannot be opened when the oven temperature exceeds 550°F. The lock usually consists of a latch and a strike. The latch is operated by either a motor or a solenoid. When energized, the solenoid or motor controls the movement of the latch. In most range ovens, at any time the door is locked and the oven is above 550°F, the clean indicator lamp will also be on.

The winding of the latch motor or coils of the solenoid which operate the latch assembly can be checked with an ohmmeter. If the motor and solenoid are good, the ohmmeter should read a small resistance. The latch switch(es) may also be checked with an ohmmeter. Most latch switches are of the single-pole double-throw type.

Oven window. The window feature of a self-cleaning model allows the user to view the inside of the oven only during cooking operations. An insulated shield is used as a heat barrier between the window glasses during the self-clean operation. This shield must be in its proper position before the door can be latched. The method of placing the shield in its cleaning position varies according to the manufacturer. This information is in the service manual.

Fan assembly. In most self-cleaning models, the fan is energized either when the clean controls are set or when the temperature switch reaches about 550°F. The purpose of the fan is to channel air up both sides of the range and exhaust it out the surface unit openings to cool the body and cooktop surfaces of the range. A portion of the forced air is directed on a fan fail switch.

Some models have a fan switch which is used as a safety device to ensure proper fan operation during the cleaning cycle. The fan switch interrupts power to the oven transformer when the voltage to the fan motor is 105 V or less. Since the fan provides air circulation for components and external surfaces, it is necessary to

Figure 9-44. Typical self-clean circuits below 500°F (top); above 500°F (bottom).

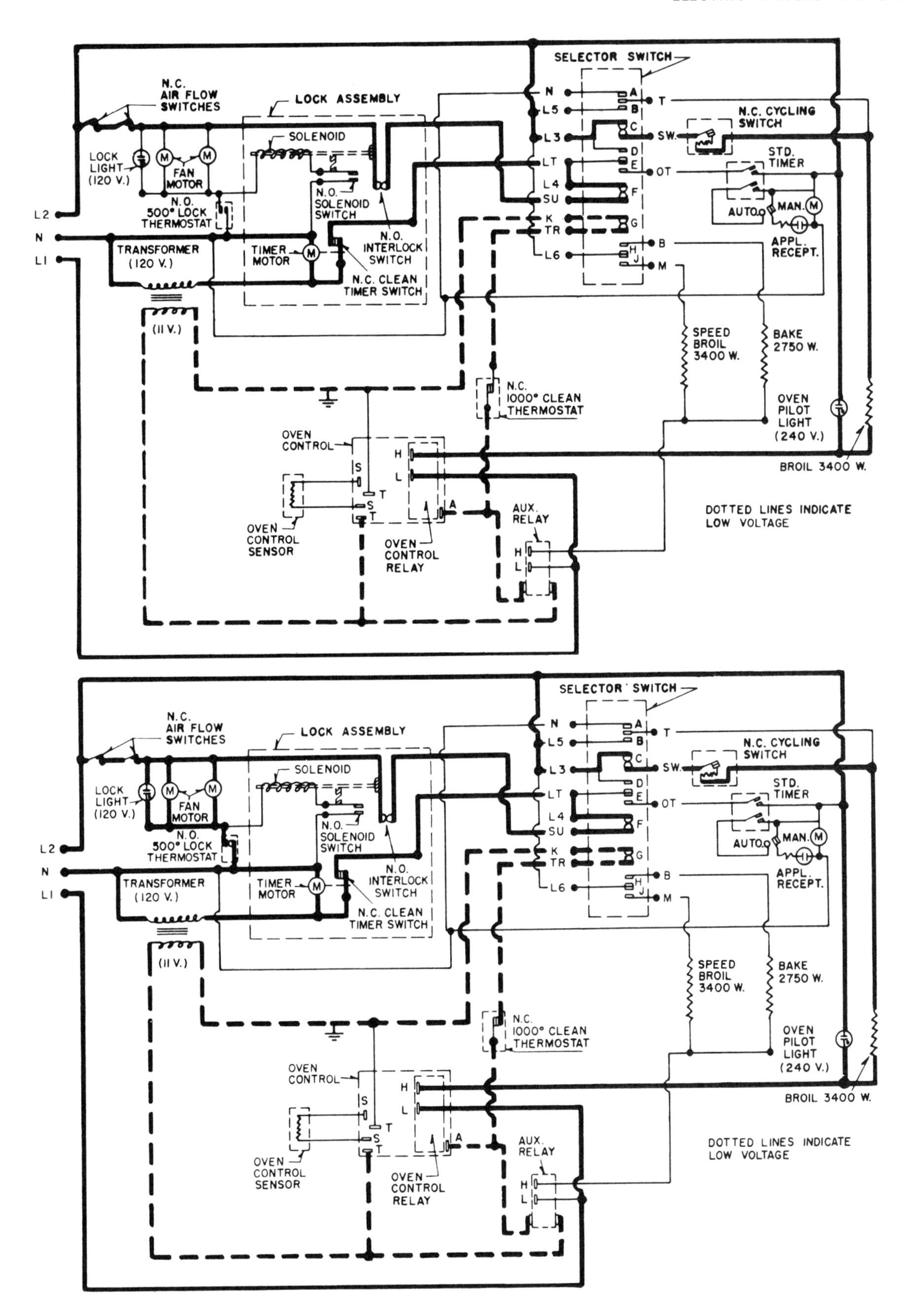
SELECTOR SWITCH
N.C. AIR FLOW SWITCHES
LOCK ASSEMBLY
SOLENOID
LOCK LIGHT (120 V.)
FAN MOTOR
N.O. 500° LOCK THERMOSTAT
N.O. SOLENOID SWITCH
N.O. INTERLOCK SWITCH
N.C. CLEAN TIMER SWITCH
TRANSFORMER (120 V.)
TIMER MOTOR
(11 V.)
L2
N
L1
N.C. CYCLING SWITCH
STD. TIMER
AUTO
MAN.
APPL. RECEPT.
SPEED BROIL 3400 W.
BAKE 2750 W.
N.C. 1000° CLEAN THERMOSTAT
OVEN PILOT LIGHT (240 V.)
BROIL 3400 W.
OVEN CONTROL
OVEN CONTROL SENSOR
OVEN CONTROL RELAY
AUX. RELAY
DOTTED LINES INDICATE LOW VOLTAGE

maintain a voltage higher than 105 V in order to prove the proper volume of air (as measured in cubic feet per minute). The switch is generally actuated (closed) by the operation of the fan damper. The switch actuator arm rests behind, but up against, the damper and closes when the fan is operating and blowing the damper open. This means that if the fan switch does not close for any reason, the oven transformer cannot be energized.

Most self-cleaning ovens that employ a fan have a thermal switch of some design. This usually is a normally closed disk-type switch, and it is in the circuit only during the cleaning cycle. As a rule, the switch is wired in series with the timer contacts, oven control transformer, latch switch, and the oven switch. Should the fan fail for any reason, the fan thermal switch will open and remove power from the oven transformer when the oven temperature reaches approximately 650 to 700°F.

Smoke eliminator. Most self-cleaning ovens have some sort of smoke elimination device to handle smoke removal during the cleaning process. Because of the high temperatures during cleaning, the smoke eliminator is needed to consume the smoke from the soil which is burned off inside the oven. It will handle normal amounts of smoke but can be overloaded if extremely large amounts of grease or soil are present. Therefore, utensils or large piles of soil must first be removed before cleaning the oven.

In general, the operation of any smoke eliminator, which is usually located in the oven vent system, is such that the end result is the conversion of food soils to CO_2 and H_2O (vapor). This is accomplished by a chemical reaction partially inside the oven, by adding heat plus oxygen, and finally inside the smoke eliminator assembly, by a catalyst plus heat. The catalyst is a coated screen mounted over the heater wire. Food soils in the oven are carbohydrates. In the presence of heat inside the oven, they break down into water, carbon monoxide, carbon, and oxygen. Then inside the smoke eliminator assembly, the oxidation catalyst plus heat complete the chemical reaction, and carbon dioxide and water in vapor form are exhausted in harmless amounts.

There are two basic types of smoke eliminators in common use.

1. *Parallel wired.* This unit consists of a single coil of a standard heating element, and is usually rated at about 150 to 200 W at 120 V.
2. *Series wired.* This unit consists of a heater wire encased in a ceramic-block assembly. It is connected in the line that feeds power to the regular oven units. All the line current, therefore, flows through the smoke-eliminator heater. The nominal resistance of the series smoke eliminator is usually about 1 Ω.

The smoke eliminator is checked by giving it a continuity test. If found open, it must be replaced.

A few self-cleaning ovens use a so-called automatic smoke eliminator rather than electrical heater types just described. This design of eliminator consists of a shallow pan located between the top heater in lower oven and top of the oven. The pan has a pattern of small holes in it which matches the outline of the broil heater.

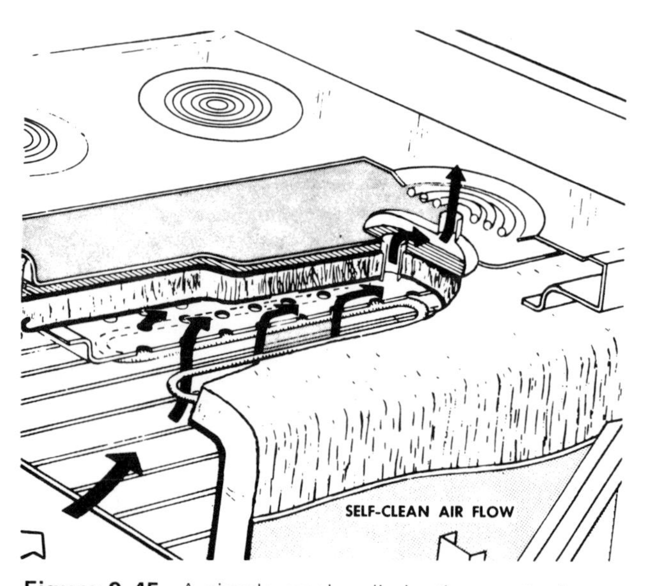

Figure 9-45. A simple smoke elimination method.

All smoke generated in the oven during the cleaning cycle must pass through the holes in the pan, which is connected to the normal oven vent, and is exhausted through a rear unit. Because of the strategic location of the holes in the pan, the smoke must pass over the red-hot heater unit, which causes the smoke to effectively change to harmless gases (Fig. 9-45). These gases are predominantly carbon dioxide and water vapor.

TROUBLE DIAGNOSIS

The secret behind any good service technician is the ability to properly analyze or diagnose the customer's problem. Many times, the service technician does not find a mechanical or electrical problem. The trouble diagnosis chart will be of no value since it is based, generally, on electrical failure when the electrical range is discussed. Since all charts are merely guides, not positive answers, they must be tempered with experience. Although not all problems can be discussed or all possible causes or remedies supplied, the charts do cover the major problem areas on an electrical range. For more detailed information, the service technician must check the service manual of the model in question.

Before any checks are made, be sure that you have the proper power to the range. Never overlook shorted, pinched, loose, or broken wires. If the customer has pinpointed the problem for you, the following causes or remedies may apply.

Problem: Meats burn on the bottom.

Possible cause	*Solution*
1. Unit on cycle too long or too frequently.	1. Adjust the oven door to reduce heat loss.
2. Oven rack overcrowded.	2. Instruct customer to select a pan that leaves at least $1\frac{1}{2}$ in of the oven rack visible on all sides.
3. Insufficient preheat time.	3. Instruct customer to heat the oven empty for 10 min or until the signal light goes out.
4. Oven door opened too frequently.	4. Instruct customer to set the clock for the minimum recommended time and check then.
5. Meat in pan without a trivet.	5. Instruct customer to elevate the meat on a trivet or small piece of crushed foil.

Problem: Meats are underdone.

Possible cause	*Solution*
1. Oven too hot.	1. Correct the oven calibration.
2. Oven rack overcrowded.	2. Instruct customer to select a pan that leaves at least $1\frac{1}{2}$ in of the oven rack visible on all sides.
3. Pan too deep, or pan covered.	3. Instruct customer that current recommendations are for open-pan roasting of meat and poultry. Meat should sit up on a trivet so all sides are exposed to oven air equally. A covered pan or foil-wrapped item calls for higher oven temperatures.

Problem: Roasting times are too long.

Possible cause	*Solution*
1. Oven too cool.	1. Correct the oven calibration.
2. Inadequate ventilation.	2. Instruct customer to be sure the vent is clear and free or to adjust the oven door for correct air intake and venting.
3. Improper use of foil.	3. Instruct customer that foil should not be

3. (*cont'd.*) used over the circulation holes, over the thermostat bulb, over an entire oven rack, or over the oven vent hole.

Problem: Steaks are too well done by the time the outside is brown enough.

Possible cause	*Solution*
1. Steak too far from heat.	1. Instruct customer to move the broil pan to a higher rack position.
2. Steak too thin.	2. Instruct customer not to expect steaks of less than 1-in thickness to be rare.
3. Cooking too slowly.	3. Instruct customer to turn the thermostat to broil (highest) setting and be sure oven door is ajar.

Problem: Steak surface is too brown by the time the steak is well done.

Possible cause	*Solution*
1. Steak too close to heat.	1. Instruct customer to move the broil pan to a lower rack position.
2. Cooking too fast.	2. Instruct customer to turn the thermostat to a lower setting (350 to 400°F) so that the unit will cycle off and on.

Problem: Cakes do not level properly.

Possible cause	*Solution*
1. Oven or racks not level.	1. With a spirit level, check the four corners of the oven rack in position, when the rack is cold and when the rack is hot.
2. Crowded baking utensils.	2. Instruct customer to check pan position and to place pans $1\frac{1}{2}$ in from the oven walls and from other pans.
3. Warped or bent baking pans.	3. Instruct customer that warped or bent pans will also cause this.

Problem: Cake does not rise properly.

Possible cause	*Solution*
1. Oven out of calibration.	1. Check the oven thermostat calibration. *Low calibration*: cakes will be flat, undercooked in the recommended time, and uneven in texture. *High calibration*: cakes will be flat and dry but even-textured.
2. Possible misuse of controls.	2. Check the customer's procedure. Cakes will not brown or rise properly unless the selector switch is turned to BAKE after the fast PREHEAT.
3. Oven thermostat probe touching heat shield.	3. Slightly bend thermostat mounting bracket so probe does not touch heat shield. (Do not bend thermostat probe.)

Problem: Cake not done in the center.

Possible cause	*Solution*
1. Oven too hot.	1. Correct the oven calibration.
2. Oven rack overcrowded.	2. Instruct customer to use second or third rack for one item.
3. Pan of incorrect size.	3. Instruct customer to select a pan that leaves at least $1\frac{1}{2}$ in of the oven rack visible on all sides.

Problem: Cookies brown too rapidly at sides.

Possible cause	*Solution*
1. Incorrect size cookie sheet.	1. Check the fit of cookie sheet in oven. One and one-half inches of air space on all sides

Possible cause	Solution
	of the sheet are necessary. A standard cookie sheet 12 × 15½ in for the 20-in oven, 14 × 17 in for the 24-in oven, or 10 × 14 in for the small upper oven is satisfactory.
2. Incorrect use of cookie sheet.	2. Cookie sheets with sides will also cause this unless cookies are placed at least 1 in from all edges.

Problem: Cookies brown too slowly.

Possible cause	*Solution*
1. Oven door leakage.	1. Check the oven door seal.
2. Oven uneven.	2. With a spirit level, check the four corners of the oven rack in position when rack is cold and when rack is hot.

Problem: Cookies are too dark on the bottom.

Possible cause	*Solution*
1. Incorrect rack placement.	1. Instruct customer to check the placement of the cookie sheet in the oven. Rack position three or four (counting from the bottom) is recommended.
2. Incorrect baking procedure.	2. For most desirable browning, only one sheet of cookies should be baked at a time.

Problem: Pies do not brown properly.

Possible cause	*Solution*
1. Incorrect rack placement.	1. Instruct customer to check the pan placement. Rack position two or three (counting from the bottom) is recommended for one rack of pies.
2. Oven out of calibration.	2. Check the oven thermostat calibration.

Problem: Fruit pies have soggy bottom crust.

Possible cause	*Solution*
1. Incorrect procedure.	1. Recommend to customer an oven temperature of 425 or 450°F for the first 10 to 15 min to quickly set the bottom crust, then to reduce oven temperature to 385 or 400°F for remainder of the baking time.
2. Incorrect procedure.	2. Instruct customer to be sure the oven is preheated thoroughly and the selector switch is reset to BAKE before the pie is placed in the oven.

Problem: Frozen pies have soggy bottom crust.

Possible cause	*Solution*
1. Incorrect procedure.	1. Instruct customer to bake frozen pies in a hot oven at 400 to 425°F for the entire baking period.
2. Incorrect procedure.	2. Instruct customer to place the pie plates on a cookie sheet when baking frozen pies.

Problem: Single surface element inoperative.

Possible cause	*Solution*
1. Burned out element.	1. Check continuity; if open, replace.
2. Wires loose or burnt off.	2. Tighten wires or replace.
3. Inoperative infinite switch.	3. Replace switch.
4. Incorrect wiring.	4. Change wiring.
5. Unit terminal contacts bad.	5. Replace terminal block.

Possible cause	Solution
6. Inoperative control switch.	6. Check and replace switch, if necessary.

Problem: All surface units fail to heat.

Possible cause	*Solution*
1. Line fuse open.	1. Check house fuse box. If it has breakers, turn them off and then back on. Check for power at range receptacle.
2. Loose, broken, or burned leads or connections.	2. Check visually and for continuity. Repair as necessary.

Problem: Infinite-heat switch does not function properly.

Possible cause	*Solution*
1. Unit inoperative.	1. Check continuity; replace, if faulty.
2. Control or switch inoperative or faulty.	2. Replace control or switch if tests prove that it is defective.
3. Utensil bottom not touching sensing unit.	3. If possible, adjust sensor or surface unit to permit contact. Otherwise, replace sensing control unit.
4. Sensing unit head dirty.	4. Clean off any baked-on deposit.
5. Calibration off.	5. Check and adjust.
6. Sensing unit head jammed, not free to move.	6. Free if possible, and if not, replace control.
7. Improper type of cooking utensils.	7. Caution user on recommendations in the cooking guide.
8. Broken or loose electric connections at unit or switch.	8. Check visually and check for continuity. Repair as necessary.

Problem: Signal light will not light or stays on constantly.

Possible cause	*Solution*
1. Faulty contact in one of the switches.	1. Check continuity; replace switch if faulty.
2. Loose or burned out lamp.	2. Check and replace lamp if necessary.
3. Poor electric connection.	3. Check visually and for continuity. Make necessary repairs.
4. Pinched leads.	4. Check visually and correct.

Problem: Fluorescent lamp does not operate properly.

Possible cause	*Solution*
1. Component faulty.	1. Check fluorescent bulb, ballast, and switch operation. If component found defective, replace.
2. Lamp not making proper contact with lamp holder.	2. Replace lamp holder or adjust holders to make contact.
3. Difficulty in electric connection.	3. Check visually and for continuity. Make necessary repairs.

Problem: Appliance or convenience outlet will not work.

Possible cause	*Solution*
1. Circuit breaker or fuse open.	1. Reset circuit breaker or replace fuse.
2. Timer inoperative.	2. Check continuity. If no continuity, replace timer.
3. Switch defective.	3. Check continuity. If no continuity, replace timer (or switch).
4. Wrong timer setting.	4. Instruct customer on how to set timer correctly.
5. Poor electric connection.	5. Check visually and for continuity. Make necessary repairs.

Problem: One or two surface units not heating (ceramic cooktop only).

Possible cause	*Solution*
1. High-limit switch open.	1. Check for continuity. Replace if open at room temperature.
2. Heating element open.	2. Check for continuity and replace as directed by manufacturer.
3. Infinite-heat switch inoperative.	3. Check for continuity and replace if necessary.
4. Loose, broken or burned leads or connection.	4. Check visually and for continuity. Make necessary repairs.

Problem: All surface units not heating (ceramic cooktop only).

Possible cause	*Solution*
1. Safety circuit open.	1. Ceramic top cracked. Replace top as needed, also fuse links.
2. Shorted triac in module.	2. Replace if necessary. Also check and replace fuse links.
3. No power to cooktop.	3. Check house fuses and wiring to terminal block on range.

Problem: Indicator or "hot-surface light" does not operate properly (ceramic cooktop only).

Possible cause	*Solution*
1. Light designed to indicate for units on one side only; defective surface thermostat on opposite side.	1. Check lamp and thermostat for continuity; if either is defective, replace.
2. Indicator light very slow to come on which indicates that it is being heated by elements rather than thermostat heater.	2. Check heater leads. Repair or replace heater as needed.
3. Indicator light does not come on at all or only glows dimly.	3. Check lamp and replace if necessary. Also check voltage and wiring to the lamp. If trouble is found, take proper action.

Problem: Oven heats but light fails to operate.

Possible cause	*Solution*
1. Bulb burned out.	1. Replace bulb.
2. Switch inoperative.	2. Check for continuity across oven light switch terminals. If no continuity, replace switch.
3. Poor electric connection.	3. Check visually and for continuity. Make necessary repairs.

Problem: Oven does not get hot enough.

Possible cause	*Solution*
1. Thermostat calibration too low.	1. Recalibrate thermostat.
2. Thermostat inoperative.	2. Check and replace, if faulty.
3. Bake or broil heater element out.	3. Check unit and replace, if faulty.
4. Selector switch open.	4. Check switch and replace, if faulty.
5. Improper door fit.	5. Check and take necessary action.
6. Altitude above sea level.	6. Temperature calibration is about 4°F low for each 1,000-ft elevation.

Problem: Oven gets too hot.

Possible cause	*Solution*
1. Thermostat off calibration.	1. Recalibrate thermostat.
2. Thermostat faulty.	2. Check and replace.
3. Stuck oven relay (if used).	3. Check and replace if faulty.

Problem: Oven will not heat (no heat).

Possible cause	*Solution*
1. Timer inoperative.	1. Check timer and replace if faulty.
2. Timer set wrong.	2. Reset timer. Instruct customer on correct use of timer.
3. Thermostat defective.	3. Check thermostat and replace if necessary.
4. Open circuit in oven elements.	4. Check for continuity and take necessary action.
5. Open line fuses.	5. Check house fuses or circuit breakers.
6. Open oven relay heater (if used).	6. Check and replace if faulty.
7. Loose, broken, or burned leads or connections.	7. Check visually and for continuity. Make necessary repairs.
8. Selector switch defective.	8. Check switch and replace if necessary.

Problem: Oven will not turn off (constant heat).

Possible cause	*Solution*
1. Timer defective.	1. Check timer and replace.
2. Selector switch faulty.	2. Check switch and replace if necessary.
3. Stuck or shorted oven relay (if used).	3. Check relay and replace if necessary.
4. Heating element ground.	4. Check and correct problem.

Problem: The oven door opens under heat.

Possible cause	*Solution*
1. Door needs adjustment.	1. Check and take proper action.
2. Pin loose or worn.	2. Replace entire hinge.

Problem: Oven signal light will not come on.

Possible cause	*Solution*
1. Contacts in thermostat inoperative.	1. Check and take necessary action.
2. Lamp burned out or resistor in lamp base open (models with neon light only).	2. Check for power to lamp when oven elements are heating. Replace bulb or lamp and base assembly as needed.
3. Broken or loose lead.	3. Check visually and for continuity. Voltage checks can be made only while elements are heating.

Problem: Oven signal light stays on.

Possible cause	*Solution*
1. Contacts in oven temperature control inoperative.	1. Check for continuity or voltage when elements are not heating. Replace if faulty.
2. Resistor in lamp base shorted (models with neon light only).	2. Replace if a low voltage (usually below 80 V) is being supplied to lamp with switch off.
3. Pinched or grounded lead.	3. Check visually and correct.

Problem: Oven signal light flickers.

Possible cause	*Solution*
1. Loose connection.	1. Tighten connection.
2. Open capacitor on cycling relay or leads reversed at terminals on relay (if relay is used).	2. Check and correct the situation.

Problem: Rotisserie skewer will not turn.

Possible cause	*Solution*
1. Switch open.	1. Check and replace switch if necessary.
2. Motor open.	2. Check and replace

Possible cause	Solution
	motor if necessary.
3. No power at rotisserie.	3. Check wiring and make necessary correction.

Problem: Oven door sweats or drips water.

Possible cause	*Solution*
1. Doors not adjusted properly.	1. Check and adjust if needed.
2. Vent plugged.	2. Clean vent.
3. Lamp shield not fitting properly.	3. Correct the problem.
4. Foil over holes in drip bowels.	4. Instruct customer that foil should not be placed over these holes.
5. Large, moist load undergoing prolonged cooking such as uncovered pot roast, custard cups in water, or baked beans.	5. Some sweating may occur normally, particularly when humidity is high.

Problem: Timer does not control oven.

Possible cause	*Solution*
1. Incorrect timer setting.	1. Instruct customer and suggest referring to user's guide book.
2. Incorrect connection.	2. Check wiring diagram and necessary connections.
3. Timer inoperative.	3. Replace timer.
4. Timer motor inoperative.	4. Check and if faulty, replace the motor or the timer.

Problem: Timer does not keep time.

Possible cause	*Solution*
1. Timer motor inoperative.	1. Check and if faulty, replace the motor or timer.
2. Burr on timer gear.	2. Replace timer.

Problem: The oven will not bake with selector set on BAKE.

Possible cause	*Solution*
1. Bake element defective.	1. Check continuity of bake element. Replace if no continuity found.
2. Thermostat faulty.	2. Check and replace, if defective.
3. Connection loose.	3. Repair connection.
4. Selector switch defective.	4. Check and replace if faulty.
5. Open interlock switch (if used).	5. Replace switch.
6. Inoperative oven relay (if used).	6. Check and replace, if faulty.
7. Inoperative pulse switch (if used).	7. Check and replace, if faulty.

Problem: The oven will not bake with selector switch set on TIMED BAKE.

Possible cause	*Solution*
1. Selector switch defective.	1. Check and replace switch, if necessary.
2. Timer defective.	2. Check and replace timer, if faulty.

Problem: The oven does not broil.

Possible cause	*Solution*
1. Broil element defective.	1. Check continuity of broil element. Replace if none.
2. Thermostat faulty.	2. Check and replace, if defective.
3. Connection loose.	3. Repair connection.
4. Selector switch defective.	4. Check and replace, if faulty.
5. Open inter-	5. Replace switch.

Possible cause	Solution
5. (*cont'd.*) lock switch (if used).	
6. Inoperative oven relay (if used).	6. Check and replace, if faulty.
7. Inoperative pulse switch (if used).	7. Check and replace, if faulty.

Problem: No heat for the cleaning operation (self-cleaning ovens only).

Possible cause	*Solution*
1. Control setting incorrect.	1. Show customer proper cleaning procedure.
2. Selector switch defective.	2. Check and replace, if faulty.
3. Oven door not locked.	3. Check and correct situation.
4. Thermostat defective.	4. Check and if faulty, replace thermostat.
5. Line fuse open.	5. Check house fuse box. If it has breakers, turn them to OFF and then back to ON. Check for power at range receptacle.
6. Oven relay defective.	6. Test and replace relay if defective.
7. Broil heating element open.	7. Check and replace, if necessary.
8. Timer switch open.	8. Check and replace, if necessary.
9. Oven thermostat open.	9. Replace thermostat.
10. High-limit switch open.	10. Replace switch.
11. Open interlock switch (if used).	11. Replace switch if found faulty.
12. Open pulse switch (if used).	12. Check and if faulty, replace switch.
13. Open bias circuit (if used).	13. Repair or replace bias control.

Problem: Incomplete cleaning (self-cleaning ovens only).

Possible cause	*Solution*
1. Setting incorrect.	1. Show user proper cleaning procedure.
2. Timer set for too short a time period.	2. Advise customer that most models require at least 2 h for cleaning.
3. Excessive soil in oven.	3. Instruct user to remove excess before starting cleaning cycle.
4. Line voltage low.	4. Check and take corrective action.
5. Heating element(s) defective.	5. Check each for operation in bake-and-broil position. Check continuity of mullion heater, if used.
6. Oven temperature control calibration for cleaning is low.	6. Check oven temperature. Adjust or replace oven temperature control if necessary.
7. Bias circuit (if used) set too high.	7. Recalibrate.

Problem: CLEAN cycle does not shut off (self-cleaning ovens only).

Possible cause	*Solution*
1. Timer inoperative.	1. Replace timer.
2. CLEAN timer switch stuck.	2. Replace switch.

Problem: CLEAN cycle shuts off but oven does not cool down (self-cleaning ovens only).

Possible cause	*Solution*
1. Oven relay stuck.	1. Replace relay.

Problem: Oven gets too hot (self-cleaning ovens only).

Possible cause	*Solution*
1. Oven temperature control calibration off.	1. Adjust.
2. Oven temperature control defective.	2. Check and if faulty, replace control.

Problem: Door locks but indicator light does not light (self-cleaning ovens only).

Possible cause	*Solution*
1. Indicator light defective.	1. Replace bulb.
2. Open lamp resistor (if used).	2. Replace resistor.
3. CLEAN thermostat defective.	3. Replace thermostat.

Problem: Oven door will not open (self-cleaning ovens only).

Possible cause	*Solution*
1. Circuit breaker or fuse open.	1. Reset circuit breaker, or replace fuse.
2. Door latch motor or solenoid defective.	2. Check continuity of motor or solenoid; if open, replace motor or solenoid.
3. Motor (or solenoid) limit switch, selector control, door lock switch, or door lock thermostat open.	3. Check continuity of switches. If open, replace the switch in question.
4. Door latch and door catch not properly adjusted.	4. Adjust door.
5. Broken or loose wires or terminals.	5. Check for broken or loose wires and repair same.

Problem: Oven door will not unlatch after CLEAN cycle (self-cleaning ovens only).

Possible cause	*Solution*
1. Cool-down period not complete.	1. Instruct customer to wait until lock light goes out.
2. Check items listed in previous problem.	2. Take appropriate action.

Problem: Lock indicator light is on and fans operate all the time (self-cleaning ovens only).

Possible cause	*Solution*
1. Lock thermostat or switch closed.	1. Check and replace lock thermostat, if faulty.
2. Lock motor or solenoid switch shorted.	2. Replace switch or solenoid or motor.
3. Cool-down period not complete.	3. Allow oven to cool.

Problem: Lock handle will latch but will not lock (self-cleaning ovens only).

Possible cause	*Solution*
1. Solenoid or motor open.	1. Check and replace, if necessary.
2. Solenoid or motor switch inoperative.	2. Check and replace, if necessary.
3. Fan motor open.	3. Replace fan.
4. Oven too hot.	4. Allow oven to cool below 450°F.
5. Open airflow switch (if used).	5. Replace airflow switch.
6. Improper power source.	6. Check power source and take proper action.

SERVICE SAFETY

The word "safety" has become common in the manufacture and sale of electrical appliances. The industry has used this word for many years, especially in teaching people how to properly service an electrical appliance. A word of caution is especially appropriate in discussing the electrical range, because of the voltage required and the transfer of heat in the range.

When the appliance is assembled at the factory, each and every component is submitted to numerous checks to ensure that it meets all national electrical standards for function, safety, and appearance. A safety committee at the factory is constantly on the alert to notice any area where a possible safety hazard may be involved and to have it corrected. The appliance design section, where safety begins, is constantly on the alert for a good safe design. The manufacturer also provides good installation and usage instructions. But in addition to all these efforts, the service technician must see that the appliance is installed properly in the customer's home and that the customer receives instructions on how it operates and how to use it intelligently. Here is a safety checklist for the service technician.

1. Always disconnect the range or throw the house circuit breaker before attempting service.
2. Use the correct tool for the job.
3. If tests are to be made with a hot range, be sure all tools have been removed and all connections are tight, or turned away from any other connection on the range.
4. Make sure that the range is grounded.
5. Be sure to use only the part specified for repair.
6. Do not leave jumpers or test cords connected after repairs.
7. Always set temperature tester on chair or stool away from range.
8. Do not use plastic or friction tape to cover or make connections.
9. Be sure all connections are secure before replacing panels.
10. Test-check the operation of the range before leaving it.

Index